# Demystifying Orchid Pollination
## Stories of sex, lies and obsession

Adam P. Karremans

Kew Publishing
Royal Botanic Gardens, Kew

© The Board of Trustees of the Royal Botanic Gardens, Kew 2023
Text © The Board of Trustees of the Royal Botanic Gardens, Kew
Images © as stated in the captions

The author has asserted their right to be identified as the author of this work in accordance with the Copyright, Designs and Patents Act 1988.

All rights reserved. No part of this publication may be reproduced, stored in a retrieval system, or transmitted, in any form, or by any means, electronic, mechanical, photocopying, recording or otherwise, without written permission of the publisher unless in accordance with the provisions of the Copyright Designs and Patents Act 1988.

Great care has been taken to maintain the accuracy of the information contained in this work. However, neither the publisher nor the author can be held responsible for any consequences arising from use of the information contained herein. The views expressed in this work are those of the author and do not necessarily reflect those of the publisher or of the Board of Trustees of the Royal Botanic Gardens, Kew.

First published in 2023 by Royal Botanic Gardens, Kew, Richmond, Surrey, TW9 3AB, UK.
www.kew.org

ISBN 978-1-84246-784-8
eISBN 978-1-84246-785-5

Distributed on behalf of the Royal Botanic Gardens, Kew in North America by the University of Chicago Press, 1427 East 60th Street, Chicago, IL 60637, USA.

British Library Cataloguing in Publication Data
A catalogue record for this book is available from the British Library.

Design: Nicola Thompson, Culver Design
Production Manager: Georgina Hills
Copy-editing: James Kingsland
Proofreading: Sharon Whitehead

Printed in Great Britain by Short Run Press

For information or to purchase all Kew titles please visit shop.kew.org/kewbooksonline or email publishing@kew.org

Kew's mission is to understand and protect plants and fungi, for the wellbeing of people and the future of all life on Earth.

Kew receives approximately one third of its funding from Government through the Department for Environment, Food and Rural Affairs (Defra). All other funding needed to support Kew's vital work comes from members, foundations, donors and commercial activities, including book sales.

IN MEMORY OF

Robert L. Dressler (1927–2019)
and
Calaway H. Dodson (1928–2020)

*Telipogon glicensteinii*, its flowers presumably mimic a female fly. © APK.

*"Much of the work of biologists is still to be performed out in the forest, deserts, mountains and lakes, in fact, anywhere in the open air. We are too often prone to think of the biologist in a white coat in a laboratory engrossed in study of a rack of test tubes or long-dead plant or animal specimens. There is yet, and will be for many centuries to come, a great deal of work to be done in direct contact with living organisms in their natural environment in an attempt to piece together the various and curious interrelationships of life"*

DODSON & FRYMIRE, 1961

A male fungus gnat attempts to copulate with the flower of *Lepanthes*.
© Sebastián Vieira Uribe.

# CONTENTS

**FOREWORD** — 11

**PREFACE** — 15

**CHAPTER 1  PLOT** — 19

1.1 **I PUT A SPELL ON YOU** — 20
Brief history of pollination studies in orchids

1.2 **US AND THEM** — 27
Orchid diversity and floral morphology

1.3 **SOMETHING TO BELIEVE IN** — 35
Pollination syndromes and strategies

**CHAPTER 2  DECEIT** — 43

2.1 **AIN'T TALKIN' 'BOUT LOVE** — 47
Sexual deception in the European genus *Ophrys*

2.2 **FIRST DATE** — 55
Sexual deception in Australian orchids

2.3 **YOUR LATEST TRICK** — 64
Sexual deception in the Neotropics

2.4 **PERFECT STRANGERS** — 68
Non-insect-like orchids and the importance of pheromones

2.5 **A MILLION MILES AWAY** — 74
Generalised food deception

2.6 **DOWN UNDER** — 83
Food deception by means of Batesian mimicry

2.7 **NEVER ENOUGH** — 91
*Epidendrum*: an ecological conundrum

2.8 **ORIGINAL PRANKSTER** — 96
The food-deceptive spider orchid and its spider-hunting pollinator

| 2.9 | FEAR OF THE DARK<br>Fungal mimicry and the pollination of *Dracula* | 102 |
|---|---|---|
| 2.10 | IT'S A LONG WAY TO THE TOP<br>Aphid mimicry, oviposition and pollination of slipper orchids | 108 |

## CHAPTER 3  REWARD  121

| 3.1 | EVERYBODY KNOWS<br>Darwin's famous prediction about the nectary length of *Angraecum sesquipedale*, and the discovery of the moth that pollinates it | 123 |
|---|---|---|
| 3.2 | LINGER<br>Pollination by fruit flies employing aggregation pheromones and nectar rewards | 129 |
| 3.3 | MISUNDERSTOOD<br>Hummingbird pollination | 137 |
| 3.4 | MADNESS<br>Pollen as a reward | 147 |
| 3.5 | TWO OUT OF THREE AIN'T BAD<br>Pseudopollen and its nutritional value | 152 |
| 3.6 | DARK NECESSITIES<br>Protein secretion and a pollination strategy based on kleptoparasitism | 157 |
| 3.7 | YOU'RE SO VAIN<br>Fragrance collection by euglossine bees | 161 |
| 3.8 | THAT SMELL<br>The pollination of *Gongora* by bees capable of recognising species by scent | 167 |
| 3.9 | WHAT HAVE I DONE TO DESERVE THIS?<br>The highly unusual ecology of the bucket orchid | 175 |
| 3.10 | HURT<br>On the sadistic means by which *Catasetum* and *Mormodes* are pollinated | 181 |
| 3.11 | WHOLE LOTTA LOVE<br>Fragrances in the fly's rectum and the pollination of *Bulbophyllum* | 190 |
| 3.12 | GIMME SHELTER<br>The cozy flowers of *Serapias* offer an unexpected form of reward | 194 |

| CHAPTER 4 | **MISFITS** | 199 |

| 4.1 | YOU CAN'T ALWAYS GET WHAT YOU WANT | 201 |
| | Ant pollination | |

| 4.2 | WITH A LITTLE HELP FROM MY FRIENDS | 210 |
| | Beetle pollination | |

| 4.3 | WHAT YOU'RE PROPOSING | 218 |
| | Aphid- and thrip-assisted pollination | |

| 4.4 | SECRET GARDEN | 222 |
| | Termite pollination in *Rhizanthella* | |

| 4.5 | IMAGINE | 227 |
| | Cricket pollination in *Angraecum* | |

| 4.6 | LADY IN BLACK | 231 |
| | Bat pollination in *Vanilla* | |

| 4.7 | WHAT IF | 246 |
| | Lizard and rodent pollination | |

| CHAPTER 5 | **REDESIGN** | 253 |

| 5.1 | COMFORTABLY NUMB | 255 |
| | Poisonous nectar and inebriated pollinators | |

| 5.2 | THUNDERSTRUCK | 260 |
| | Mass synchronic flowering in orchids, and the gregarious flowering of *Dendrobium crumenatum* | |

| 5.3 | TRIGGER | 267 |
| | Touch-mediated triggered movement and its relation to pollination | |

| 5.4 | PATIENCE | 277 |
| | Protandry and pollinaria transformations | |

| 5.5 | NOTHING ELSE MATTERS | 286 |
| | Self-pollination | |

| 5.6 | RIDERS ON THE STORM | 295 |
| | Rain-assisted pollination | |

| 5.7 | BLOWIN' IN THE WIND | 300 |
| | Wind-mediated pollination | |

## CHAPTER 6    FALLACIES                                         309

**6.1** STRAY CAT STRUT                                           311
The deadly mimicry of the orchid mantis

**6.2** EVERYWHERE                                                315
Anti-pollinators and their effect on orchid pollinators

**6.3** WORTH FIGHTING FOR                                        321
Pseudoantagonism as a pollination strategy

**6.4** PARANOID                                                  332
Pseudoparasitism as a pollination strategy

**6.5** NICE GUYS FINISH LAST                                     335
The relative importance of bees, flies and others as pollinators

**6.6** SOMEBODY TOLD ME                                          340
Specificity of orchid–pollinator relationships

**6.7** DIAMONDS AND RUST                                         346
The evolution of orchids inferred from pollinator fossils

**6.8** YOU ONLY LIVE ONCE                                        355
Ghost orchids and the importance of field studies

## CHAPTER 7    CHANGE                                            365

**7.1** UNDER PRESSURE                                            366
Climate change and its effect on orchid pollination

**7.2** IT'S THE END OF THE WORLD AS WE KNOW IT                   374
The effect of forest fragmentation and habitat loss

**7.3** BEAUTIFUL DAY                                             378
Is there hope for orchids and their pollinators?

**ACKNOWLEDGEMENTS**                                              387

**GLOSSARY**                                                      389

**REFERENCES**                                                    395

**INDEX**                                                         431

# FOREWORD

'Orchid' is derived from the Greek *orchis*, meaning testicle. With a name like that, it is no wonder that stories of sex, lies and obsession abound in the orchid literature. In this book, Adam has dived into the lurid details of the various contrivances through which orchids are pollinated. Some of the stories are true and some he reveals to be fanciful. For others, the jury is still out, as reaching a verdict can be difficult when the evidence is incomplete and contradictory.

Carl Linnaeus, who called this group of plants *Orchis*, was fond of using sexual references in his names and descriptions, which got him into trouble with some members of the clergy and their followers. According to Linnaeus, *Orchis* were plants that had underground parts reminiscent of paired testicles, one slightly larger than the other. Legend has it that they grew in pastures where bulls and cows had sloppy sex, with semen seeding the soil to produce lovely flowering orchids in the spring. Adam might be sceptical of such details and would point out that definitive experiments have not yet been done. You may have heard this story, or perhaps another that entails outlandish claims of orchids involved in sex, lies and obsession. Whatever the story may be, there is a good chance that Adam will have something to say about it in an entertaining way, while keeping the science tight.

Unfamiliar with orchids? Is your memory sketchy on the process of pollination? No problem! Before venturing into the amazing lengths to which orchids go to gain the services of pollinators, Adam explains what makes an orchid, how the parts work to attract pollinators, how certain traits indicate who the pollinators might be, and occasionally how misleading these same traits can be.

All of this is fun, but there are serious underpinnings to the fantastic stories revealed in this book. The means and mechanisms by which orchids are pollinated have been interpreted by Charles Darwin, and biologists ever since, as adaptations to increase the probability of cross-fertilisation, which is a fundamental tenet of Darwin's evolutionary theory. Cross-fertilisation produces variable, often robust offspring, whereas self-fertilisation produces less variable, often weak offspring, if any at all. Therefore, variants that enhance the probability of outcrossing are more likely to be represented in generations to come. This process is natural selection, and the outcome over time is evolution.

Darwin wrote his 1859 magnum opus, *On the Origin of Species by Means of Natural Selection,* as an abstract of some 400 pages. The numerous treatises that followed provided details of the studies and experiments that he conducted to bolster his arguments. And the first to be published after *Origin* was *The Various Contrivances by which British and Foreign Orchids are Fertilised by Insects*. He argued that the amazing array of morphological adaptations to enhance the probability of outcrossing in orchids was evidence that the variation produced by cross-fertilisation must be extraordinarily important, providing the raw materials by which natural selection could occur in the 'struggle for existence'. Orchids became his prime model system and, according to his letters, they provided as much intrigue as he could handle — from simple joy to abject frustration. So, the stories of orchid pollination not only allow us to wonder at the intricacies of nature, but also provide evidence for the theory that unites all life.

We tend to lionize Darwin, but he was not perfect in his observations. Although many of the orchid pollination mechanisms he described were fantastical, even Darwin could not believe all that he witnessed, a prime example being nectarless orchids. He considered the insects that visit flowers to be far too intelligent to be duped; they must be getting something from those flowers. Yet Darwin turned out to be as duped as the pollinating insects as numerous studies since have shown that such flowers are indeed rewardless and their pollinators get nothing. Darwin's failure to recognise what was under his nose is a weakness that occasionally hinders scientific progress. Our preconceived notions often blind us to what is really happening. Ironically, it was Darwin (and Alfred Wallace) who had broken the intellectual and cultural shackles of the day to propose the theory of natural selection. Adam's demystification efforts often revolve around phenomena that biologists had expected, only to find that with acute observations something novel was before them.

A drosophilid fly in the process of pollinarium removal of *Masdevallia zahlbruckneri*. © APK.

*Vanilla costaricensis*, a rare species with extraordinary green and white flowers. © APK.

CHAPTER 1

# PLOT

*"We tend to assume that flowers are here to gladden our hearts. They're not, of course. Plants produced flowers long before humanity appeared, to summon mammals, birds and, above all, insects: and although they sometimes are given some payment, they are the servants. The masters are the plants."*
SIR DAVID ATTENBOROUGH in *The Private Life of Plants*

Before embarking on the subject matter of this book — the secrets of orchid pollination — you will need some background knowledge. In this chapter, we first travel through a brief historical overview of the science of orchid pollination, followed by a description of the orchid flower and its parts, and finishing with the theories behind floral syndromes and pollination mechanisms in orchids. But bear with me, it will be worth the effort.

## 1.1 I PUT A SPELL ON YOU

*"They have a magic whose secret will always elude us, a blend of witchcraft and romance, that sets them apart, above all other plants — a sorcery that conjures up our deepest passions, from idolatrous worship through the pure spirituality of aesthetic veneration to fear, revulsion, and mindless hatred."*
LUIGI BERLIOCCHI in *The Orchid in Lore and Legend*

There is a distinct lack of written historical evidence about the interactions between orchid flowers and their pollinating agents. Early naturalists considered ecological interactions to be a mere intellectual curiosity rather than an academic priority. Botanists initially concentrated their efforts on describing the morphological features of plants, classifying them in order to establish their potential use as foods, medicines or tools. Even today, most popular literature on plants revolves around folklore and herbalism. In his book *In Defense of Plants*, Matt Candeias points out that "whether for economic gain or some purported medicinal benefit, we only seem to care what plants can do for humans". Plant biologists, much more than their zoological counterparts, typically find themselves having to justify the study and conservation of plants in terms of their potential use, whereas their ecological function has been largely neglected. A case in point is the recently discovered ecological function of the world-famous aromatic compound vanillin, which is extracted from the fruit of the *Vanilla* orchid.

In fact, the idea that flowers actively attract their pollinators was revolutionary at the end of the 18th and beginning of the 19th centuries. It was not until 1793, with the publication of Christian Konrad Sprengel's book *Das Entdeckte Geheimnis der Natur im Bau und der Befruchtung der Blumen* [Discovery of the Secret of Nature in the Structure and Fertilisation of Flowers] that the pollination strategies employed by plants were first recognised and described in a systematic fashion. The book was one of the first steps towards a more rational approach to natural history, with guesswork slowly giving way to critical and reasoned research on orchid pollination. However, a review by Robert Brown published in 1833 reveals that there were still a significant number of authors 'who — from certain peculiarities in the structure and relative position of the sexual organs in this family — have regarded the direct contact of these parts as in many cases difficult or altogether improbable, and have consequently had recourse to other explanations of the function'. These authors, including renowned botanists, believed that the direct application of pollen onto the stigma of orchid flowers was simply too difficult and therefore was unlikely to be necessary for fertilisation. What little faith they had in the resourcefulness of orchids!

The astonishing complexity of the schemes used by orchid flowers to ensure fertilisation was put under the spotlight many years later when Charles Darwin dedicated an entire book to the subject: *On the Various Contrivances by which British and Foreign Orchids are Fertilised by Insects*, first published in 1862. In his book, Darwin presented detailed accounts of his personal observations on the fertilisation of British orchids and also discussed the possible pollination mechanisms of a number of tropical orchids. At the time, his book was largely ignored, but for Darwin it provided the strong support needed for his theories on natural selection and speciation through interspecific interactions. The second edition of the book, published in 1877, presented a "remodelled" account with additions and corrections, in particular regarding tropical orchids. Darwin never witnessed the interaction between tropical orchids and their floral visitors first hand. He had to rely on the accounts of multiple local collaborators, and of course on a very keen sense of deduction, speculating on the possible purposes that particular floral traits could serve. Many of the floral mechanisms that he described as having developed to ensure fertilisation were completely unheard of at the time. His studies of orchid flowers captured the imagination of many readers, fascinating and inspiring several generations of floral ecologists and enthusiasts who from then on looked at orchids in a new light. One of his biggest fans was Calaway Homer Dodson.

Cal Dodson was born in California in 1928. At the age of 18, he enlisted in the US Army, serving as a paratrooper in the Korean War. Upon his return home, he earned a bachelor's degree in botany from Fresno State College. For his dissertation, Dodson conducted fieldwork in Ecuador, working on Andean orchids. It was in the Andes that Dodson, no doubt stunned by the overwhelming diversity of these plants, caught 'orchid fever'. After obtaining his PhD at Claremont College, Cal returned to Ecuador as professor at the University of Guayaquil, where he founded the Institute of Botany. Dodson had a dramatic influence on botanical studies in the Andes, especially in Ecuador, raising academic awareness of orchids and biodiversity in general.

In 1960, Dodson relocated with his family to Saint Louis, Missouri, where he was appointed Taxonomist and Curator of Living Plants of the Missouri Botanical Garden. It was at Missouri Botanical Garden that Cal met Robert Lee Dressler — whom most of us knew simply as Bob. That same year, Cal and Bob published a very important paper under the title 'Classification and phylogeny in the Orchidaceae'. The paper was the first major attempt to classify the orchid family since the beginning of the century. Dodson graciously named the orchid genus *Dressleria* in Bob's honour — though he later admitted that the smell of the plant reminded him of Bob's field socks (see Figure 1.1.1).

Both botanists became very interested in orchid pollination, but were somewhat limited by the lack of a representative collection of living orchids at the Missouri Botanical Garden, which would ultimately prompt them to leave Saint Louis. In 1963, Bob applied for a job at the Smithsonian Institution's Tropical Field Station in Panama (which later became the Smithsonian Tropical Research Institute), becoming their first hire. A year later, Cal became Assistant Professor and Curator at the University of Miami in Coral Gables, where his academic legacy began. In 1966, he and Leendert van der Pijl coauthored *Orchid Flowers: Their Pollination and Evolution*, a book that came to be considered a 100-year update of Darwin's orchid book.

*Orchid Flowers* was the most comprehensive treatise on orchid pollination for half a century, laying the groundwork for many of the modern studies on the subject. It included a large number of novel observations on the interactions between tropical orchids and their pollinators, at a time when relatively little was known on the subject — despite orchids in the tropics being much richer in species numbers and floral diversity than their temperate counterparts. Ecological studies in the tropics were virtually unheard in those days, and that opened a large window of possibilities for discovering and understanding new and exciting relationships among plants and animals. The authors' accounts were nothing short of mind-blowing: orchids that mimic spiders and are pollinated by an enormous parasitic wasp, bees that pollinate orchids by means of frantic territorial aggression, orchid

**Fig. 1.1.1.** *Dressleria dilecta*, Calaway Dodson honoured his friend Bob Dressler in the name of this orchid genus.
© APK.

flowers offering fragrances that male bees collect to use as cologne, orchids imitating excrement and rotting flesh to attract flies, birds pollinating orchid flowers, and many more extraordinary strategies.

Robert Louis Dressler was born in 1927 in rural Taney County, in the Ozark mountains of Missouri. His family later moved to California, where Bob graduated from Gardena High School in Los Angeles. After briefly serving as a finance clerk in the US Army, he attended the University of Southern California, graduating with honours in 1951. In 1957 he received his PhD in biology from Harvard University, moving back to Missouri the following year to work as editor of the *Annals of the Missouri Botanical Garden*. Bob got his first taste of the Neotropics in 1961 when working in Panama, and after meeting Cal Dodson at Missouri, they often accompanied each other in the field. Once in the Neotropics, Bob and Cal became especially interested in the relationship between male Euglossini bees and orchid flowers, which offered fragrances as a reward for pollination. Cal published several papers based on his field observations in Ecuador and Peru, while Bob

**Fig. 1.1.2.** Dressler's resting place at Lankester Botanical Garden in Costa Rica. © APK.

continued with his studies in Panama, and later also Mexico, Guatemala and finally Costa Rica. He worked at the University of Costa Rica's Lankester Botanical Garden, where he now rests peacefully under a large specimen of the orchid *Sobralia* (see Figure 1.1.2). Both Bob and Cal became obsessed and enchanted by orchids, an everlasting spell that would prove more powerful than any army. They both impacted orchid research on a global scale and their work on pollination, as well as the work of their students, is still influential today.

In 1969, Joe Arditti prepared a list of key references on orchid pollination that had been published since Darwin's orchid book. The list was not meant to be exhaustive, but it is revealing. First, it tells us something about how prevalent the works of Dodson and Dressler were: together or on their own, they authored more than half of the references cited by Arditti. Second, it shows how little research was being done on the interaction between orchids and their floral visitors, given that only 24 works made the list. Research on pollination has skyrocketed since the 1960s. As of January 2021, *Bibliorchidea* — the most complete database on orchid literature in the world — held 3,000 works with the keyword 'Pollination' in its records, of which only the first 1,000 (3%) were published prior to 1960. Jim Ackerman and his colleagues have estimated that the average number of works on orchid pollination grew from fewer than one per year before 1966, to more than 20 a year since.

To prepare a complete account of everything that is known about orchid pollination worldwide is a colossal endeavour, given the body of literature that is available today. Although I have enjoyed this project, analysing the available research for this book has been overwhelming at times. There is simply too much loose information. The most recent effort to summarise our knowledge on the subject of orchid pollination was carried out by Nelis A. van der Cingel. In 1995, van der Cingel published *An Atlas of Orchid Pollination: European Orchids*, which was complemented six years later by *An Atlas of Orchid Pollination: America, Africa, Asia and Australia*. Van der Cingel was trained in the study of animal behaviour, and despite warning of the weaknesses he saw in the interpretation of the ecological interactions between certain orchid flowers and their floral visitors, he loved to document and publish observations on pollination.

More recently, comprehensive works dealing with the pollination of orchids in particular regions have surfaced. The terrestrial orchids (orchids that grow on the ground) of Europe have received great attention from scientists since the time of Darwin, and a majority of pollination papers today still focus on these members of the orchid family. An extraordinary compilation was published in 2011 by Jean Claessens and Jacques Kleynen, entitled *The Flower of the European Orchid: Form and Function*. Claessens and Kleynen's book showcases the orchids of Europe in a way that has never been done before, with breathtaking photographs of the floral details and the insect visitors.

The pollination of orchids of Australia — which is certainly the most-researched continent after Europe — has been explored in great detail in many recent papers. It is also discussed by John Alcock in his wonderful *An Enthusiasm for Orchids: Sex and Deception in Plant Evolution*, published in 2006. Alcock presents the terrestrial orchids of Australia and shines a new light on the evolution of their pollination syndromes. Finally, a comprehensive compilation of the state of knowledge on the pollination of orchids found growing north of Mexico and Florida was published in 2012 by Charles L. Argue in *The Pollination Biology of North American Orchids.*

Insightful as these works are they have one thing in common with the majority of papers being published on orchid pollination: a strong bias towards terrestrial orchids from temperate regions. Research on tropical epiphytes (which grow on other plants) lags decades behind. The gap is certainly worsened by the policy of scientific journals to discourage the publication of field observations on the pollination of tropical orchids in favour of experimental laboratory and computational work. Although basic ecological insights were presented by European, North American and Australian naturalists a century or more ago, they are completely lacking for the tropics, where myths and the unknown consequently still abound.

The reader will find that this book has a slight bias towards describing the interactions between neotropical orchids and their pollinators. This is partly because the author has witnessed these interactions first hand in the field. But it is also motivated by the fact that studies on orchid ecology come predominantly from temperate regions, despite the significantly greater diversity of orchids growing in the tropics. The foundations for much of what we know about tropical orchid pollination were laid by researchers such as Dodson and Dressler. Several of their ideas are discussed here and, although some are contested in the light of new findings, many still hold true today. Their contributions are not marked by big eureka moments. Rather, they worked hard and gradually accumulated observations and data, with time transforming these into ideas, some of which matured and eventually became hypotheses. They published their findings one step at a time, and continually refined initial interpretations in the light of new evidence.

Much like Sprengel, Darwin and the naturalist Alfred Russel Wallace before them, Cal and Bob did something that today is undervalued by many in academia and funding agencies: they spent a lot of time patiently observing nature in action; they went out to the field and untiringly examined the interactions between plants and animals; they smelled, felt and sampled; they took notes and measurements; they tried and failed, and then tried again. It is thanks to this painstaking and often seemingly fruitless perseverance of our predecessors that we have something to build upon today. Ecology requires the investment of a precious and increasingly scarce resource: time.

## 1.2  US AND THEM

In his novel *The Picture of Dorian Gray*, Oscar Wilde alludes to orchids as monstrous metaphors of subtle colour. Like many of us, Wilde was stricken by the exotic beauty and complexity of the orchid flower. The famous Irish poet and playwright was a spokesperson for the Aesthetic movement in England, which espoused the doctrine that art exists for the sake of its beauty alone. He was a strong believer in the creation of art for art's sake. But unlike the autonomic beauty of art that aestheticism advocates, the orchid flower's appeal is bound to its function and purpose.

In 1993, Dressler estimated that there were about 19,500 named species of orchids. In his book *Phylogeny and Classification of the Orchid Family*, he pointed out that the discovery of new species had significantly accelerated in recent years, and that trend has continued. Today, we estimate that there are between 310,000 and 320,000 species of flowering plant worldwide, and the family Orchidaceae accounts for around 30,000 of all recognised species. The orchid family is so ubiquitous and species-rich that one out of every ten flowering plant species on Earth is an orchid. Take a minute to really think about that. Of all the different flower forms this world has to offer, 10% are orchids!

Even though all orchids share a basic floral morphology, each species has a more or less unique flower. This means that in the orchid family alone there are 30,000 different floral possibilities. In addition to those unique individual species, there are thousands of subspecies, forms, varieties, cultivars and hybrids that increase the diversity of colour, shape, size and smell of orchid flowers to a few hundred thousand variations. Not to mention that previously unknown species are being discovered almost on a daily basis in all corners of the globe — every year between 150 and 200 new orchid species are discovered by scientists. Many of them come from poorly explored, remote tropical areas, but others may be discovered after carefully inspecting someone's backyard. There are so many different orchids that, you could spend an entire lifetime trying to see them all.

When we talk about a species, we are normally referring to a group of individuals that share a common ancestry — but this is not the only definition in use. Species typically form populations that are interconnected. They are able to mate with each other and not with

Fig. 1.2.1. **A**. A fully grown, flowering plant of the minute *Platystele tica* fits entirely on a one-cent coin. **B**. *Grammatophyllum speciosum*, possibly the most massive orchid in the world, growing in Singapore
© APK (A) and Andre Schuiteman (B).

other species, and the offspring they produce are fertile. This is a very broad definition of what a species is, and it doesn't, and can't be, applied equally to all organisms. But it is good enough for the purpose of this book. Many orchid species are in fact able to cross with others, forming what are known as hybrids. Orchids that are commonly sold as house plants are hybrids created through a process of artificial selection. Hybrid orchids also occur in nature. Closely related species may hybridise naturally under certain conditions, producing offspring with intermediate features. Hybrid individuals may not always be able to reproduce themselves, so their line ceases to exist when they die. However, hybrids that are able to reproduce may form a whole new lineage that is different from both parents. That novel lineage may eventually become stable, having its own population of individuals that breed only among themselves, producing fertile offspring. A new species can evolve in this way. There are thousands of hybrid orchids, either spontaneous or artificial, but we will not go further into that here.

This apparently unending diversity is one of the most appealing features of orchids. Contrary to what occurs in other plant groups, the orchid enthusiast's desire for new variants may never be sated. Orchids are not only rich in species numbers, they are also highly diverse in floral shape and functionality. Their plant structures can be as varied as their flowers, but I will concentrate on the latter. It may come as a surprise to the reader that the flower of every orchid represents only a slight variation of the same basic floral scheme. From the pouch flower of a lady's slipper to the star-shaped Darwin's orchid, from the rotten flesh smell of the *Bulbophyllum* flower to the delicately scented *Vanilla*, from the insect-like *Ophrys* to the soldier-like *Orchis*, from the giant *Grammatophyllum* to the tiny *Platystele* that fits on a US one-cent coin (Figures 1.2.1), from the exotic bucket orchid growing on an ants' nest to those *Phalaenopsis* hybrids sold at your local store — all orchids have the same essential parts (Figure 1.2.2).

Orchid flowers are easily recognised by having one outer 'whorl' or layer composed of three 'sepals' and an inner whorl made up of three petals (see Figure 1.2.2). Sepals and petals are modified leaves that have changed function from being an exclusively photosynthetic organ. Photosynthesis, the process by which sunlight is used to synthesise nutrients from carbon dioxide and water, is probably still taking place in some sepals and petals, but in general terms, these organs serve

**Fig. 1.2.2.** A typical orchid flower, with three sepals and three petals. One petal, known as lip, is modified to accommodate visitors. It subtends a central column that bears the stigma and anther cap, within which we find a pollinarium. © APK.

another purpose. The main function of the sepals is to protect the delicate floral bud from physical factors such as rain, sunlight and wind, as well as from the attack of other living organisms, including herbivores and disease-causing pathogens. The main role of petals is to attract pollinators. They are usually larger and more colourful than the sepals. In essence, the function of these modified leaves has changed from energy production to the reproduction of the plant. Naming these organs sepals and petals rather than leaves is useful to recognise their unique function, but it is important to keep in mind where they come from. The leafy origin of flowers is still recognised in some languages, such as Dutch and German, which call petals *bloemblad* and *blütenblatt*, respectively, literally meaning flower leaves.

In orchid flowers, one of the petals differs substantially from the other two. This modified petal is known as the lip or labellum. The lip is generally the largest and most ornamented segment of the flower and it gives orchids their characteristic appearance. The floral segments of many plants are equally arranged around a central axis, resulting in flowers that look the same from any side. This circular arrangement means that you get two identical halves from any plane crossing the central axis, known as radial symmetry. Several everyday flowers, such as daisies, lilies, roses and sunflowers, are radially symmetric, with any plane crossing the central axis resulting in two identical halves. However, the presence of a single petal with a shape that differs from that of the other two means that orchid flowers have no radial symmetry. There is only one imaginary line that, when crossing the exact middle of the flower, will result in two identical halves. This arrangement is known as bilateral symmetry and it immediately distinguishes the orchid flower from that of relatives such as the iris. There are of course some asymmetrical orchid flowers, such as those of *Ludisia* and *Mormodes*, but we will discuss those later on (see Figure 1.2.4).

So far we have learned that orchid flowers typically have six floral segments: three exterior sepals and three interior petals. One of the petals, known as the lip, is conspicuously different from the others, giving orchids their characteristic bilateral symmetry. The innermost whorls of the orchid flower constitute the sexual organs. Beneath the six floral segments, just below where they are inserted, is the ovary of the orchid. An ovary with this positioning is known in the botanical world as an inferior ovary, as the seeds will sit under the flower parts.

**Fig. 1.2.3.** Orchid flowers, like this *Vanda*, typically show bilateral symmetry, meaning there is a single plane that when crossing the central axis results in two identical halves.
© APK.

When the orchid is flowering, the ovary is more or less undifferentiated from the other stalks of the inflorescence, but as it becomes fertilised it slowly swells, transforming into a thick fruit. Most flowers typically have a central 'style', which is the female organ, surrounded by several 'stamens', which are the male organs. In orchids, the style and stamens are fused into a single central structure known as the 'column' or 'gymnostemium'. The column is usually cylindrical and elongate, bearing a single 'anther' at the tip and with the stigma just below that, facing the lip. The anther is the male organ and contains the orchid's pollen, while the stigma is the female organ and receives pollen in order for the flower to be fertilised (see Figure 1.2.2).

**Fig. 1.2.4.** *Ludisia discolor*, an unusual orchid flower with organs naturally twisted to one side. As a result, the flower lacks bilateral symmetry. © APK.

Orchids also have characteristic pollen. Rather than producing loads of dust-like pollen, as many other plants do, the pollen grains of orchids are compacted into larger packets called 'pollinia'. In more primitive orchids, the pollen masses are soft and easily disrupted, and they are transferred to the pollinator in fragments rather than as a single unit. Powdery pollen is only found in species of the very primitive and little-known subfamily Apostasioideae. In subfamilies Cypripedioideae and Vanilloideae, to which lady slippers and vanilla belong, respectively, the pollen masses are sticky, but not always aggregated enough to be defined as pollinia. For example, in Costa Rica, I observed the cluttered lumps of *Vanilla* pollen masses being carried on the back of a bee. This granular pollen is removed and deposited only partially. In this case, the pollen masses have no discernible shape, a feature that is uncommon in the family.

Advanced orchids, including the main subfamilies Orchidoideae and Epidendroideae, typically have two, four, six or eight distinct pollinia that are compacted, have a definite shape and configuration, and are charmingly adorned with accessory structures — which facilitate their removal, transport and deposition. The most common accessory structures are the caudicles (a granular process to which the pollinia are attached), stipe (a slender stalk often connecting a viscidium to the caudicles), and viscidium (a sticky pad or viscid drop). The single, well-defined unit formed by the orchids' compacted pollen, plus the accessory structures, is called the 'pollinarium' (see Figure 1.2.2). The packing of pollen into a single discrete unit is a key feature of orchids, with more than 95% of species having a well-defined pollinarium that is not easily disrupted by touch. This arrangement means that the whole pollen content of a flower is removed during a single visit by a pollinator, and likewise is deposited as a single unit in one flower. The number, texture and shape of the pollinarium parts are a characteristic features for many orchid groups. In short, you may be able to identify the orchid based only on the features of its pollinarium. This can prove very useful when one finds a pollinator carrying pollinaria and would like to know what orchid it visited.

Flowers are woven into our social life. They are a symbol of romance, sentiment and thoughtfulness. In temperate regions of the world, they announce that winter is over, wheras in the tropics, they inform us that we have entered the dry season. Flowers bring joy to the downhearted, hope to the despairing, and serenity to the distressed.

Their delicate beauty inspires the artist and motivates the passionate horticulturist. We adorn our houses and clothes with them, and they provide company for our departed loved ones. And yet, we rarely question why they stir our emotions. Flowers exist for a purpose, but it is certainly not to be beautiful for us humans. The features that we regard as attractive, including their symmetry, contrasting colour patterns, exotic aromas, and fleeting, exuberant nature are all intricately related to the survival of the species. We will discuss this role further in the following chapters.

## 1.3 SOMETHING TO BELIEVE IN

Like the sexual displays of animals, floral displays increase reproductive success. Flowers are enormously diverse, and yet similar floral patterns are found in distant plant families. These recurrent flower types suggest the existence of 'pollination syndromes' that guarantee reproduction. A pollination syndrome is the set of characters — such as colour, shape, pattern and scent — that together represent adaptations to particular types of pollinators. Broadly speaking, this means that certain features are shared by flowers that exploit the behaviours and preferences of the same kind of visitors. These similar patterns in floral characteristics can be a powerful predictor for the pollination syndrome of a particular flower.

The most important pollination syndromes among flowering plants are: 'melittophily' (literally, 'bee-loving') or bee pollination; 'phalaenophily' or moth pollination; 'psychophily' or butterfly pollination; 'myophily' or fly pollination; 'ornithophily' or bird pollination; and 'cantharophily' or beetle pollination. All of these pollination syndromes and some of their subdivisions — especially 'sapromyiophily' (pollination by flies that visit decaying substances, dung or carrion) and 'sphingophily' (pollination by hawkmoths) — have been found to occur extensively in orchids, with the exception of cantharophily.

Melittophilous flowers are those adapted to pollination by bees, a syndrome often shared with wasps (both order Hymenoptera). Given that most of these insects are active during the day, the flowers of orchids with this syndrome are said to open during the day, to have bilateral symmetry and a prominent landing platform, to have bright blue, violet, purple, green, yellow or white colours, conspicuous guides and nectar, and to smell fresh and sweet.

Nevertheless, many bee-pollinated orchids lack the floral features associated with hymenopterous flowers, such as having bilateral symmetry, or having nectar and nectar guides, or producing sweet odours. Flowers pollinated by bees through fragrance collection, such as those belonging to the genera *Coryanthes*, *Gongora* and *Mormodes*, are among the many exceptions. Their flowers are typically dull, heavily spotted or stained in brownish or purplish colours, and they lack the nectar rewards and nectar guides that are characteristic of melittophily. At the same time, several orchids pollinated by bees or wasps have features that are associated with other syndromes. For example, some have trap flowers (they trap insects), others are tubular and lack a landing platform, some are smooth and tiny, others are dull and warty.

'Lepidopterous' flowers have a well-developed nectar tube that is adapted to the nectar-foraging behaviour of moths and butterflies (from the order Lepidoptera). Moths have a nocturnal lifestyle and so phalaenophilous flowers — those with the moth-pollination syndrome — are suggested to flower and produce fragrance at night. Their bilateral symmetry is said to be weak, their colours are white, cream or greenish, their nocturnal scent is strong, their landing platform tends to be turned back out of the way or upwards, and abundant nectar is produced but it is hidden in deep cavities or tubes.

By contrast, butterflies have a diurnal (daytime) lifestyle, and as such the psychophilous flowers that are pollinated by these insects open during the day, lack bilateral symmetry, are erect with a horizontal landing plane, have vivid colours, have fresh and agreeable odours, produce nectar and have nectar guides. However, most species of the genus *Epidendrum* — the largest group of orchids pollinated by butterflies and moths — have a well-developed labellum that clearly differs from the petals and sepals, which gives them strong bilateral symmetry. Their flowers are dominated by whitish and greenish colours, though bright red, orange and fuchsia-coloured flowers are not unknown. Even though *Epidendrum* flowers have nectaries, their nectar production varies from copious to none at all. In this large genus, therefore, there are exceptions to most of the features associated with the Lepidopterous pollination syndrome.

'Dipterous' or myophilous flowers are adapted to pollination by flies (order Diptera) that are specialised visitors of bright, sweet-smelling flowers, and feed mainly on nectar. Many myophilous flowers are not exclusively pollinated by flies, but are visited by several other insect groups, including bees and butterflies. Orchid flowers that are pollinated by flies have mostly been associated with a specific subgroup of myophily known as sapromyiophily. Sapromyiophilous flowers imitate the chemistry and shape of decaying substances, dung or carrion, and the flies visit them either for food or to lay their eggs. These flowers have less pronounced bilateral symmetry. Their colours are often superfluous, dull or greenish, occasionally with brown-purple spots, they smell putrescent, lack nectar or other foods and have no nectar guides. In addition, they may have slits, holes and windows to guide the insect, brown, furry hair (often with vibrating hairs), and trap devices.

Once again, there are several exceptions to the rule. A major subtribe of orchids known as Pleurothallidinae specialise in fly pollination, and are therefore typically associated with sapromyiophily in the literature. However, among the pleurothallids, we find some that have strong bilateral symmetry while others have radial symmetry, some that are extremely glandular and hairy while others are not at all, some that have long tails while others are tailless, some that are dull-coloured, while others have bright yellow, red and orange or pearly white flowers. Some appear to deceive pollinators into believing that they provide nectar, whereas others certainly do provide nectar rewards, some have a putrid smell, while others smell very sweetly, some are closed and have windows, whereas others are completely open and flat, and yet others have trap devices. So despite being specialised in fly pollination, members of the Pleurothallidinae have highly varied pollination strategies, and many are not sapromiophylous by any definition.

Ornithophilous flowers are those pollinated exclusively by birds and generally feature vivid colours, especially bright red, orange and fuchsia. These flowers typically open during the day and have weak bilateral symmetry. They are tubular, without a landing platform, produce abundant nectar, and lack a strong odour or nectar guides. They tend to be sturdy. Neotropical orchids pollinated by hummingbirds are indeed typically bright red, orange or fuchsia, nectar-rich, and tubular. However, each of those features is not exclusive to ornithophilous flowers and may also be found in orchids that are pollinated by other creatures. A few examples are the bright red *Cattleya* species, which are pollinated by bees, and the bright fuchsia *Masdevallia* species, which are pollinated by flies.

The fact that hummingbirds are attracted to and will inspect bright red, orange or fuchsia-coloured flowers has often led to the false impression that these birds pollinate the flowers. This is discussed in more detail in Chapter 3 (section 3.3, 'Misunderstood'). Conversely, at least one species in the genus *Angraecum* with flattish, greenish-white flowers that are typical of the lepidopterous syndrome, is actually pollinated by birds.

Floral syndromes have become widely accepted by orchid researchers and hobbyists alike, and may indeed be useful in predicting the pollinator group of certain orchids. However, they suffer from an extreme oversimplification of the very complex evolutionary histories of flowers and their adaptations to pollinators. The exceptions are not simply a few odd species here and there, but large portions of the Orchidaceae that simply do not fit into the very broad categories of floral syndromes. As is often the case with broadly defined stereotypes, pollination syndromes foster mistaken beliefs. In fact, a large field test of the frequency and predictive power of pollination syndromes among plants in general found that only a small proportion could be categorised into a particular syndrome, and only 30% could be predicted correctly. This is also likely to be true for highly specialised orchid flowers. Flowers that are pollinated by the same pollinator groups may indeed look alike, but several other factors are in play at the same time. An extremely important factor in floral similarity is common descent. The closer one species is to another genetically, the more likely it becomes that the flowers they produce are similar; especially if they share the same pollinator groups. Another key factor is pollination mechanism, namely how the pollinator interacts with the flower, which is extremely important in shaping the floral features.

In a study on South African orchids that are pollinated by the oil-collecting bee, *Rediviva peringueyi*, Anton Pauw found that all of the orchids shared the same syndrome of floral features and were extremely similar to each other, due to the fact that they share the same pollinator. His study exemplified how floral syndromes could accurately predict pollination by this specialised oil-collecting bee. However, the 15 orchids

**Fig. 1.3.1. A.** *Bombus terrestris* pollinating *Goodyera repens*, a non-stereotypical melittophilous orchid. **B.** The bright red-orange flowers of *Masdevallia veitchiana* suggest hummingbird pollination, but they are rather myophilous. **C.** Flowers pollinated through sexual deceit, such as *Chiloglottis trapeziformis*, rarely exhibit any of the expected floral features. **D.** *Epidendrum piliferum* is a butterfly-pollinated orchid that lacks most of the typical features of the syndrome. **E.** *Eulaema cingulata* visiting *Catasetum maculatum* while carrying a pollinarium of a Zygopetalinae. **F.** *Eulaema cingulata* visiting *Vanilla pompona* while carrying the pollinarium of *C. maculatum*.

© Jean Claessens (A), Stig Dalström (B), Rod Peakal (C), APK (D), Ernesto Carman (E), Charlotte Watteyn (F).

species in what the author called the *R. peringueyi* pollination guild not only are pollinated by the same pollinator and through the same mechanism, but also are members of a single subtribe — the Coryciinae — and thus are very closely related. When you look at distantly related orchids that are pollinated by oil-collecting bees — such as the Oncidiinae — you find a completely different floral morphology.

In fact, some orchids that share the same pollinator species, but are unrelated and use a different pollination mechanism, have quite different floral morphologies. There are many examples, but one that we have observed in Costa Rica is that of the bee *Eulaema cingulata*, which visited the flowers of *Vanilla pompona* while carrying a pollinarium of *Catasetum maculatum* (see Figure 1.3.1). This particular bee species is known to be the effective pollinator of both of these orchids. However, the flowers of *V. pompona* are much more similar to those of other orchids pollinated by bees through food deception (mimicking a food source), such as *Cattleya* and *Sobralia*, than to *C. maculatum*. The latter has unique floral features more similar to those of close relatives that are also pollinated through fragrance collection. These two orchid species share the same pollinator species and yet look completely different.

Broad pollination syndromes do not sufficiently explain floral features in the Orchidaceae. Being pollinated by a bee, wasp, butterfly, moth, fly, hummingbird or beetle certainly conveys certain information about a flower. But to be a useful diagnostic tool, the complexity of specialisation in the features of orchid flowers requires a much higher level of refinement. Some authors have suggested using functional groups, that is animals that share morphological and behavioural features, instead of simply the group to which the pollinator belongs. Indeed, this proves to be much more accurate because functional groups add more levels of complexity.

I would suggest that there are three main factors driving floral features in orchids: 1) evolutionary history, or genetic predisposition; 2) pollinator group, defined as the kind of animal that the flower attracts, and; 3) pollination mechanism, namely the animal behaviour that the flower is exploiting. It is the combination of these three factors that shapes the features of the orchid flower. Therefore, predictions regarding pollinators become more robust when, in addition to floral features, we consider how floral visitors actually interact with the orchid flower. To address the latter, we must first discuss pollination strategies in the Orchidaceae.

In terms of strategy, orchid flowers have mainly been divided into two large categories: reward and deceit (though the distinction between them is not always clearcut). In general terms, rewarding flowers are those that provide a benefit for the pollinating insect in exchange for a service; whereas deceptive ones falsely advertise a prize but provide no benefit at all. In the next chapter, I discuss the different mechanisms of deception that orchid flowers are known to employ. This will be followed by a chapter on the different kinds of rewards on offer. Some orchids may use more than one mechanism from either category to attract and interact with their pollinators and thus ensure reproduction.

The exotic flowers of *Coryanthes horichiana*, the bucket orchid, lacks features typical of the melittophilous floral syndrome.
© APK.

The deceptive *Diuris punctata* growing as a terrestrial west of Longford, in Victoria, Australia © Ron Parsons.

CHAPTER 2

# DECEIT

*"Darwin's bafflement over the phenomenon he had glimpsed suggests that perhaps he had not fully realised the consequences of his own botanical revolution"*
JIM ENDERSBY (2016)

The concept of pollination by means of deception has always been — and continues to be — controversial. It was first proposed by Christian Konrad Sprengel and described in his 1793 book *Discovery of the Secret of Nature in the Structure and Fertilisation of Flowers*. Sprengel was the youngest of 15 children, born in Brandenburg, Prussia, in 1750. He studied theology and philology before being appointed director of the Great Lutheran Town School of Spandau, close to Berlin. After complaining of a serious irritation of his eyes, he was advised to spend as much time as possible in the open and to avoid indoor work. It was then, at the age of 30, that his interest in botany began. Sprengel became fascinated by the world of flowers, a passion that consumed him to the point that it conflicted with his duties as a school teacher and eventually cost him his job. A self-proclaimed philosophical botanist, in *Discovery of the Secret of Nature* he explained the construction, conformation and purpose of floral organs.

Sprengel's work was undervalued and during his lifetime his ideas did not receive the appraisal and support he had hoped for. Over the years, however, many of his tenets have been shown to be correct, and today we consider most of them self-evident. But at a time when people thought of flowers as merely casual objects of beauty, Sprengel's ideas were revolutionary. His proposal that flowering plants possess floral organs that are specifically designed to attract the insects that pollinate them was groundbreaking. Sprengel found that flowers produced — and carefully concealed — nectar with the sole purpose of luring pollinators. He noticed that flowers had 'nectar guides', special colour patterns and fragrances that were used to signal the visitors. He also theorised that by using those signals, some nectar-less flowers could cheat their pollinators through false advertising.

Long after Sprengel's death, praise from Charles Darwin would give his work the publicity among botanists that it had previously lacked. After all, many of the observations in *Discovery of the Secret of Nature* could only be explained through Darwin's theory of evolution by natural selection. Nevertheless, Darwin did not believe that orchid flowers could deceive insects, and certainly not bees, moths and butterflies, which he regarded as being highly intelligent. "I cannot swallow the belief of such roguery" he wrote in a letter to the botanist George Bentham. And in his personal copy of Sprengel's book, Darwin underlined the word *scheinsaftblumen* (sham nectar flowers) — in two places. And in blue crayon in the top margin of page 404, he scribbled "cannot be deceptive for insect has at once to fly to other flower". Despite his disbelief, many of the orchid flowers that Darwin inspected lacked apparent rewards. But they still required visiting insects to become pollinated. Could orchids somehow be 'outsmarting' their pollinators after all? Sprengel postulated that flowers could attract visitors using colour patterns and fragrances alone, and even though he lacked experimental proof at the time, his observations were later confirmed in experiments using insects. Darwin looked for other explanations. Again to Bentham he wrote: "Do you know whether nectar is ever secreted and reabsorbed promptly? — I am utterly puzzled: I have watched the nectaries till 11:30 at night and all as dry as a bone."

Darwin's bafflement over this phenomenon partly stemmed from something that is still very true today: we see plants as passive, stationary and unconscious elements of biodiversity, inferior to more advanced and intelligent living beings such as insects. But plants are not as altruistic and guileless as we believe them to be. Their benevolent reputation is a romantic human construct. Plants — like every other organism on Earth — are in a competitive struggle to survive and reproduce at any cost. For many of us, nature is an escape: it brings peace and serenity during stressful times. But we should not forget that amid the superficial tranquility, every plant is selfishly looking to thrive at the expense of other organisms. To do so, it will cheat and deceive. We now know that orchids typically exploit preexisting animal behaviours, and there is overwhelming evidence that deception is in fact a widespread strategy among them. Orchids are not the only deceivers among flowering plants. Species belonging to more than 30 plant families are known to have flowers that are pollinated through deception. However, failing to give pollinators an expected reward is far more common in orchids than in any other flowering plant family. Out of every ten deceiving flowers in the world, nine are orchids.

When we think about deception, we generally think of the flower promising pollinators a reward in the form of food, which ultimately is not provided. But there are other types of deception in orchids and many rely on mimicry. The phenomenon was first discovered by the British naturalist and explorer Henry Walter Bates in 1862, when he was studying butterflies in the Amazonian rainforest. Even though it is frequently associated with animals, floral mimicry is quite common and key to the deceptive pollination strategies in orchids. From flowers that chemically and visually mimic a pollinator's sexual partner, to those that imitate a place for pregnant females to lay their eggs, pollination strategies involving deception are extremely diverse. The most common involves flowers that use the shape, colour, scent and other visual or tactile guides to advertise the presence of food. In most cases, it has been determined that the orchid flower is not mimicking a particular rewarding flower, but instead relies on the use of general cues for the instinctive foraging behaviour of naive or exploring insects.

Deceptive pollination mechanisms in the Orchidaceae can be divided into two main classes: sexual deception and food deception. They have been further divided into different categories based on the nature and behaviour of the visiting animal. One of the most commonly used systems defines six broad categories: 1) Sexual deception, when flowers mimic the female insects' mating signals to fool males into attempting to copulate with them; 2) Rendezvous attraction, when pollination is ensured or enhanced by the presence of mating males and females on a flower or inflorescence; 3) Generalised food deception, which exploits pollinators' innate food-foraging behaviour through general floral signals; 4) Batesian floral mimicry, which is similar to generalised food deception except that the flower is mimicking a specific rewarding species; 5) Brood-site imitation, which refers to flowers that attract insects that are looking for a place to lay their eggs; and 6) Pseudoantagonism or territorial defence, when orchids exploit the territorial behaviour of certain animals, which attack flowers in the belief that they are intruders. Examples of all of these deceptive categories will be presented in the remainder of this book.

But let's begin by asking ourselves, are orchids truly deceptive? Can we prove that certain flowers are able to deceive their pollinators consistently? Darwin had his doubts, and there are still skeptics among enthusiasts and scientists alike. But whether this skepticism is due to a perceived lack of evidence, or stems from a deep-seated dogma, the fact remains: belief only constitutes knowledge if it is true. We will explore what is known in the pages ahead.

*"In many of the highly evolved orchids it seems as if the orchid often takes complete advantage of the insect."*
C.H. DODSON and G.P. FRYMIRE (1961)

*Bulbophyllum renipetalum*, a fly-pollinated flower lacking evident rewards.
© Kurt Metzger.

## 2.1 AIN'T TALKIN' 'BOUT LOVE

*'There is something definitively awe-inspiring in the pollination-history of the orchids when it is understood that to ensure their sexual success there has been developed subserviency to two of the dominating instincts of animals: the urge of hunger and the sexual impulse.'*
OAKES AMES (1937)

It is a calm, warm day in spring. The gentlemen in French-occupied Algeria eagerly wait for their ladies. Each is trying his best to stand out from the others. Suddenly, they perceive a lady in the distance. Her striking red hair and shiny blue ornament make her even more enticing. The anxious lovers incessantly dance around in seduction. They clash among themselves to get a chance with her. The most valiant gentleman prevails. The extent of his conquests is revealed in the opulence of the garments on his head.

What reads like a scene from a romance novel is in fact a slightly exaggerated reinterpretation of Maurice-Alexandre Pouyanne's first description, in 1917, of the pollination of *Ophrys* flowers through the mimicry of female insects — as communicated by Louis Henry Correvon to the National Horticultural Society of France.

Sexual deceit is one of the most fascinating and intriguing pollination strategies employed by orchid flowers. Like other sexual relationships, it too relies heavily on chemistry. As humans — contrary to perfume manufacturers' advertisements — none of us really expects to become more successful at finding a mate by spraying our bodies with aromatic products alone. However, certain odours can be overwhelmingly powerful for some animals, and on occasion may bring together the most unlikely of partners. Many orchid flowers rely on mimicking these odours to stimulate insects, and other animals, to visit and interact with them.

Sexually deceiving flowers dupe male insects into believing they are females of their own species, and pollen transfer occurs while the unsuspecting male tries to copulate with the flower. It's always the boys, and not the girls, that are fooled by sexual deception — did you really expect otherwise? They may seem gullible, but the strategy is highly precise and very convincing. It needs to be, because otherwise the plant would not be able to ensure that the same male makes the mistake at least once more, in order to deposit the pollen it picked up the first time. So the plants usually make sure they smell, look and feel like the real thing. Sexual deceit is not unique to orchids. There are a few other plant species, in the daisy and iris families, that also take advantage of this unusual floral specialisation. However, in no other plant family is sexual deception as commonly found and in such diversity as in orchids. It is thought that more than one thousand different species of orchids are sexually deceptive. The strategy has appeared independently in orchid groups from Africa, Asia, Australia, Europe and Central and South America.

The insects that are fooled by the orchids' fraudulent sex signals are likewise highly diverse. The sexually deceptive terrestrial orchids of Australia are mainly pollinated by solitary wasps, but at least one species is pollinated by male social ants. In Europe, species of the well-known terrestrial genus *Ophrys* are mostly pollinated by solitary bees, but pollination by predatory and parasitic wasps, as well as by beetles, has been documented in a few cases. In Central and South America, the epiphytic genera *Mormolyca* and *Trigonidium* are pollinated through the sexual deception of social bees, while the genera *Lepanthes* and *Telipogon* have been shown to sexually deceive flies. In each of these cases, reproductive success depends on the orchid's capacity to persuade a male insect that the flower it visits is a female. However, each deception is accomplished in unique ways.

The resemblance of some orchid flowers to insects — as perceived through human eyes — has long been a matter of intrigue. Today it seems unquestionable that the flowers of most species of the European orchid genus *Ophrys* have an insect-like look, but how that related to their pollination remained unclear until quite recently. In 1833, the botanist Robert Brown noted that bug-like orchid flowers always resembled insects native to the same country, but given that fertilisation in *Ophrys* "is frequently accomplished without the aid of insects", he speculated that "the remarkable forms of the flowers in this genus are intended to deter not to attract insects". Darwin did not agree with Brown. He was obsessed with *Ophrys* and their pollination, requesting help from his many correspondents and spending countless hours in the field meticulously observing and analysing them. He never witnessed *Ophrys* species being pollinated, and being unable to explain the flower's resemblance to an insect, considered it nothing more than "fanciful" —

a product of our imagination. "Who ever saw a Bee with violet wings like the petals of the Bee Ophrys?" the father of evolution by natural selection wrote in a letter in 1870. Even for the revolutionary naturalist, the idea of a flower taking the form of an animal to somehow fool it into a behavioural response was too far-fetched.

It was not until the early 20th century that Louis Henry Correvon, a prominent Swiss botanist with a special interest in Alpine plants, and Maurice-Alexandre Pouyanne, a scholar of the judicial system and amateur botanist based in Algeria, carefully explained how species of the genus *Ophrys* were fertilised through sexual mimicry. They revealed how the flowers combine chemical, visual and tactile cues, to get the pollinating insect to agitate its abdomen against the orchid's lip in an attempt to copulate. It took Pouyanne over 20 years of diligent observation and experimentation in Algeria, but after extensively documenting male wasps visiting the flowers of *Ophrys* species, he was able to establish conclusively that they were pollinated through pseudocopulation.

Pouyanne's studies focused on the native *Ophrys lutea*, *O. fusca* and, especially, *O. speculum* (see Figure 2.1.1). His findings were published with Correvon, who highlighted the extreme difficulty of carrying out such pollination studies, pointing out that they require not only a high density of plants and flowers for study, but also great patience and perseverance. *Ophrys* species — the authors note — are pollinated by male hymenopterans that emerge from the soil a month or so before the female. It is during this window of time, between the emergence of the male and the emergence of the female, that the *Ophrys* get their chance to be pollinated. As the male insects incessantly patrol and hover over the areas where the females are to emerge, they rush onto the flowers. There is a precise concordance between the emergence of the males and the flowering period of *Ophrys*, both being retarded or even halted by adverse climate conditions.

The flowers are mostly visited by their pollinators at the beginning of their flowering period. *Ophrys speculum* is pollinated by the male of the species *Dasyscolia ciliata* (Scoliidae). Upon detecting the flowers, the males desperately try to copulate with them. Up to four individuals may aggressively engage each other for a spot on the lip of a single flower. The successful male aligns itself longitudinally with the lip and places its head between the base of the lip and column — where it

**Fig. 2.1.1.** The sexually deceptive *Ophrys* species first studied by Pouyanne. **A.** *Ophrys speculum*, Mallorca, Spain. **B.** *Ophrys lutea*, Sicily, Italy. **C.** *Ophrys fusca*, Rhodes.
© Ron Parsons.

contacts the pollinarium. The abdomen is inserted between the long hairs on the lip and once in position, it will then tremble, making the convulsive movements typical of copulation behaviour. The male is completely absorbed by this, and is undisturbed by the researcher's close inspection. During the copulatory movements, the male wasp gets the pollinaria stuck onto its head. Insects carrying these are easily spotted due to the relatively large size and bright colours of the pollen sacs.

Pouyanne found no evidence of nectar or any other rewards on the flowers, and given their general morphological resemblance to the insects, he suspected that the males believe them to be females. To prove his hypothesis, he performed a series of experiments. First, he offered the *Dasyscolia* males diverse floral bouquets, including those of other *Ophrys* species, and noticed that they did not react. This led him to conclude that there was a certain level of specificity between *O. speculum* and *D. ciliata*. He also offered female *Dasyscolia* the *O. speculum* bouquet, and to his astonishment, not only were they not attracted by the flowers, they even seemed disturbed.

Pouyanne then shaved the flower's long hairs and noticed that the male insects would take a bit longer, but still visit the flower. Then he removed the lip, after which the amputated flower was no longer of any interest to the wasps. Instead, he noted, they would visit the lip on its own, even when it was turned upside down or cut into pieces. Finally, even when he covered a bouquet of 10 flowers with newspaper, the *Dasyscolia* males would inspect them and attempt to copulate. From this, Pouyanne concluded that, although imperceptible to the human nose, the lip of *O. speculum* produced an odour that incited the males to copulate, even when the visual and tactile stimuli were absent. The insect-like appearance and touch probably played a lesser role than the strong chemical attraction.

His hypotheses were met with scepticism, but further research has consistently proven that male insects do indeed attempt to copulate with *Ophrys* flowers in the apparent belief that they are females of their own species. What Pouyanne had discovered — and meticulously described — is a unique pollination mechanism. In 1982, Georges Pasteur classified this type of mimicry as 'Pouyannian mimicry', in honour of its discoverer. Today it is mostly referred to as 'pseudocopulation'.

*Ophrys* species are now known to be pollinated almost exclusively by male bees or wasps. The male insects are lured to the orchid's flower through chemical, visual and tactile signals, and it is those cues that elicit the sexual behaviour. *Ophrys* flowers not only look like a female insect, they feel, and especially smell, like one too. The orchid produces abundant and complex volatile compounds, some of which are detectable by the human nose, as part of a scent bouquet that attracts its pollinator. By physically hiding the flowers from the insects, researchers have now established beyond doubt that the main attractant is indeed chemical. Studies on the chemistry of those compounds show that similar volatile chemicals are produced by the female insects. Using scented dummies, the substances have been shown to trigger the approach of patrolling males in a similar manner to when they approach virgin* females. The odour from the flowers also triggers copulation attempts. Its composition has been found to be specific enough that minor differences will attract males of different species, allowing for closely related species of *Ophrys* to be consistently pollinated by different insects. *Ophrys sphegodes*, a species native to Europe and the Middle East, is known to be pollinated through pseudocopulation by males of the bee *Andrena nigroaenea* (see Figure 2.1.2). Using a technique called electroantennographic detection — where a living bee's response to an odour is read using electrodes connected to its antennae — researchers showed that the receptors in the antennae of *A. nigroaenea* males responded to several compounds shared by the fragrance of this particular *Ophrys* species and the cuticles of virgin female *A. nigroaenea*. In the field, a bouquet of compounds extracted from the lip of *O. sphegodes*, presented to flying males of *A. nigroaenea* using dummies, elicited copulation behaviour.

Further behavioural tests have confirmed that the relationship between the bee orchid and its pollinator is highly specific, with each *Ophrys* species attracting one or a few pollinator species. Different *Ophrys* species offer a particular aromatic blend that mimics specific female bees and wasps. But the mimicry goes further than chemistry. The flowers of some *Ophrys* species also imitate the shape and colouration of the female insects. One of the more striking cases is that of *Ophrys speculum*, the mirror orchid, which mimics the blue

---

\* It is relevant that they are virgins because female bees modify their scent after mating, releasing anti-aphrodisiac compounds that inhibit the male's mating behaviour.

Fig. 2.1.2. *Ophrys* species pollinated by male *Andrena* bees through pseudocopulation. **A.** *Ophrys sphegodes*, showing its insectiform flower. **B.** *Ophrys splendida*, with an *Andrena* male removing pollinaria with its head. **C.** *Ophrys lojaconoi*, pollinated by *Andrena* sp. in reverse.
© Ron Parsons (A), Jean Claessens (B), Matteo Perilli (C).

shiny wings — including the way they reflect ultraviolet light — the reddish brown body hairs and the middle and hind legs of the female scoliid wasp *Dasyscolia ciliata*. Tactile stimuli come into action once a male has landed on a female, or a flower mimicking the female. Competition for females is fierce, so the males must be quick. Prior to copulation, males orient themselves on the female body, or flower, using hairs as guides. During pseudocopulatory events, the male bee touches and removes the pollinarium, either with his head or the tip of his abdomen.

As if being fooled into copulating with the wrong organism isn't embarrassing enough, male insects occasionally prefer the orchids over the real females. In some *Ophrys* species, the flowers mimic female insects imperfectly rather than copying their exact scent. Behavioural field tests show that male bees remember the odour bouquets of individual flowers during mating attempts, just as they do when interacting with actual females. This allows them to recognise these individuals in later encounters and, in this way, they avoid mating with the same one again. The plant takes advantage of the learning ability of the bee by varying the relative proportions of the aromatic compounds. Slight variations between the flowers in a single inflorescence, and larger variations among flowers of different individuals in a population of orchids, stimulate the males to search continuously for new mates. In addition, the male insect is adapted to prefer mating signals that are slightly different from those of its female siblings, in order to avoid inbreeding. This results in males having a tendency to prefer the imperfect smell of the orchid over the more commonly encountered signals produced by the females of their own species. When presented with the choice of real females and the orchid flowers, therefore, the male will choose the orchid.

This chemical, tactile and visual variation is thought to be one of the main drivers for the high speciation and diversification in *Ophrys*. Some authors have gone so far as to suggest it may represent a true case of coevolution — where the plant not only relies on a specific pollinating insect, but also influences the spatial dynamics of the insect population.

When it comes to the sexually deceptive flowers of *Ophrys*, we sure ain't talkin' 'bout love.

**AIN'T TALKIN' 'BOUT LOVE**
*Ophrys* pollination through sexual deception

https://youtu.be/6fZJ-o3sytA

The insect-like appearance of the flowers of the European orchid genus *Ophrys* has intrigued naturalists since before Darwin. The reason for this resemblance remained a mystery until the 20th century, when it was revealed that the flowers of *Ophrys* were pollinated through copulation attempts by male bees and wasps that mistake them for a female of their own species. The flower not only looks and feels like a female bee or wasp, but it smells just like one too. The sexual pheromones that the orchid flower mimics are so convincing that the insects are unable to distinguish them from those of a real mate. Males fight for a place to mate on the *Ophrys*, not realising that it's a trick to get them to transfer the orchid's pollen.

The videos and photographs shown here are owned by Matteo Perilli (https://www.youtube.com/channel/UCpr4ef82Gl13KgUMKRNZ1CQ) and have been reproduced and edited by the author with permission.

## 2.2  FIRST DATE

*"It may be that those who would reject the evolutionary approach to an understanding of life and who prefer to regard the world as the product of Special Creation will lean a little more lightly on human weakness when they discover moral turpitude among the insects. And it may be that entomologists, who see for insect societies parallels in human institutions, will become Freudian in their outlook when discussing the sexual vagaries revealed by symbiotic phenomena and introduce such terms as Lissopimplan behaviour or Ophrydean complex. Perhaps even the poet will have to reconsider whether 'Only man is vile'"*

OAKES AMES (1937)

Charles Darwin was not impressed with Australia. To the continued dismay of many local biologists, he described it as rather dull. Local naturalists have of course done their best to prove Darwin wrong, calling our attention to the numerous biological wonders that are unique to the continent 'Down Under'. One of the most fascinating interactions between Australian plants and animals is the sexually deceptive pollination strategies of orchids in the Diurideae tribe. Had Darwin known about the "moral turpitude among the insects" that pollinate these orchids, he'd surely been blown away.

There are several sexual deceivers among Australian members of the Diurideae, but the most famous are probably the species of genus *Drakaea* — popularly known as hammer orchids. There are ten or so currently known species of hammer orchid, and all of them rely on sexual deception of male thynnine wasps for pollination. Each species of these orchids is unique in the shape, size and colouration of their flowers and it has now been established that there is high specificity in their pollination. Every hammer orchid species seems to be pollinated by a specific species of thynnid wasp in the genus *Zaspilothynnus*. Although hybrids between different species of these orchids do occur on occasion, the usually high specificity has helped to reveal that there are more species of *Drakaea* than was previously thought. Like the sexually deceptive bee orchids of genus *Ophrys*, the flowers of *Drakaea* mimic the female sex pheromones that are sensed by patrolling males. But there is a striking difference between the two strategies. The wingless female thynnine wasp — which the *Drakaea* flower mimics — typically perches on a blade of grass where she is swept off her feet by the mate to partake in a romantic midair consummation. As a tiny plant with an unassuming inflorescence, this orchid has perfectly disguised itself as a greenish reed with a female wasp perched on top (see Figure 2.2.1). Unlike the static flower of the bee orchid, the lip of the hammer orchid bears a wasp-like decoy, delicately hinged to the rest of the flower. As the male wasp lands on the flower, it briefly attempts to copulate before grasping the decoy and trying to fly away. The eager male wasp's takeoff is impeded by the lip's hinge, however, and instead it forcefully bangs away repeatedly against the column of the orchid. It is during this struggle that the orchid's pollinia are transferred. How did such an odd system come about?

**Fig. 2.2.1.** The pollination mechanism of the hammer orchid. **A.** *Drakaea livida*, ready to receive the male thynnid wasp *Zaspilothynnus nigripes*. **B.** The male *Z. nigripes* grabs the lip of *D. livida* and tries to fly away. **C.** Its flight is impeded by the hinged orchid's lip and the wasp is banged against the column. **D.** A female *Zaspilothynnus* perching on a blade of grass, next to its mimic *Drakaea glyptodon*. © Rod Peakall.

Thynnine wasps are among the most common orchid pollinators in Australia. They are parasitic wasps that lay their eggs in burrowing beetle larvae. The wasps emerge from the ground after hatching from their cocoons in spring. The females, which are adapted to burrowing in the soil to hunt the beetle larvae, are wingless and much smaller than the males. Thynnine wasps have a unique mating behaviour that involves males transporting the wingless females and copulating with them mid-air, then offering them gifts in the form of regurgitated food. This feeding has been interpreted as a form of paternal investment in the offspring.

After hatching underground, the males dig themselves out and fly away. On sunny mornings, they fly close to the ground in search of receptive females. After hatching, by contrast, the females construct underground exit tunnels to the surface, where they stick their heads out of the ground — unless they are blocked by tall grass or shrubs. If that is the case, they crawl up onto bushes or grass stems to heights ranging from 5 cm to 2 m. There, they perch and wait to be literally swept off their feet. The posture varies according to the species: they can be perched head-up or head-down, and may be perfectly straight or with the abdomen bent at a particular angle. Receptive females release a pheromone that attracts passing males of the same species, which fly in a weaving, zigzag pattern, before alighting to search for the motionless female. When a male comes into close contact with a female, he places his antennae on her head and is then permitted to climb onto her back and fly away with her. The copulation starts shortly before or immediately after the male becomes airborne. The males always fly away from the initial pickup point, presumably fleeing from other possible suitors that may also have located the female. Fighting for possession of the female does occur. Having copulated, the female eventually releases her grip and falls to the ground. After remaining motionless for a few seconds, she rapidly walks off and begins to burrow into the soil to look for new beetle larvae in which to lay her eggs. And the whole cycle starts again.

The mimicry of the *Drakaea* orchid is highly effective — and it needs to be if it wants to keep up with the wasp. Field experiments show that hammer orchid flowers placed in areas patrolled by males are visited within the first minute by up to seven wasps, but then the interest declines. The excess of males in relation to receptive females suggests that there is intense reproductive competition. This explains the rapid response. But first come is not always first served. The wasps are selective in their mate choice and so the orchid's deceit has to be very precise. Both males and females may be subject to rejection by their mates. The orchid's sex signals need to be highly stimulating to the males for pollination to be successful, and at the same time, the wasp's stimulus threshold needs to be relatively low. A very picky male could learn to avoid the orchid, but may miss out on the real female as a result.

Wasp couples remain together for about an hour, and the males are able to mate several times in the course of a day. Visiting the orchid does not prevent the males from successfully mating, and they have been observed with real females after being fooled by the flowers. This suggests that the failed copulation attempt does not impact the reproductive success of the individual wasp significantly — especially given that the male does not waste sperm on the flower.

The successful mimicry in *Drakaea* requires both visual and chemical stimuli that originate from the same point. Curiously, despite the continuation of the stimulus, male wasps don't revisit the orchid flower. Wasps are able to learn precise locations and they may keep away from females (and flowers) that are in a location that they have already visited in order to avoid wasting their time. Some males are even able to recognise the odour of a female with which they have already mated. At any rate, after visiting a flower the male leaves the area immediately, which leads to increased pollen flow and may ensure genetic diversity in the orchids.

Hammer orchids are not the only sexually deceptive orchids in Australia. In fact, sexual deceit is a well-known feature of species belonging to the Diurideae tribe, a group that comprises exclusively terrestrial orchids with a distinct peak of diversity in Australia. Sexual deception in the Diurideae evolved several times independently from ancestors that used food deception to lure pollinators. Most of these sexually deceptive orchid species seem to have a highly specific relationship with males of different hymenopteran families, and require only so-called pre-copulatory behaviour to attain pollination. That is, the insect removes and subsequently deposits the pollinia while preparing to copulate, rather than during attempted copulation itself. Members of the genus *Chiloglottis* (see Figure 1.3.1C) harm the reproductive success of their pollinators by making it harder for the female wasp that it mimics to attract mates. The flowers have been found to produce compounds, known as chiloglottones, as pheromone mimics. After their initial visit, the males learn to avoid areas where the orchids are present, and thus neglect the females in those areas. There is strong evolutionary selection pressure on the insects to learn to avoid the orchid, and on the orchids to perfect their mimicry in order to continue to fool the insects.

One of the most remarkable aspects of the sexually deceptive orchids of Australia is that some 'sympatric' species (which occupy the same geographical area without interbreeding) have independently evolved to be pollinated by the same insect species. Species of *Drakaea* and *Caladenia* (Figure 2.2.2), which are members of Drakaeinae and Caladeniinae subtribes, respectively, not only share the same sexually deceptive pollination strategy but also the same pollinator species. Incredibly, this convergent evolution between these orchids boils down to having developed the very same biosynthetic pathways to produce the chemicals necessary to attract the pollinators.

Even more fascinating is the fact that sexually deceptive orchids can stimulate their pollinators to such a degree that the males ejaculate on them. Yes, you've read that right. This is the case for flowers of the genus *Cryptostylis* — commonly known as tongue orchids. In 1929, the Australian naturalist Edith Coleman described how her daughter had observed males of *Lissopimpla excelsa*, an ichneumonid wasp, visiting and pollinating the flowers of the tongue orchid under the impression it was a female wasp (Figure 2.2.3). The male insects backed into the flower, clasping a fold at the base of the lip with their abdomen, ejecting seminal fluid, and carrying off the pollinaria on the tip of their abdomen. Sperm wastage in these orchids is not a rare event. It occurs in about 70% of all interactions between the male *Lissopimpla* wasps and the *Cryptostylis* flowers. Typically, however, sperm wastage is rare in sexually deceiving orchids. This makes sense given that it may impose a high cost on the pollinator. One critical factor affecting male and female reproductive fitness is sperm availability, so sexual deceit that causes ejaculation can reduce female mating success. In addition, sexual deception can make it more difficult for a pollinator to find females among the orchid's false signals.

The evolutionary biologists among you may be wondering how such a pollination scheme — in which an insect loses not only precious time

**Fig. 2.2.2.** Sexually deceptive species of *Caladenia* with their male bee and wasp pollinators. **A.** Arrowsmith spider orchid, *Caladenia crebra*. **B.** Boranup spider orchid, *C. ambusta*. **C.** The white spider orchid, *C. longicauda*. **D–E.** Cherry spider orchid, *C. gardneri*.
© Rod Peakall (A), Jeremy Storey (B-E).

**Fig. 2.2.3.** The sexually deceptive Australian orchid *Cryptostylis* causes its male wasp pollinator to ejaculate on the flower. **A–B.** Male *Lissopimpla* pseudocopulate with the flowers of *C. ovata*, removing pollinaria. **C–D.** Pollinaria are transferred a male *Lissopimpla excelsa* attempts to copulate with *C. erecta*.
© Mark Brundrett (A–B), Rod Peakall (C-D).

and energy, but also sperm — could persist over time. Pollinators that prefer orchids over real females or are unable to find a real female on account of the orchid would tend to die out. Conversely, insects that learn to avoid the orchid would have higher chances of reproducing and transmitting those advantageous, orchid-avoiding genes to their progeny. It has now been well established that sexually deceptive orchids rely on insects that have a fitness advantage over more discriminating males in responding promptly to female signals. This is because if a male is too picky, it may not be able to get to the real female in time either. Nevertheless, pollinators do learn to avoid sexually deceptive orchids, which in turn puts strong selective pressure on the orchids to be increasingly persuasive in their mimicry. This causes an arms race in the reproductive success of the orchid and its pollinator.

But how could the fascinating pollination strategy of *Cryptostylis* species be so successful and persistent, given the sperm wastage of the pollinator? Australian researchers have proposed an intriguing explanation. They found that the pollinating wasps of sexually deceptive orchids are frequently 'haplodiploid'. What this means is that the females can still bear offspring from unfertilised eggs. In haplodiploid insects, population dynamics depend on female sperm management. Unfertilised eggs become male offspring (they are 'haploid', with only one set of chromosomes), while fertilised eggs become female offspring (which are 'diploid', with two sets of chromosomes, one from each parent). Therefore, despite the waste of sperm in the sexually deceptive orchid, the pollinator species will still be able to reproduce. Haplodiploidy also implies that selection imposed on male insects by orchids cannot be inherited directly by their sons, because any eggs that are fertilised by a male's sperm always become daughters. This means that males that learn to avoid the orchid are not likely to pass this trait on to the next generation of males. In fact, the orchid could promote the production of naive males that would act as pollinators during subsequent flowering periods. At the same time, they reduce the number of females and therefore increase competition among males for those females. This means that the males are also more likely to compete for access to the orchid's flowers. Incredibly, *Cryptostylis* species seem to maintain their pollination success precisely because they interfere with the pollinator's reproduction by generating additional males (see Figure 2.2.3).

**FIRST DATE**
The sexually deceptive Australian orchids

**SCAN ME**
https://youtu.be/GMZ-EujMIhc

Shortly after pollination through sexual deception was confirmed in the European genus *Ophrys*, a similar strategy was discovered in terrestrial orchids from Australia. Species of the genus *Drakaea*, known popularly as hammer orchids, are among the most famous sexual deceivers Down Under. What makes the hammer orchid extraordinary is its unusual flower, which mimics a female thynnid wasp perching on a blade of grass. Males typically search for the females and locate them using pheromones. Upon finding a perched female, they grab her and try to fly off with her. The insect-like lip of the *Drakaea* flower is delicately hinged to the flower and as the wasp tries to fly away with it and copulate, the wasp is thrown back towards the flower, banging against the column and transferring the orchid's pollinarium. The closely related genera *Caladenia* and *Chiloglottis* also mimic female insects and sometimes share the same pollinator with *Drakaea*. However, their flowers look quite different, and it is the weight of the male insects that tilts the lip in such a way that it is pressed against the column. It removes the pollinarium as it exits.

The videos and photographs shown here are owned by Rod Peakall and have been reproduced and edited by the author with permission.

## 2.3 YOUR LATEST TRICK

Before the 1960s, sexual deception was only known to occur in a handful of terrestrial orchids from Europe and Australia, and was mainly associated with solitary bees and wasps. This changed radically when botanists reported sexual deception in neotropical epiphytes involving two groups of insects not previously associated with sexual deceit: social bees and flies. The new ideas on sexual deception of epiphytic orchids from the New World were consolidated in 1966 with the publication of *Orchid Flowers: Their Pollination and Evolution*. Sexual deceit was later reported to occur in the genus *Trichoceros* and its allies in subtribe Oncidiinae, and in the genus *Trigonidium* and a few of its relatives in subtribe Maxillariinae. However, reports of the pollination of sexually deceptive neotropical orchids remained mostly anecdotal, and the belief that sexual deception occurred mainly in Australian and European orchid species, through male solitary bees and wasps, lingered well into the beginning of the 21st century.

Members of the highly diverse and species-rich genus *Telipogon* are among the most visually striking neotropical orchids. Given their close relationship and overall similarity to *Trichoceros*, and their insect-resembling flowers, botanists have assumed that *Telipogon* is also pollinated through sexual deception. *Telipogon* flowers frequently have very large, spreading, bright yellow petals which are intricately adorned with reddish bands and stripes. In sharp contrast, the centre of the flower is dark and hairy. The result is a flower that looks something like a large fly stuck in the middle of a bright spider's web (see Figure 2.3.1). This notorious resemblance to a true fly — in our eyes — has fuelled the belief that *Telipogon* flowers are sexually deceptive. The idea was that the black hairy bulge in the centre mimics a female tachinid fly and that pollination occurs when male tachinids attempt to copulate with it. But until recently, this was unproven.

A field study in Peru by Carlos Martel and his colleagues showed that the black and hairy tachinid flies are indeed visually and chemically attracted to the flowers of *Telipogon*. The flowers — which appear scentless to our noses — produce compounds such as alkanes and alkenes, which may have a role as sex pheromones. But the visual and tactile cues also play an important role in this pollination strategy. When a male tachinid locates a receptive female resting on a flower or leaf, it first reaches for her abdomen and thorax, and then, while clenching them, tries to make the female drop to the ground. If the male succeeds, it positions itself behind the female in order to copulate. This was exactly the kind of foreplay observed when the tachinid male approached *Telipogon* flowers. The orchid has 'figured out' the flies' mating behaviour and deceives the male by looking, and probably smelling, like a receptive female ready for action. The male fly tries to fly off with the orchid's flower, but is unsuccessful and has to find a new lover. From the tachinid's perspective, it has all been in vain. But, it is during this pre-copulatory struggle that the *Telipogon* pollinia are transferred to the male insect (see Figure 2.3.2).

**Fig. 2.3.1.** The insect-like floral design of *Telipogon costaricensis*. © APK.

*Telipogon* is a species-rich genus with an incredible floral diversity. They are endemic to the mid- to high-elevation mountain ranges of Central and South America, where many await discovery. Given how little we currently know about their ecological interactions, there surely remains much to be said about their pollination. An exciting follow-up study by the same group of researchers suggests that *Telipogon* orchids have one more trick up their sleeves. It turns out that their very large, bright yellow petals may not only help to highlight the hairy black centre, but the petals themselves may mimic the bright yellow daisy flowers on which the insect perches. The flies are known to feed on the daisies, and the high contrast between the daisy inflorescence and the female allows the males to locate them more easily. So it seems that *Telipogon* flowers mimic not only the female insect but also one of its favourite places to eat.

## YOUR LATEST TRICK
*Telipogon* pollination

SCAN ME
https://youtu.be/c6tNO4UZ3bQ

Flowers of *Telipogon* are among the most visually striking of all neotropical orchids. Most *Telipogon* species have yellow flowers with intricate brownish-purple markings and a dark, hairy centre. Studies on their pollination show that male tachinid flies mistake the *Telipogon* flowers for a perching female. In an attempt to copulate with the perceived female, as a pre-mating behaviour the male tachinid approaches the flower and tries to grab and carry off his mate. Unable to do so, the fly leaves, but not without first getting the orchid's pollinarium attached to the side of his body. Ongoing studies in Peru suggest that some species of *Telipogon* may in fact elicit pseudocopulation of the tachinid male as part of their pollination strategy.

The videos and photographs shown here are owned by Manfred Ayasse, Heiko Bellman and Adam Karremans and have been reproduced and edited by the author with permission. Materials published in Martel *et al.* 2016 (https://doi.org/10.1371/journal.pone.0165896) are open access and have been reproduced with the kind permission of the authors.

Fig. 2.3.2. The sexually deceptive neotropical genus *Telipogon* is pollinated by tachinid flies. **A.** A male visits *T. peruvianus*, trying to take away what it believes to be a perched female. **B.** Male *Eudejeania* species (Tachinidae) carrying a pollinarium of *T. peruvianus*.
© Manfred Ayasse (A), Heiko Bellman (B).

## 2.4 PERFECT STRANGERS

In 1999, while Mario Blanco was working in the pleurothallid greenhouse at Lankester Botanical Garden in Costa Rica, Gabriel Barboza — an experienced nature guide — wandered in with a group of tourists. As the visitors enjoyed their exclusive glimpse of orchids grown 'behind-the-scenes' at the garden, Gabriel talked to Mario about the pollinators he had witnessed regularly visiting orchids in his private garden. "You must come to Monteverde, so you can see the pollinators of *Lepanthes* in action," he said casually. Mario could not quite believe what he was hearing. The pollination of *Lepanthes* — his favourite orchid genus — was a mystery to science. Several researchers had unsuccessfully spent numerous hours in the field hoping to see the magic happen. Mario scrambled to gather his equipment and a few days later he was seated on a bus to Monteverde. The next morning — after an obligatory cup of coffee — the two Costa Ricans carefully observed a tiny fungus gnat in the Sciaridae family visit and inspect the flowers of a native *Lepanthes* species. Before their eyes, the gnat then attempted to copulate with the flower. As they patiently sat there for hours with their camera equipment and notebooks, they became the first scientists to document pollination by pseudocopulation in *Lepanthes*.

Sexual deceit has now been observed in about a dozen *Lepanthes* species, and is likely to be widespread in the genus, given the similarity in the floral morphology of its members (see Figure 2.4.1). With its more than 1,200 species, found throughout the Neotropics, this would make *Lepanthes* the most species-rich, sexually deceptive genus of plants in the world. If confirmed, this would dramatically shift the received wisdom that sexual deception is better represented among temperate terrestrials pollinated by bees and wasps, towards it being more common in tropical epiphytes pollinated by flies.

Interestingly, and somewhat surprisingly, none of the *Lepanthes* species look like the flies that pollinate them. Unlike the sexually deceptive orchids discussed in previous stories, *Lepanthes* flowers do not even look insect-like. The fungus gnats that pollinate them are blackish, mosquito-like insects, with long legs and a slender abdomen, whereas the flowers are typically bright red, orange or yellow and not particularly slender. To attempt copulation, the male fungus gnat grabs the lateral petals from the front and locks its genitalia to the 'appendix' on the lip

**Fig. 2.4.1.** *Lepanthes* are pollinated by fungus gnats (family Sciaridae) attempting to copulate with the flowers. **A.** Sciaridae male and female mating on a leaf. **B.** Fungus gnats visit *Lepanthes* species to pseudocopulate by locking their abdomen onto the flower's appendix. **C.** A male fly locks its genitals onto the flower of *L. sabinadaleyana*. © Brian Valentine (A), Nicolás Gutiérrez (B), Sebastián Vieira Uribe (C).

of the *Lepanthes* flower. The appendix is a narrow, central lobe of the lip that most *Lepanthes* species have. Each one differs slightly in length and ornamentation, but the majority are not even half a millimetre long, which allows the minute fungus gnat to grasp the flower. However, a few species, such as *Lepanthes telipogoniflora*, have an appendix that can reach 3 mm long. Besides the lip lobes, no other floral structure resembles any insect body parts. It is this central area of the flower that has been shown to be biochemically active. Given that sciarids mate by clasping their abdomens while facing away from each other, visual and tactile mimicry may be less important than chemical mimicry in this instance.

In the Neotropics, especially at mid and high elevations, *Lepanthes* species are relatively common and diverse. Anyone who has gone looking for them has probably noticed that some species are abundant when present, and they are typically found in bloom. The inflorescence of *Lepanthes* bears dozens of small flowers that open one at a time in very slow succession. Unlike many other orchids, the inflorescence of *Lepanthes* also keeps producing new branches over time and thus seems to be endlessly producing floral buds. As a result, large *Lepanthes* plants give the impression of flowering continuously. An ongoing experiment, using living plants cultivated at Lankester Botanical Garden in Costa Rica, shows that individual *Lepanthes* plants may bloom for up to 60 straight weeks or 14 months under greenhouse conditions. In the 365 days between 1 June 2021 and 1 June 2022, three individual plants have had at least one open flower for 349 to 362 days. Meaning these *Lepanthes* plants had a flower available 95 to 99% of the time during this one year period, and have continued blooming.

Despite their ubiquitous presence and prolific nature, pollination seems to be extremely rare. Curiously, therefore, certain *Lepanthes* species are frequently observed with fruits, despite there being no indication that they are self-pollinating. This led my colleague Diego Bogarín to investigate more carefully, and what he discovered is truly remarkable. Diego found that most of the so-called fruits of non-selfing species of *Lepanthes* are actually not fruits at all. They are galls. A plant gall is an abnormal growth in the external tissue which is caused by a parasite. Certain wasps commonly induce plant galls for the development of their larvae. The gall protects the wasp's larva during the most vulnerable stages of its life cycle, and after feeding inside the bulge it eventually emerges as an adult. In the case of some *Lepanthes* species, it seems that the wasps induce the unfertilised ovary of the orchid to develop into fruit-like galls. In one population in Costa Rica, every single fruit inspected was in fact a gall. They all contained a parasite and none contained seeds. This supports the observation that pollination events are rare.

The scarcity of fruits, and a high degree of specificity in their host trees, are said to be among the main causes of rarity and high risk of extinction among *Lepanthes* species. One may then wonder how *Lepanthes* became so widespread and diverse in the first place. Has pollination by sexual deception been key in the spectacular radiation of *Lepanthes* species? Interestingly, studies in Puerto Rico suggest that low gene-flow and the low number of breeding adults may actually be one of the main drivers of speciation in the genus, through an evolutionary process known as genetic drift. The rarity of pollination events, and therefore the limited gene-flow through populations, may play a part in causing genetic isolation and the speciation that follows.

But how are *Lepanthes* species able to maintain the sexually deceptive mechanism? One explanation may be 'paternal genome elimination', which is a very odd feature of Sciaridae flies. In flies with this characteristic, the males only transmit chromosomes that they inherit maternally, while chromosomes inherited from the father are eliminated. Perhaps in a similar way to the haplodiploid wasps discussed earlier, male fungus gnats that learn to avoid *Lepanthes* flowers do not transmit the genes that make this possible to their offspring. This remains to be studied in the future.

*Lepanthes* species are not the only sexually deceptive epiphytic orchids that do not resemble insects. Contrary to what we may expect, several others have flowers that are not insect-like — at least to our mammalian eyes. Flowers of the genera *Mormolyca* and *Trigonidium*, for example, employ sexual deception as part of their pollination strategy and look nothing like insects (see Figure 2.4.2). *Trigonidium* also differs from most sexual deceivers in that the removal of pollen does not occur during the copulation attempt. The sepals and petals of this orchid form an upward-pointing tube, and the bees that pollinate them are attracted to the edges of that tube. They expose their genitalia and contract their abdomens, repeatedly attempting to copulate with the edges of the floral tube. It is during this endeavour that they fall into and are briefly trapped in the slippery tube. As the bees try to exit the tube, they find their way to the base, where they encounter and remove

Fig. 2.4.2. Not all neotropical sexually deceptive orchids are insect-like. **A.** A *Mormolyca ringens* flower showcasing its odd lip. **B.** Tubular flower of *Trigonidium egertonianum* with its alien-like blueish petals. **C.** A male stingless *Scaptotrigona* bee attempts to copulate with the lip of *M. ringens*. **D.** A male *Plebeia* bee emerges from the tubular flowers of *T. obtusum* carrying a pollinarium.
© APK (A-B), Rodrigo Singer (C–D).

the pollen. To avoid self-pollination, the flower's pollinia are initially too large for the stigmatic cavity, and it is only after dehydration that they will fit the stigma. By then, the bee has already left its disappointing lover in search of another *Trigonidium* flower. After slipping into the tube during a posterior copulation attempt, the bee finds itself at the base once again where the dehydrated pollinarium can now be deposited in the stigmatic cavity, successfully pollinating the flower.

Sexual deceivers that do not resemble insects are also found outside the Neotropics. The pollination of two sister species belonging to the terrestrial genus *Disa* has also been suggested to be through sexual deceit. Both are exclusively found growing in the Cape region of South Africa, where their flowering is strongly stimulated by fire*. Like those of *Lepanthes* and *Trigonidium*, the flowers of *Disa* do not resemble their pollinators morphologically (see Figure 4.2.3). Rather, based on studies of their light absorbance and reflectance, the whole inflorescence may act as a single attractive unit that optically stimulates males that are searching for mates. Researchers found that patrolling male wasps visited the flowers repeatedly despite the lack of any rewards. The two orchid species are able to coexist despite their similar reproductive strategies because each is pollinated by a different male wasp species.

Several other orchids are presumed to be sexually deceptive because they have insect-like features. Indeed, it is likely that additional sexual deceivers will be discovered in the future, especially among poorly studied tropical orchids. Unfortunately, it is difficult to assess the pollination mechanism without actually observing how the insect interacts with the flowers. It seems that the most important element in the female mimicry of sexually deceptive flowers is neither visual nor tactile, but rather chemical. The effect of sexual pheromones may be so overwhelming for the insect that these orchids need little more than the right scent to secure a 'mate' among perfect strangers.

Sexual deceit is certainly one of the most fascinating orchid pollination strategies, with much still to be learned. There are, of course, other types of deception in the orchid family and we will now turn our attention to those that involve food.

---

\* The largest number of individuals of these orchids are found flowering on sites that were burnt the previous summer or autumn.

**PERFECT STRANGERS**
Sexual deception in non-insect-like orchids

SCAN ME
https://youtu.be/qiyZOmq6aTY

Not all sexually deceptive orchid flowers look like female insects. When sexual deception by male fungus gnats was discovered to occur in the neotropical genus *Lepanthes*, it put many of our preconceived notions about this form of pollination to the test. Pollination through sexual deceit had been mostly known to occur in terrestrial orchids from Europe and Australia and to involve bees and wasps. With the discovery of the strategy in genus *Lepanthes*, neotropical epiphytes suddenly became an important player in the sexual deception arena. But the most important paradigm change was the realisation that flowers don't need to look much like female insects as long as they smell like them. Sexual deception was later also confirmed to occur in other neotropical epiphyte orchid genera, such as *Trigonidium* and *Mormolyca*, extending the limits of this extraordinary pollination syndrome even further.

The videos and photographs shown here are owned by Álvaro Salazar Méndez (Jardín de Orquídeas de Monteverde), Sebastián Vieira Uribe (https://www.flickr.com/photos/kligo/) and Rodrigo Singer (https://www.youtube.com/@rbsinger70) and have been reproduced and edited by the author with permission.

## 2.5  A MILLION MILES AWAY

*"Sprengel calls these flowers 'Scheinsaftblumen,' or sham-nectar-producers; — he believes that these plants exist by an organized system of deception… But… we can hardly believe in so gigantic an imposture. He who believes in Sprengel's doctrine must rank the sense of instinctive knowledge of many kinds of insects, even bees, very low in the scale."* -
CHARLES DARWIN in *Fertilisation of Orchids* (1877)

Despite calling Sprengel's book "a curious old book full of truth with some little nonsense", Darwin was in fact very fond of *The Secret of Nature* and referred to — and debated — Sprengel's work continuously in his letters with diverse correspondents. He thought of Sprengel as a most careful observer who had thoroughly studied orchids. Darwin was inspired by many of the ideas of the German naturalist and theologian, but he was also sceptical about others. Perhaps most notoriously, he did not believe in the existence of flowers with empty nectaries that achieve pollination through food deception. Darwin explained in *Fertilisation of Orchids*: "I still suspected that nectar must be secreted by our common Orchids, and I determined to examine O. morio rigorously. As soon as many flowers were open, I began to examine them for twenty-three consecutive days: I looked at them after hot sunshine, after rain, and at all hours: I kept the spikes in water, and examined them at midnight, and early the next morning: I irritated the nectaries with a bristle, and exposed them to irritating vapours: I took flowers which had quite lately had their pollinia removed by insects, of which I had independent proof on one occasion by finding within the nectary grains of some foreign pollen; and I took other flowers which from their position on the spike would soon have had their pollinia removed; but the nectary was invariably quite dry."

Sprengel was right. About a third of all orchid species are known to be deceptive. Even more strikingly, of the 10,000 flowering plant species on Earth that are known to use deception as a pollination strategy, 90% are orchids. Let that sink in. Given the large number of orchid species that use it, deception must be considered a highly successful evolutionary strategy. Deceptive strategies have evolved many times and are also highly diverse in the family. But the question then arises: if some orchids produce rewards, why do others rely on deception instead? One obvious answer is that it is cheaper not to feed your pollinator. But saving resources on nectar production does not seem to outweigh the risk of reduced pollination success. The most widely accepted hypothesis is that deceptive pollination strategies discourage pollinators from visiting many flowers on the same plant. In a nutshell, deception increases the chances of cross-pollination by discouraging loyalty. This works well in orchids because all of the flower's pollen can be removed or deposited in a single visit, and therefore a couple of visits to the flower may be enough.

The previous stories dealt with different types of sexual deception, but by far the most common type of deception in orchids involves tricking animals that are searching for a meal. Food deception is said to occur when the flower advertises — through guides, colours, shapes or smells — that it contains nectar, but in fact does not. Authors divide food-deceptive strategies into two broad types: 'generalised food deception', which refers to the use of general floral signals such as showy floral displays, spectrally pure flower colours, sweet scents, nectar guides and nectaries — to exploit the pollinators' innate food-foraging behaviour; and 'Batesian floral mimicry', where the plant mimics the flower of a specific species on which the pollinator feeds. Essentially, the difference is that flowers using Batesian floral mimicry are mimicking a nectar-producing species so well that pollinators are unable to distinguish between them. Flowers using generalised food deception, on the other hand, don't have a specific model. Instead, they advertise general floral signals that the pollinator associates with rewarding plant species.

There is a fine line between generalised food deception and Batesian floral mimicry. In essence, the difference depends on what the flowers are modelling. When there seems to be no specific model we talk about generalised food deception, and when there is a specific model we talk about Batesian floral mimicry. However, finding a potential model may be challenge, and proving that this is truly the model goes far beyond simple superficial similarity. Consequently, most food-deceptive orchids are thought to use generalised food deception and the proven cases of Batesian floral mimicry remain relatively few, with several suspect cases failing when put to the test.

One example of generalised food deception is that of the beautiful orchid species *Calypso bulbosa*, which grows in the thick litter of the coniferous forests of Asia, Europe and North America (see Figure 2.5.1). Theodore Mosquin suggested that *C. bulbosa* flowers specifically mimic those of the shooting star plant, *Dodecatheon radicatum*. However, most authors agree it is a more generalised food deceiver, and support for specific mimicry has not been found in later studies. Instead, these studies have found evidence that the deception of *Calypso* flowers relies on deceiving a large number of naive pollinators. Recruiting these less experienced new individuals counters the effect of avoidance learning. *Calypso bulbosa* benefits from synchronising its flowering period with

**Fig. 2.5.1.** The beautiful circumboreal *Calypso bulbosa* is morphologically quite variable along its widespread distribution. **A.** Humboldt County, California. **B.** Huanglong, Sichuan Province, China. **C.** Mount Tamalpais State Park, Marin County, California. **D.** Humboldt County, California.
© Ron Parsons.

the emergence of bumblebee queens, thus relying on the initial attraction and deception of these naive bumblebees for pollination.

Theories about generalised food deception propose that the larger the population size of the deceptive species, the less likely their flowers will be pollinated, because the bees are more likely to learn to avoid the rewardless species. Yet, no significant difference in pollen removal between small and large populations of *Calypso bulbosa* has been found. Recent studies also suggest that rewarding plants that share the same general area with *Calypso*, such as willows and bilberry, may serve as 'magnet species' to attract the bumblebees that also pollinate the orchid — provided their flowering coincides. Interestingly, there are exceptions to this rule too. The lack of competing food sources in the immediate vicinity of the *Calypso* flowers can actually make them more successful.

Several of the more common European terrestrial orchids are food deceptive. Among the most well-known, generalised food-deceptive orchids are the members of genus *Dactylorhiza* — in the past frequently known as *Orchis*. The flowers of most *Dactylorhiza* are completely nectarless, despite having a well-defined nectary spur. The spur, as Darwin put it, is 'bone dry'. *Dactylorhiza* flowers do not resemble any particular food flower, but are rather pollinated by inexperienced bees that follow general optical cues. But how does the deception actually work? One would imagine that after probing a few flowers, bees would learn to avoid *Dactylorhiza*, and yet these species keep reproducing year after year. Studies have suggested that for the deception to work, *Dactylorhiza* need to be isolated and remote, having exceptionally attractive floral displays and few competitors for the pollinators. In this scenario, the bees will visit the orchid's flowers because there are few rewarding species available to them. However, there are several alternative hypotheses. Some researchers find that *Dactylorhiza* in fact require other food flowers to be abundant in the vicinity. This suggests that the orchids benefit from a portion of the large number of insects spilling over from their nectar-rich neighbours. Profiting from co-occurring food plants that help to attract pollinators is known as 'magnet species theory', and may be essential to keep the deceptive mechanism of *Dactylorhiza* working. Another key element in this mechanism seems to be floral variation.

Every plant species has its own natural variation, with each individual member of the same species being more or less similar to the next. This morphological variation may be the result of the genetic or environmental circumstances of an individual. The flowers of many orchid species are notoriously less variable, in shape, size, colour and smell, than other plant organs such as the leaves and stems. This is most likely a result of the constraint that particular shapes, sizes, colours and aromas have in attracting particular pollinators and their ability to place the pollinaria on them. Striking morphological differences between orchid flowers are generally considered to be evidence of reproductive isolation and therefore speciation. Broadly speaking, when diagnosing different species, taxonomists give a much higher value to consistent differences in the floral morphology over those in other plant organs. However, generalised food deception is usually associated with higher than expected flower colour variation. This is because colour variation may make it much more difficult for the pollinator to learn that a specific colour is a predictor of the absence of rewards. In other words, colour variation impedes insect learning, and thus reduces avoidance of the deceitful orchid flower. Variable colour forms in orchids may come in the form of a 'discrete polymorphism', with two or more clearly distinct morphological types — known as morphs — or 'continuous polymorphism', with a continuum of variation among the morphs.

Pollinators can learn to associate floral traits with the quantity and quality of rewards, eventually learning to avoid deceptive orchids. Sexually deceptive orchids, or those pollinated through Batesian mimicry, avoid being detected by becoming virtually indistinguishable from their model. Because the actual model rewards the pollinators, if the orchid flower is similar enough, pollinators will never learn to avoid it. In the case of generalised food deception, where a specific model is absent, this is not possible. Such orchids have therefore resorted to polymorphism as a means to avoid detection. Rather than trying to mimic something very closely, they try not to fit any single pattern very precisely. The more dissimilar they are, the longer it takes the pollinators to learn to avoid them. European botanists are familiar with the striking colour variation in *Dactylorhiza* (see Figure 2.5.2). Species such as *Dactylorhiza maculata* and *D. sambucina* have individuals of two discrete, strikingly different morphs: their flowers are either intense yellow or deep purple. And the two — or more — morphs live side by side, intermingled within the same population. Researchers have found that the dramatic colour polymorphism in these rewardless orchids is not

**Fig. 2.5.2.** The food-deceptive European genus *Dactylorhiza* is well-known for having discrete floral morphotypes (morphs) growing intermixed within a single population. **A.** *D. romana*, Sicily, Italy. **B.** *D. sambucina*, Monte Baldo, Italy. © Ron Parsons.

accidental — it plays a key role in their deceptive pollination strategy. But how do the different types persist over time? It has been suggested that the frequency of each morph in a population may determine its reproductive success, with a rare morph being more likely to reproduce than a common one. This phenomenon of preference for the rare floral type is known as negative frequency-dependent selection. It is the pollinator, through its own preferences, that selects the morphological types, and maintains an equilibrium between the different morphs.

But not all studies find evidence that pollinators prefer the rarer morph. In fact, a recent paper by researchers from Naples suggests that perhaps the pollinators are not as capable of perceiving morphological variations as we humans are. There may therefore not be a direct link between reproductive success and high floral variation, as was previously believed. And yet, the fact remains that all across Europe *Dactylorhiza sambucina* populations have individuals with different colour morphs. The morphs are also unevenly distributed. The yellow type seems to be more common in some countries, including Italy, France and Austria, while the purple type has been reported to be more common in the Czech Republic and Sweden. One study showed that pollen movement among the different morphs is not random either. Pollen was more likely to be transferred between morphs of the same colour than between the different morphological types. Pollen from yellow flowers placed on the stigmas of yellow flowers overwhelmingly dominated over any other possible combination. The disproportionate yellow dominance in both the number of individuals and pollen movement contradicts the negative frequency-dependent hypothesis. Likewise, a study on *D. maculata* in Finland showed that there were no differences in the abundance of the morphs, but the bumblebees that pollinated them preferred the purple over the yellow. These studies show that there is a definite selective influence of the bees over the morphs, but how the balance works precisely is not yet clear.

Darwin noticed that many orchid flowers withered without ever being visited by pollinators. He examined *D. incarnata* (then known as *Orchis latifolia*), *D. maculata* (as *O. maculata*) and *O. purpurea* (as *O. fusca*), and was surprised to find how few pollinia had been removed — and few fruits formed — in relation to the number of flowers produced. He noted: "From these facts the suspicion naturally arises that *O. fusca* is so rare a species in Britain from not being sufficiently attractive to our moths, and consequently not producing a sufficiency of seed." He called this phenomenon of relatively low pollen removal and fruit set "imperfect fertilisation". Indeed, despite bees being available in large numbers, there is low visitation of bees to *Dactylorhiza* flowers. This suggests that bees do eventually learn to avoid the deceptive flowers.

Interestingly, a study in which a sugar-rich solution was manually injected daily into the nectary spur of flowers of the nectarless *D. sambucina* did not increase pollen removal. However, this manual nectar supplementation did result in a significant increase in the proportion of flowers pollinated. One may then wonder why, if the orchid is able to produce more fruits and therefore seeds by being rewarding instead of deceptive, has it developed and maintained these deceptive systems? The answer is likely to be that there are fitness advantages to being deceptive. Yes, perhaps there is lower seed production, but the seeds that are produced are of higher quality. Deception may be more cost-effective, reducing the expenditure of resources while increasing cross-pollination, because a deceived pollinator will more quickly leave a non-rewarding flower and fly to another. Given that orchids are long-lived plants that produce enormous amounts of dust-like seeds, a lower frequency but continuous high-quality fruit set may represent the best evolutionarily stable strategy to ensure reproductive success during the orchid's lifetime.

Orchids give the impression of having planned it all so well. The idea that high floral diversity reduces avoidance learning by the pollinators seems obvious. But botanists are still in the dark about how these morphological types are established and how they are maintained. Under certain conditions, researchers have found evidence for negative frequency-dependent selection — where rare forms are more reproductively successful than common ones. However, these seem to be the exception rather than the rule. An extraordinary hypothesis stemming from one study is that inexperienced bees initially visit the *Dactylorhiza* morphs randomly. As the bees gain experience with the nectar-rich floral models, however, they become more selective and will prefer the morphs that are more similar to rewarding flowers. Therefore, the bees may in fact be the selective force acting on the orchid morphology. This pressure is weak when pollinators are very diverse and naive, but increases as the insects become more experienced and selective. Maybe floral plasticity allows these deceptive orchids to always have — in their genetic arsenal — the capacity to produce morphs that will fool insects under certain circumstances. Alternatively, there may be fewer natural selection constraints on orchid flowers that use generalised food deception and so the broad morphological diversity may simply be a result of these lax selective pressures. There probably isn't a single answer. But what is undisputed is that — despite what Darwin thought — certain orchids find success with food-deceptive pollination strategies. They may not produce large numbers of fruits, but they produce more than enough quality seed to maintain large and highly diverse populations. Like many of us, when it comes to reproduction, orchids choose quality over quantity.

## 2.6 DOWN UNDER

> *"It would be a great mistake to assume that there is any purpose, sense or reason behind what orchids 'do'. It is all a result of adaptation and developments the outcome of which is the survival of the fittest — i.e., evolution. But, orchid 'behaviour' is sometimes so much like that of people that I can't resist asking the question... Are orchids people?"*
> JOSEPH ARDITTI (October 1981)

The first time I ever set eyes on a blue orchid flower was back when I was a teenager. I grew up at an agricultural institute in Turrialba, a small town on the Caribbean lowlands of Costa Rica. Turrialba was known for its harsh heat, intense rain, restless creepy-crawlies, and — obviously — lots of exotic tropical orchids. I began my own personal orchid collection in my early teens, initially picking them up around the neighbourhood. But eventually, as my curiosity for the unknown grew, I started bugging my parents to take me out to look for plants further away. My mom has always loved plants and among the many in her garden thrived some of the local orchids. Hanging in baskets under a shaded tree were *Cochleanthes aromatica* and *Stanhopea wardii*. Every year I waited patiently for their exquisite large flowers to appear. They bloomed only for a few days, but their scent was so overpowering and unforgettable that it still transports me to my childhood today. My dad had no particular interest in orchids — or plants for that matter. But he craved adventure about as much as he enjoyed indulging my brother and me. He used to take us on orchid hunts in his old Toyota 4Runner. On one occasion, he took us on a roadtrip through Central America, driving all the way from Costa Rica to Guatemala. Our car was broken into in Honduras, and at the Nicaraguan border, we saw duffel bags being flung over the wall — closely followed by their owners. Another time he rented a little boat to take me into the mangroves, where I learned the hard way how inseparable *Myrmecophila* species and their ant tenants really are. Those orchids we encountered together are still imprinted in my mind.

The orchid fever eventually led me to Lankester Botanical Garden. It was the turn of the century and specialised orchid websites were still at a very early stage of development. Back then, we knew much less about Costa Rican orchids, and no one was thinking of using social media to identify plants. Orchid fanatics — like myself — had to rely on the available magazines and books. Floyd S. Shuttleworth's pocket-sized *Orchids A Golden Guide* was an absolute must-have at the time. The golden guide featured orchids from all around the globe, many of which I had never heard of before. It was there that I first learned about *Thelymitra*, the mythical blue-flowered Australasian sun orchid. I was blown away. Depicted in the tiny book was a flower as blue as my dad's old car. I had never seen anything like that and since then have always dreamed of one day visiting Australia to see it.

The name *Thelymitra* comes from the Greek *thelys* meaning female, and *mitra*, a headband, referring to the presence of a hat-like crest on top of the column of many species. *Thelymitra* species are quite the oddball among orchids. Not only do their flowers have a very unusual colouration, but their shape is unlike that of other orchids. The lip in *Thelymitra* species is flat and plain, lacking the characteristic lobes and complex ornamentation of most orchids. It looks just like the other two petals. The sepals are not much different from the petals either. The result is a flower that is more or less radially symmetric with six floral segments that are virtually identical. It looks more like a lily than any other orchid flower. The very short column has another odd feature. It is highly ornamented with intricate lobes, sometimes forming a brightly coloured crest of hairs or trichomes that stand out in the middle of the flower.

Back in the day, I believed there were only a few species of *Thelymitra* and that they were all blue. However, more than 100 different species are recognised today. The highest diversity is found in Australia and

New Zealand, but a few are known to occur in New Guinea, Timor, Indonesia and the Philippines. Many species have flowers that are the typical sky blue, but they also range in colour from red to violet, passing through brown, orange, pink, purple and grey (see Figure 2.6.1). Some have spots, others streaks, but many have a solid colour. And they can be strongly scented. They are popularly called 'sun orchids', and the reason for this is that the perianth (the sepals and petals) responds to sunlight and temperature. Rather than opening early in the morning and staying open until the flower withers, like most other flowers, sun orchids open up when it's hot and the sun is high in a cloudless sky. On cool, cloudy or rainy days, *Thelymitra* refuse to open their flowers and simply close up shop.

Early naturalists noted that small-flowered *Thelymitra* species are 'cleistogamous' — a type of self-pollination that does not require the flower to open at all. Other species would open occasionally, despite having their own pollinia already germinating in place and self-pollinating the flower. For about a century, authors believed that *Thelymitra* species relied exclusively on self-pollination and therefore did not require insects to do their bidding. Australian botanist Robert David Fitzgerald — who corresponded with Darwin — is credited with having made the first observations about self-pollination in *Thelymitra* species. Darwin noticed that the flowers had specialised "structures adapted for cross-fertilisation" but, relying on Fitzgerald's observations, regarded the genus in its entirety as cleistogamous self-pollinators that would only open their flowers 'on fine days'.

It wasn't until the second half of the 20th century that the first interactions between the *Thelymitra* flowers and insects were recorded, and hypotheses were made about the function of their peculiar floral ornaments. Speculations about *Thelymitra* flowers abounded, from imitating a pollen source to the mimicry of enemies or mates. The mechanism of sun orchid pollination only became clear in the 1980s. By studying the behavioural patterns of insects visiting the flowers, researchers were able to establish that they are mimicking other flowers that offer pollen as a reward. Bees that collect pollen from a wide range of unrelated plants to feed their larvae are known as 'polylectic'. What researchers discovered was that *Thelymitra* are pollinated by polylectic bees attempting to harvest pollen through thoracic vibration. Female bees manipulate the intricate lobes of the sun orchid's column

Fig. 2.6.1. The unorchid-like flowers of *Thelymitra* are pollinated by polylectic bees through the mimicry of irises and daisies. **A.** *T. ixioides* at Grampians National Park, Victoria. **B.** *T. crinita* in Walpole, Western Australia. **C.** *T. nuda* in Campbelltown, Victoria. **D.** *T. epipactoides* in Golden Beach, Victoria. **E.** *T. rubra* in Tasmania. **F.** *T. antennifera* in Frankland, Western Australia.
© Ron Parsons.

while looking for pollen to feed their larvae. During this process they inadvertently get the orchid flower's pollinarium stuck to their bodies instead. They also discovered that the reason for the unusual blue colour and odd perianth is that the flowers are mimicking the floral models that these bees regularly visit: lilies and irises.

Within the genus *Thelymitra* we may find different types of food-deception: 1) generalist food mimics that lack a model and attract generalist foragers; 2) Batesian mimics that imitate a specific co-blooming species and are pollinated by specialists; and 3) guild mimics, a lesser known third type of food deceivers that have characteristics in common with several flowers. In practice, it may be quite a challenge to tease apart the latter two types and they may grade into each other. True Batesian mimics are visually and morphologically similar to the model, share pollinators with it, flower simultaneously and have a higher fitness when the model is present. The population size of the mimic is typically smaller than that of the model. We talk about guild mimesis when the orchid lives in the same area and blooms at the same time as several unrelated plants with very similar flowers that share the same subset of pollinators.

The large, metallic blue-flowered *T. epipactoides* seems to be pollinated through a combination of visual and olfactory cues, suggesting general food deception. The beautiful yellow or cream-flowered *T. antennifera* appears to be a general mimic of co-blooming flowers. The blue flowers of *T. nuda* — possibly a guild mimic — are buzz-pollinated by bees that harvest pollen from blue-flowered lilies through thoracic vibration. Meanwhile, *T. crinita* and *T. macrophylla*, another pair of bright blue species, are very likely to be Batesian mimics of the blue iris *Orthrosanthus laxus* (see Figure 2.6.2). These fraudulent orchids fool the polylectic female bees through scent and floral display, producing flowers that visually resemble those where they regularly harvest grains of pollen. Studies show that although not identical to the model, the scent and shape of *T. macrophylla* closely resembles that of the rewarding *O. laxus*.

The mimicry of lilies and irises certainly explains why *Thelymitra* flowers are so unorchid-like, but one mystery remains. Why do these orchids open and close their flowers over and over again? Researchers in Australia have come up with an interesting theory. The 'new again, more again hypothesis' is an intriguing — yet unfortunately named —

**Fig. 2.6.2.** Certain *Thelymitra* flowers are likely Batesian mimics of the blue iris *Orthrosanthus laxus* shown here.
© Mark Brundrett.

proposal that suggests that *Thelymitra* flowers open and close several times during their flowering period to enhance their visual and olfactory display. What I would have called the 'intermittent-accumulation hypothesis' argues that because *Thelymitra* individuals bloom almost in synchrony and their flowers last for more than two weeks. The accumulation of these large blue flowers during this period, in addition to their opening and closing daily, provides an exaggerated floral display. The explosive and increasingly exuberant floral stimulus that is then interrupted and renewed, continues to entice the resident bees over a longer period of time than would otherwise have been the case.

Down Under, we find yet another group of food-deceptive orchids that take floral mimicry to the limit. Species of the Australian genus *Diuris* are well known for having a pair of large, ear-like petals that have earned them the popular name of donkey orchids — now, don't get ahead of yourself, they are not mimicking donkeys! The flowers of most *Diuris* species are bright yellow with various reddish-brown markings that differ in style and intensity. Because of their shape and colour, botanists suspected that they mimic flowers of the legume family, Fabaceae (see Figure 2.6.3). Unlike everyday legumes such as peas, chickpeas and beans, which regularly self-pollinate, these particular Fabaceae need bees to pollinate their flowers. Researchers later found

**Fig. 2.6.3.** The food-deceptive orchid genus *Diuris* mimics the nectar-rich flowers of pea plants in the Fabaceae family. **A.** *D. magnifica* at Wireless Hill, Western Australia. **B.** *D. brumalis* at Serpentine National Park, Western Australia. **C.** *Bossiaea eriocarpa*, a *Diuris* floral model. **D.** *Daviesia horrida*, another *Diuris* model.
© Ron Parsons (A–B), Daniela Scaccabarozzi (C-D).

additional support for the suspected mimicry when it was shown that the orchids and the legumes share the same pollinating insects. In a study led by Daniela Scaccabarozzi, researchers showed not only that the flowers of *D. brumalis* overlapped in colour and morphology with those of several species of genus *Daviesia* — a genus in the legume family — but also that the pollinating bee showed the very same foraging behaviour when visiting both model and mimic. In fact — as one would expected — the orchid is more successful when the *Daviesia* plants are present because they increase the presence of the bees, as explained through the magnet species theory that was, discussed earlier.

The colour patters of the *Diuris* flowers are strikingly similar to those of the legumes, but you may be wondering whether the morphology itself couldn't use a bit of fine-tuning. Indeed, orchids and their models tend to have almost identical colour patterns, but not necessarily the same exact shape. This phenomenon is known as imperfect mimicry and it may in fact allow mimicry of a more general search pattern of the pollinators. Rather than mimicking one model exactly, the orchid may mimic a group of similar species. Such is the case of *D. maculata* and *D. magnifica* which — using evidence from the colour reflectance of their flowers — have been shown to be guild mimics of several nectar-rich species of pea plants that live in the same neighbourhood. *Diuris magnifica* is primarily pollinated by a colletid bee in the genus *Trichocolletes*, which forages nectar and pollen on the yellow-red Fabaceae flowers. It manipulates the orchid flower in the very same way it does the real thing. As expected in such mimicry systems, a higher density of *D. magnifica* plants reduces their fruit set. But the orchid has one more trick up its sleeve. Rather than having a restricted population size, it begins flowering late in the season, well after the model has already bloomed. This guarantees that when the orchid flowers finally show up, the bees are already familiar with the 'pea-flower image' and are therefore keen to visit the mimics too.

Just as in the *Thelymitra* discussed above, there are multiple different food-deceptive pollination strategies within the genus *Diuris*. Species such as the white-flowered *D. alba* and the purple-flowered *D. punctata* may not mimic pea plants at all. In fact, they appear to be pollinated through generalised food deception, by generalist bees. Unlike specialist bees, generalists visit many different kinds of rewarding flowers without having a preferred species.

Among closely related species of *Thelymitra* and *Diuris*, switching from self-pollination to being pollinator-dependent, and from having highly specialised to more generalised pollination strategies, is a clear indication of evolutionary plasticity. Given how recent most of the discussed studies are, it also speaks to how much we still have to learn about the complex ecological interactions in orchids. I haven't managed to realise my childhood dream of looking for orchids in the Australian bushland, but my expectations have grown enormously now that I see these extraordinary plants in a new light.

## 2.7 NEVER ENOUGH

With an estimated 2,400 species, *Epidendrum* is possibly the most species-rich genus in the Orchidaceae family, and one of the top three richest genera among vascular plants. Despite their diversity and ubiquitous presence in the Neotropics, *Epidendrum* remains ecologically quite mysterious. Flowers of species belonging to the genus are typically known to have nectarless nectaries and are mostly regarded as being pollinated by moths and butterflies through deception. However, there have been relatively few pollination studies on *Epidendrum*, and as I dug into the literature — for the purposes of writing this book — I came to the realisation that most commonly held notions regarding pollination in the genus have not been well established.

*Epidendrum radicans* is one of the more common roadside reed orchids in most of Central America. Its pollination strategy is often cited as a classic example of Batesian mimicry — where the deceitful flower mimics a specific rewarding species with which it shares its habitat. This species with its bright orange flowers, and a few of its close relatives, are famously known for being rewardless mimics of the nectar-rich flowers of common 'weeds' in the genera *Asclepias* (family Apocynaceae) and *Lantana* (family Verbenaceae) (see Figure 2.7.1).

In 1980, Thomas Boyden published a paper about the floral mimicry by *E. radicans*\* of the common weeds *L. camara* and *Asclepias*

---

\* Boyden cites *Epidendrum ibaguense* instead of *E. radicans*, but that species, which is florally extremely similar, is not known to occur in Panama and has been extensively confused with *E. radicans*.

Fig. 2.7.1. The food-deceptive *Epidendrum radicans* is believed to be a Batesian mimic of two common weeds, *Lantana camara* (Verbenaceae) and *Asclepias curassavica* (Apocynaceae). **A.** Common lantana, *L. camara*. **B.** A roadside reed-orchid, *E. radicans*. **C.** The tropical milkweed *A. curassavica*.
© APK.

*curassavica*, also known simply as lantana and tropical milkweed, respectively. The author noted that *E. radicans*, *L. camara* and *A. curassavica* were frequently found growing together near his study site in Boquete, Panama. While the flowers of both weeds are rich in nectar, the *Epidendrum* is notoriously nectarless. Some of the monarch butterflies caught and examined by Boyden carried pollen and pollinia of all three species, which led him to suspect that the pollination of all three species was interrelated. Based on their evident floral similarity, sympatric occurrence and shared pollinators, the author hypothesised that the nectarless *E. radicans* was a *bona fide* mimic of the nectar-rich weeds *L. camara* and *A. curassavica*.

Unfortunately, ecological relationships are rarely as straightforward as one would hope. A virtually simultaneous publication by Paulette Bierzychudek, concluded that in Costa Rica the elevational range and microhabitat of *E. radicans* in fact does not coincide well with that of either the lantana or the tropical milkweed. Furthermore, she pointed out that the origin of both weeds is widely debated, and it remains uncertain whether they were as widespread in the recent past as they are today. Contrary to the earlier research, her study found no conclusive evidence of mimicry. While the three lookalikes are indeed pollinated by butterflies, cross-visitations — a single individual butterfly visiting more than one species of plant — were rarely observed. In addition, the *Epidendrum* species were found to be equally successful well outside the area of overlap with *Asclepias* and *Lantana*, and no significant differences were found in pollen removal between populations of *E. radicans* when standing alone or accompanied by the two weeds.

So, can it be proven that *E. radicans* is truly a Batesian mimic of *L. camara* and *A. curassavica*? Despite establishing that the three species are similar, may live in the same area, and may share pollinators, the studies from the 1980s are inconclusive as to the nature of the deception. Twenty years later, two more studies on the pollination of *E. radicans*, published in 2000 and 2004, once again put the mimicry to the test — and hopefully, the case to rest. Unlike the Bierzychudek study, both these reports found support for the Batesian floral mimicry hypothesis involving the three plant species. An increased fitness was found for *E. radicans* as a result of proximity to the rewarding models *A. curassavica* and *L. camara*, and the authors observed that butterfly visitation and pollen removal from the orchid flower was higher than when the plant was growing on its own. They proposed that in patches consisting solely of *E. radicans*, the pollinator was less likely to visit multiple flowers and was more likely to leave. Case closed. Or so we thought.

A Brazilian study published in 2010, this time involving another reed orchid, the florally similar *E. fulgens*, once again set out to test the possible floral mimicry of *Asclepias* and *Lantana* by *Epidendrum* flowers. This team showed that the flowering times of the three species do sometimes overlap, but are certainly not synchronised. The researchers also found butterflies carrying the *E. fulgens* pollinaria and pollen from the weeds in places where the orchid species was absent, which definitively demonstrated that the butterflies do travel among areas and cross-visit the flowers of the different plant species. But despite the correlation between visits to their flowers, the authors found that a closer presence of *Lantana* did not increase the probability of pollinia removal from the *Epidendrum* flowers, as would be expected if the latter mimics the former. Once again, the authors were unable to demonstrate conclusively that a mimicry system exists between the three plants.

Another assumption that is not met in this particular case is that mimics should in theory be less frequent than their models for the mimicry to work, and this is certainly not true of either *E. fulgens* in Brazil or *E. radicans* in Costa Rica and Panama. Both are extremely common, living in high density populations of approximately 20 plants per square metre. In fact, they may have ten or more times the number of individuals in an area than their alleged models.

Despite numerous studies, we are not yet able to say with certainty whether the flowers of *E. radicans* and closely related reed orchids represent a true case of Batesian floral mimicry. However, it has been well established that, unlike the nectar-rich lantanas and milkweeds, they are nectarless deceivers. Or has it? In 2018, two papers dealing with the presence/absence of nectar in the nectaries of *Epidendrum* surfaced. After carefully studying the nectaries of *Epidendrum* species, one study estimated that "approximately 10.3% of *Epidendrum* species do not produce nectar" while the other stated that "all investigated species produce nectar or nectar-like secretion to varying degrees."

So despite appearances and previous beliefs, it seems that both studies agree that most *Epidendrum* species are in fact secreting nectar and thus rewarding. But our old friend *E. radicans* was examined in both studies. In one it was treated as a definite nectarless deceiver, whereas in the other, despite the lack of obvious nectar, the nectary was found to be highly metabolically active and the presence of some nectar residue at the surface indicated that it was actually secretory. The authors of the second paper caution that the reported lack of food rewards in *Epidendrum* requires careful microscopic reexamination as the presence of meagre amounts of nectar may be sufficient to maintain the interest of pollinators. They postulate that most, if not all, species in the genus may secrete varying amounts of nectar and may in fact be rewarding. But I wonder if the presence of residues or meagre secretions in the nectaries of *Epidendrum* — as well as those of other orchids — is rather indicative of a deceptive strategy that involves the reduction of nectar to traces, stimulating the pollinator's probing behaviour without encouraging them to linger on the flowers. This would reduce the energy cost for the plant while at the same time increasing the chances of outcrossing.

Are nectar-less nectaries always completely dry and residue-free? Does the presence of any trace of nectar, meagre as it may be, constitute a reward? Or are these small offerings simply never enough? In an engaging study of how widespread deceit is in orchids, Australian researchers propose that "trace rewards may succeed as a pollination strategy by exploiting neural mechanisms in insects' brains that code rewarding experiences" and as such "pollinator manipulation by trace rewards might still be considered largely deceitful." It is clear that the presence of varying amounts of rewards and its consequences need to be tested further in orchids. Only the study of the pollinator's behaviour when visiting the flowers will tell us with absolute certainty whether it is being fooled or not. From my own observations of butterflies visiting and pollinating the flowers of *E. radicans* (see Figure on p. 394), I find it hard to believe that — like several other species that contain no observable amounts of nectar in their nectaries — they could really be rewarding. This remains to be tested of course. But what is certain is that the hypothesis about a *Asclepias-Epidendrum-Lantana* floral mimicry complex requires further study.

To compound our uncertainties regarding the ecology of *Epidendrum* species, it seems that they may not be strictly lepidopterous (pollinated by species of Lepidoptera, namely butterflies and moths), as was generally believed. It has now been established that some species are pollinated by flies. While most literature asserts that this mammoth genus of over 2,400 species is mostly butterfly- and moth-pollinated, the fact is that only 24 species — or one out of every 100 species of *Epidendrum* — have been confirmed to be pollinated by Lepidoptera. It remains to be determined how widespread pollination by flies, or myophily, really is in *Epidendrum*. Perhaps it is merely a fraction of all the species. But a possible transfer of a few hundred species from the pollinated-by-Lepidoptera-through-food-deception category to the pollinated-by-Diptera-through-nectar-rewards really tells us something about the importance of being careful with broad analyses and conclusions based on relatively few hard ecological data.

## 2.8 ORIGINAL PRANKSTER

Arachnophobia — or fear of spiders — is among the most common phobias. The mere sight of an arachnid may trigger feelings of overwhelming fright and distress in millions of people. Even though the vast majority of spiders are completely harmless to humans, the thought of their touch alone may make your skin crawl. With their dark, hairy bodies and multiple long legs, it may be understandable that some people find these creepy-crawlies terrifying. The same, however, cannot be said about the magnificent flowers of the oddly named spider orchid. *Brassia* flowers elegantly parade in rows on either side of the densely flowered raceme (an inflorescence with flowers attached by short stalks along a central stem) (see Figure 2.8.1). They are typically greenish or yellowish and heavily flecked with brownish-maroon spots. The long and narrow sepals and petals give the flowers their characteristic elongated appearance. The narrow floral segments may be reminiscent of long spidery legs, but besides that, they look quite unlike any actual spider. So why are *Brassia* known as spider orchids?

The name *Brassia arachnoidea*, or spider *Brassia*, first used in 1877, proves that the association between these orchids and spiders can be traced back at least to the 19th century. I like to believe there is more to the story behind the peculiar nickname than just mere superficial resemblance. In fact, it is persistently claimed that *Brassia* flowers are pollinated by wasps that supposedly attack the flowers in the belief that they are spiders. These fearsome wasps belong to the Pompilidae family and are popularly known as spider wasps because of their predatory feeding behaviour. Once an adult, female wasp has been fertilised, she

**Fig. 2.8.1.** The slender flowers of the spider orchid *Brassia arcuigera*. © APK.

sets out to hunt for spiders, specifically tarantulas. Upon locating the arachnid, the wasp will determine its suitability using her highly sensitive antennae. The wasp, which is relatively large, then seizes the tarantula with her mandibles and flips it on its back. She stings the spider to paralyse it, then drags it to her burrow, where she lays an egg in its abdomen. The egg later hatches and the carnivorous larva feeds on the insides of its live host. This horrifying feeding behaviour makes me wonder why 'spheksophobia', or fear of wasps, isn't as common as arachnophobia.

The tale of how spider orchids are pollinated by spider wasps that mistake them for tarantulas has been repeated so many times that it has taken on a life of its own. It shows up recurrently in literature and even Richard Dawkins refers to it in his magnificent book *The Greatest Show on Earth*, as evidence in favour of evolution. There is, however, not a single published study supporting the all-too-famous claim. So where does it come from? In their groundbreaking book *Orchid Flowers: Their Pollination and Evolution*, published in 1966, van der Pijl and Dodson note that wasps of the genus *Pepsis* have been observed carrying the pollen of *Brassia* orchids. *Pepsis* wasps, which belong to the aforementioned spider-hunting wasp family Pompilidae, are particularly well known for their spider predatory behaviour. In their book, however, the authors never suggest that the flowers could be mimicking spiders, nor do they mention having ever observed spider wasps attacking the spider orchids. Norris Williams, in 1972, was possibly the first to propose that spider wasps visit spider orchids in search of prey. When discussing the pollination of *Brassia* (*Ada*) *allenii* he states "since no nectar is available for the [Pompilidae] wasp, and food in the form of pollen is not available, it is possible that the wasps are visiting the flower in search of prey." But no further explanation or proof is given.

It was Calaway Dodson who definitively secured *Brassia*'s reputation as the most original prankster. In 1990, Cal provided a detailed description of how the orchid benefits from the wasp's spider-hunting behaviour, suggesting that it pollinates the flower while attempting to paralyse what it believes to be a tarantula. In his imaginative account, he reports that female wasps first drag themselves along the sepals in search of a spider. They then grasp the base of the column with their mandibles, and sting the surface of the lip. They finally attempt to take their 'prey' away, without succeeding. During the struggle with the flower, the wasps supposedly remove the pollinaria with their heads. No evidence, other than the anecdote itself, was ever presented in support of the incredible tale. Repetition over the years, nevertheless, ensured that it came to be regarded as an established fact. This is known as an illusory truth effect. To quote Sir David Attenborough in *Plants Behaving Badly*: "Like all good fiction, it contained a germ of truth."

In a recent study, led by my colleague Melissa Díaz-Morales, we were finally able to demonstrate how *Brassia* flowers are pollinated. As previously pointed out by both Dodson and Williams, large spider wasps belonging to the genus *Pepsis* are indeed pollinating the spider orchids. The enormous black and orange wasp approaches the flowers in the warm morning hours, landing on the large platform that is provided by the orchid's lip. It quickly explores the flower, soon shifting its full attention to a thick structure at the base of the lip. The thickened structure, known as the callus, is strategically placed just below the anther and stigma of the flower. As the insect meticulously inspects the callus, it gets the pollinarium attached to the frontal part of its head just below the antennae. The wasp then briefly scrutinises a few more flowers before heading off. From the instant the insect lands on

Fig. 2.8.2. The food-deceptive *Brassia arcuigera* is pollinated by large spider-hunting wasps. **A–B.** Spider-hunting wasps inspect the flowers. **C–D.** A female *Pepsis atalanta* (Pompilidae) carries a pollinarium of *B. arcuigera* on its head.
© APK (A), Juan José Zúñiga (B), Melissa Díaz-Morales (C–D).

the inflorescence to the moment it removes its first pollinarium, only seconds will have passed. In fact, on average the entire visit takes no more than 15 seconds. Some of the elements in our observations are certainly reminiscent of the interaction between spider orchids and spider wasps originally portrayed by Dodson (see Figure 2.8.2). But, are the wasps truly being fooled by the orchids into believing they are spiders? Not exactly.

Nothing in the behaviour of the insect while visiting the flowers suggests spider mimicry. Why, then, do spider wasps show an interest in spider orchids? At this point it is important to consider other aspects of the wasp's behavioural ecology. The adult females of some *Pepsis* species are known to have two peaks of activity: they search for nectar in the morning and hunt for spiders in the afternoon. Their ephemeral visits to *Brassia* flowers are by no means aggressive and are confined to the warm morning hours. These two factors are consistent with the spider wasps' nectar-feeding, rather than spider-hunting behaviour. When approaching the orchid, the wasp probes the flower using its mouthparts, not the stinger. Such a head-first inspection is to be expected when a wasp searches for an immediate food source such as nectar. Contrary to what was previously believed, today we can infer that to be pollinated, the orchids have tapped into the spider wasps' nectar-searching behaviour.

Interestingly, after careful inspection of the flower parts, we were unable to detect the presence of nectar, or any other reward. This means that spider orchids are profiting from a food-deceptive pollination strategy. The wasps quickly learn that the *Brassia* flowers offer no food, and swiftly move on. Incidentally, the exceptionally brief visits may explain why it was so difficult to observe and document the interaction between these orchids and their pollinators in the past. We can safely conclude that spider orchids are pollinated by the spider wasps, and that their pollination strategy is food deception rather than spider mimicry.

Spider orchids are not the only plants that use the menacing pompilid wasps as pollinators. In fact, they are not even the only orchids to do so. Flowers of species belonging to several unrelated plant families are pollinated by these wasps, and many have the dull greenish or brownish-white flowers with purple markings that we also find in *Brassia* flowers. Among them is a species belonging to the genus *Disa*, a terrestrial orchid from South Africa. In this particular case the orchid emits a sweet, spicy fragrance that can be perceived from several metres away and serves as the initial attraction. Unlike *Brassia* flowers, the relatively small, yellowish-green flowers of *Disa* have a distinct spur and offer visible nectar to the visiting wasps. The pollinia get stuck to the feet of the insect as it holds on to one flower in order to reach for the nectar of the adjacent one. The presence of nectar in some of the flowers pollinated by spider wasps may explain why they do not learn to avoid food-deceptive species. Yet another species of *Disa* is also pollinated by spider wasps, but in this case through sexual deception, as explained in the 'Perfect strangers' story above (section 2.4). Spider wasps are thus not only pollinators of different unrelated plant families, but are also involved in syndromes as diverse as food reward, food deception and sexual deception. None, however, has been proven to be fooled by spider mimicry.

Before we move on to the next subject, it is important to mention that pollination through prey mimicry does occur in orchids. Using chemical and electrophysiological analyses, a group of German and Chinese researchers showed that flowers of *Dendrobium sinense* mimic the alarm pheromones produced by honey bees to attract their pollinators. Alarm pheromones, which are released by bees to signal danger when they sting another animal, are also used by hornets to locate and hunt the bees. *Dendrobium sinense* is a rewardless orchid bearing white flowers with a red centre. On the island of Hainan in China, the plant is visited and pollinated by the hornet *Vespa bicolor*. The authors observed that rather than landing on the lip, the hornets very briefly pounced on the red centre, as if they were attacking prey. Hornets are social wasps that feed their brood with insects, and are often known to capture honey bees. Foraging hornets use a combination of visual and olfactory cues to locate their prey. By emitting volatiles that indicated the presence of prey, the flower of *D. sinense* is able to attract its pollinator.

**ORIGINAL PRANKSTER**
Pollination of *Brassia* by spider-hunting wasps

SCAN ME
https://youtu.be/BEWn6zSKAec

*Brassia* species are known as spider orchids on account of the long, slender sepals and petals of their flowers. Even though the flowers are visited by several bee and wasp species, only very large wasps are able to remove their pollinaria. However, visits by large parasitic wasps belonging to the Pompilidae and Scolidae families occasionally occur and are extremely difficult to observe and document. Given the flower's superficial resemblance to a spider and the swiftness of the wasps' actions, authors have suggested that *Brassia* may actually mimic the spiders that the female spider-hunting wasps are looking for to lay their eggs. However, this is not the case. Rather than attacking the flower, the wasps swiftly look for nectar at the base of the lip, leaving in a matter of seconds as they become aware of the deception. Fractions of a second are enough for the large wasps to drop the anther caps and remove the pollinaria.

The videos and photographs shown here are owned by Melissa Díaz, David Villalobos, Juan José Zúñiga, and Adam Karremans and have been reproduced and edited by the author with permission.

## 2.9 FEAR OF THE DARK

It's a chilly, damp afternoon in the mountains south of the city of Cartago in Costa Rica. We are in the middle of the rainy season, and here in the forest it starts to get dark at around 5pm. Karen Gil — one of our Master's students — and I are carefully inspecting the bases of the large trees searching for orchids. Not just any orchid, but *Dracula*.

Few legends stir up our fear of the dark as much as that of Dracula. The mere mention immediately transports our suggestive minds to an unholy world of bloodsucking vampires and frightening evil. In Latin, from which the word originally comes, *dracula* means little dragon. Even though we don't necessarily associate Dracula with little dragons today, it is this meaning which the botanist Carlyle Luer used when naming this orchid genus. *Dracula* flowers look as tenebrous as their name suggests, but these little dragons were not named to honour the lord of darkness.

*Dracula* orchids mostly grow as epiphytes in the dense jungles of tropical Central and South America. They may be found low on the trunks of large trees — in contrast to other orchids, they prosper in the darkness (see Figure 2.9.1). Many orchids are eccentric and flashy when displaying their flowers, whereas *Dracula* are quite the opposite. These cryptic plants produce an inflorescence that creeps and crawls within the rich layers of moss. Their large flowers typically face the ground and are produced one at a time in slow succession. To an untrained human eye they may be difficult to locate, but *Dracula* are very appealing and immediately apparent to the minute flies that thrive on them. The flowers are made up of three large, long-tailed sepals which are covered with prominent hairs and warty glands. In the middle of this large flower, dangling straight down like a pendulum, is the mushroom-like lip. The lip is lighter in colour, has gill-like keels, and produces a musty, fungal odour. This unusual combination of lifestyle, morphological features and scent has inspired the belief that *Dracula* flowers mimic fungi. We know today that this is in fact true.

Field studies in the Neotropics have established that the flowers of *Dracula* are pollinated by flies of the family Drosophilidae, and that these visits are the consequence of a unique kind of mimicry. Some drosophilid flies, mainly those of the genera *Drosophila* and *Zygothrica*, are known to spend part of their life cycles on mushrooms, where they feed, mate and lay their eggs. In the rich tropical forests of the New World, you may find dozens of drosophilids on the gills on the underside of mushrooms. Field tests have shown that both the mushroom-like appearance and the mushroom-like smell of *Dracula* flowers play a decisive role in attracting the insects. It is unclear whether the orchids are offering a reward or not. They may simply pretend to be a site for reproduction or breeding, but they may also be advertising food. No obvious food source has been identified. Nevertheless, several parts of the *Dracula* flower have been shown to have secretory activity, and yeasts have been found growing on the orchids. Upon visiting *Dracula* flowers, for hours at a time, the drosophilid flies can be seen probing the floral parts with their mouthparts extended. It remains to be proven whether these secretions, or the yeasts, are actively involved in the attraction of the insects, either as part of the flower's fragrance or by offering an edible delicacy as reward.

The pollination of *Dracula* has typically been associated with 'brood-site imitation', which is a special case of deception involving the attraction of female insects that seek an appropriate place to lay their eggs. The corpse flower, *Amorphophallus titanum*, and corpse lilies from the genus *Rafflesia*, mimic the smell of decaying corpses and are other examples of pollination through brood-site imitation. Eggs or larvae have never been observed on *Dracula* flowers — despite the

**Fig. 2.9.1.** The cryptic *Dracula* flowers are typically found oriented towards darker, humid places where fungi grow.
© Karen Gil.

Fig. 2.9.2. The curious fungal mimicry pollination mechanism of *Dracula erythrochaete*. **A.** *Zygothrica* flies thrive on the gilled fungi found around the *Dracula* plants. **B.** The fungi-mimicking *D. erythrochaete* flowers are visited and pollinated by the *Zygothrica*.
© Karen Gil.

number of incredulous eyes looking for them. Occasional mating events have been known to occur on *Dracula* flowers. But they are not more common than in any other orchid, and given the large numbers of males and females visiting *Dracula* flowers — and the extent of those visits — such events are bound to occur eventually.

In Costa Rica, flowers of the common *Dracula erythrochaete* attract flies on the first day they open, when they start to emit a very characteristic fungus-like smell (see Figure 2.9.2). Upon alighting on the flowers, the flies continuously inspect the lip and sepals with their extended proboscis. They may be found on their own or in small to very large groups. Up to 24 individual flies having been counted on a single *Dracula* flower at any one time and more than 100 flies on a single plant. Their visit may last only a few seconds, but typically the flies will spend a few hours on the flowers. The vast majority of the flies found on *Dracula* flowers belong to the genus *Zygothrica*. In the forest where Karen and I had been looking for *Dracula* flowers, the *Zygothrica* flies may be found by the hundreds on the underside of mushrooms. Those mushrooms belong to the order Agaricales — commonly known as gilled fungi — and many different species may be found growing in the same area as a population of *Dracula* orchids. Previous field studies have failed to establish confidently that the very same fly species visit both *Dracula* flowers and the surrounding gilled

**Fig. 2.9.3.** Brood-site imitation in the genus *Bulbophyllum*. **A–B.** The scavenger *Chrysomya megacephala* blowfly visits and pollinates the hairy, dark red flowers of *B. lasianthum* that smell of decaying flesh. **C.** The fly *Liopygia ruficornis* pollinates the flowers of *B. mandibulare* while laying its live maggots on the flower.
© Ong Poh Teck.

mushrooms, but Karen's research proves it beyond any doubt. Flies collected on the gilled fungi in the *D. erythrochaete* population proved to be genetically identical to those collected on the orchid's flowers — and both carried the *Dracula* pollinaria.

The presence of *Dracula* pollinators on the gilled fungi, in addition to the fungus-like appearance and the fungus-like smell, establishes that the flowers are indeed fungal mimics. But why are the flies not seen laying their eggs on the flowers? Studies on plants that use brood-site mimicry suggest, that in most cases, the flies and beetles that are duped typically don't go as far as to actually lay their eggs on the flowers. The reason seems to be that they require a delicate interplay between smell, taste and mechanical signals to stimulate them to do so. The mushroom-like smell and appearance are powerful attractants that surely incite the flies to inspect the flowers, but the insects will not lay their eggs if the other conditions are not just right.

Brood-site imitation has also been closely associated with 'sapromyiophily', whereby plants deceive their pollinators by producing odours of decay and mimicking the decaying flesh in which flies normally lay their eggs. Such plants are generally pollinated by carrion or dung flies, which visit the flowers expecting to find rotting protein. The set of features that describe the sapromyiophilous pollination syndrome include radial flowers (often with great depth) with dark colours (most often brown, purple and greenish), and scents characterised by sulphur compounds, which give the flowers a putrescent odour. The orchid genus *Bulbophyllum* is well-known for its dark, foul-smelling flowers that are pollinated by flies. Our understanding of the reproductive biology of most species in the genus remains quite rudimentary, and sapromyiophily has not been amply proven to occur. However, studies on the *Bulbophyllum* on La Réunion in the Mascarene Archipelago confirm that at least one species, *B. variegatum*, is pollinated through sapromyiophily. This species displays large, reddish flowers that emit an unpleasant, urine-like scent. Researchers have shown that the pollinators are flies belonging to the Platystomatidae family, which are attracted to the scent emitted by the orchid during the day.

A group of *Bulbophyllum* species in Peninsular Malaysia share having spirally arranged flowers, a mobile lip, and yellow, red or purplish flowers that emit a 'fishy' or 'musty' odour. The smell, colour and morphology are all believed to play an important role in the reproductive strategy of these orchids. Pollination is only effected when a fly of the right size and weight triggers the tilting mechanism of the lip. *Bulbophyllum lasianthum*, another species from the same region, bears hairy, dark red flowers that smell like decaying flesh and is pollinated by *Chrysomya megacephala* (see Figure 2.9.3). These blowflies belong to the Calliphoridae family and are well-known scavengers. The same flies have also been observed visiting the flowers of the carrion-scented *B. virescens*. As soon as the flowers open, the flies become very agitated. They land on the lip and cause it to tilt towards the column. A fly of the right size and weight will become pressed onto the column in such a way that it removes the pollinarium as it attempts to pull itself out of that position.

*Chrysomya megacephala* is often found on garbage and is attracted to decomposing meat and human excrement. Its larvae have been known occasionally to infect the tissue of humans and domestic animals. In a stomach-churning turn of events, one study found that the flesh flies that visit and pollinate the unpleasantly scented flowers of *B. mandibulare* are 'ovoviviparous'. That is, they may deposit hatched maggots instead of eggs on carrion, dung or the open wounds of animals. The flesh flies have been observed to drop their live larvae on the flowers of this orchid as they pollinate it — where they of course eventually starve to death (see Figure 2.9.3C). Whether this was merely a chance event or part of a gruesome deceptive pollination strategy of this orchid remains to be proven. The fact is, some orchid flowers are in the business of mimicking decaying flesh.

**FEAR OF THE DARK**
*Dracula* pollination through fungal mimicry

**SCAN ME**

https://youtu.be/nL8bJzOiT1M

*Dracula* plants are commonly found growing in the darker parts of the forest, where the more or less pendent flowers creep and crawl among the mosses. They are visited and pollinated by tiny flies in the genus *Zygothrica* (Drosophilidae). Both males and females may be found in large groups of more than a dozen individuals on a single *Dracula* flower. *Zygothrica* spend part of their life cycle on gilled-fungi (Agaricales), and it is those fungi that this orchid is mimicking. *Dracula* flowers use brood-site imitation, smelling like fungi and having a lip that looks like gills to fool the *Zygothrica* flies. Once the fly reaches the base of the lip, it is pressed against the column and trapped. Struggling to grasp the movable lip, it is finally able to escape several minutes later with the orchid's pollinarium attached to its scutellum.

The videos and photographs shown here are owned by Karen Gil and Adam Karremans and have been reproduced and edited by the author with permission.

## 2.10 IT'S A LONG WAY TO THE TOP

*"You have hit on the same very idea which latterly has overpowered me, viz the exuberance of contrivances for same object … how curiously difficult it is, to be accurate, though I try my utmost."*
CHARLES DARWIN in a letter to Asa Gray on 10 June 1862

As he examined the very particular floral structure of slipper orchids for the first time, Darwin suspected that the insect that pollinated the flowers reached into the pouch-like lip by means of a long proboscis. The insect would then have pollen smeared on the tongue as it reached through one of the two side slits of the lip, where the anthers are. It was his friend, the renowned American botanist Asa Gray, who instead suggested that small insects, such as flies, could easily crawl into the lip and transfer pollen as they crawled out. Darwin had asked Gray to try to make observations on the pollination of North American species of *Cypripedium* slipper orchids. In a letter to the botanist Sir Joseph Dalton Hooker, Darwin wrote: "I must not indulge with *Cypripedium*: Asa Gray has made out pretty clearly that at least in some cases the act of fertilisation is effected by small insects being forced to crawl in & out of flower in a particular direction; & perhaps I am quite wrong that it is ever effected by proboscis."

In the second edition of Darwin's *Fertilisation of Orchids*, his initial misinterpretation was rectified. He dutifully added Asa's findings and described the results of an experiment of his own. "Accordingly I first introduced some flies into the labellum … but they were either too large or too stupid, and did not crawl out properly," Darwin explained. He then tried again using a bee, noting that the lip acted as a conical trap that impeded the insect from crawling out easily. "The bee vainly endeavoured to crawl out again the same way by which it had entered, but always fell backwards." Ultimately the bee succeeded in forcing itself out through one of the small basal orifices where the anthers are, successfully removing pollen. Darwin had little sympathy for the bee that had managed to find an escape route from the bottom of the trap. He put it right back, and again it crawled out through the small orifice. Science requires replication and Darwin did just that. To be certain of his conclusions, he repeated the test five more times — always with the same result. No doubt it's a long way to the top for the insect trapped by the slipper orchid — certainly one in the hands of the diligent English naturalist.

Slipper orchids have remarkable flowers, with a unique and exotic shape that immediately sets them apart from all other orchids. They are easily recognised by the large, pouch-like lip, with a pair of small, lateral openings that are formed by the interlocking bases of the lip and the column. Most orchids have a single anther, whereas slipper orchids have three. One has lost its pollen-bearing function and has been transformed into what is known as the 'staminode': a large, shield-like structure in the middle of the flower. The two pollen-bearing anthers are positioned inside each of the two side orifices at the base of the lip. These two side openings are the only exits from the pouch-like lip. To reach them, the insect must crawl from the bottom of the pouch all the way to the top using the stairway of hairs that the plant has provided for this purpose. To remove and deposit pollen successfully, the insect needs to be of a precise size. The right insect will get pollen from one of the anthers smeared onto its back as it squeezes through one of the lateral slits. If it is too small, it will exit without transferring any pollen, whereas a very large insect may get stuck and die. This trap mechanism is found in all insect-pollinated slipper orchids around the world, from the *Cypripedium* species in temperate North America, Asia and Europe, to the *Phragmipedium* species in the cloud forests of Central and South America, from the *Paphiopedilum* species of tropical and subtropical Southeast Asia, to the heat and drought-tolerant *Mexipedium* of Mexico. They differ only in how each manages to lure and fool pollinators, which is something that remained a mystery until very recently.

In 1985, the American botanist John Atwood proposed an extraordinary hypothesis about the pollination of slipper orchids. After studying dozens of flowers in different populations of the beautiful and rare *Paphiopedilum rothschildianum* in Sabah, East Malaysia, he discovered that they are pollinated by the syrphid fly *Dideopsis aegrota* through a very unusual strategy (see Figure 2.10.1). The flowers emit an enticing — if you are a syrphid hoverfly that is — peppery or spicy fragrance. After their initial approach, the flies are guided to the central staminode. Some of them exhibit only a passing interest, but others actively insert their abdomens into the space formed by the staminode. The hoverfly may get stuck and it is while they are attempting to launch into flight once again that some of them fall into the sac-like lip. The insect takes about a minute to emerge from one of the two lateral exits and in doing so it transfers pollen. This is not very unlike what we observe in many other orchids, but what makes these slipper orchids eccentric is that the pollinating hoverflies lay their eggs on the flowers.

Atwood found as many as 76 syrphid eggs on a single flower of *Paphiopedilum*. The eggs appeared on different floral parts, but the vast majority were located behind the central staminode. The author also found live syrphid larvae, though fewer than you would expect based on the number of empty eggs. He did not find larvae on any of the withered flowers, nor did he find them feeding on the fresh flowers. So, what could be happening here? Atwood proposed that *P. rothschildianum* is pollinated by female hoverflies through a grisly case of brood-site deception. Syrphid larvae are known to feed on aphids — small sap-sucking insects that are commonly seen infesting plants. But because larvae are born blind, pregnant syrphid flies will look to lay their eggs close to where the larvae can feed. What the orchid appears to have done is to mimic the aphid colonies where the females regularly deposit their eggs, and on which the larvae feed once they hatch. The staminode of *P. rothschildianum* has transformed into an aphid colony lookalike to fool the pregnant hoverflies. In the process of laying her eggs, the female gets caught in the lip, eventually transferring the pollen as she exists through one of the side slits. She leaves and the orchid has got its way. But eventually those larvae will hatch, only to realise that their mother has been duped. They starve to death blindly searching for the food source that was never there to begin with.

**Fig. 2.10.1.** The deceptive slipper orchid genus *Paphiopedilum*. **A.** *P. rothschildianum*, the extraordinary flower on which brood-site mimicry was discovered. **B.** *P. villosum* mimics a floral food source attractive to hoverflies. **C.** *P. insigne* with a fly stuck at the basal exit of the lip.
© Ron Parsons (A–B), APK (C).

Not all *Paphiopedilum* species use brood-site mimicry as their pollination strategy. Some species are food-deceptive. The difference between these two strategies boils down to whom the food is meant for. Food-deceptive orchids promise food to the visiting insect itself, whereas brood-site deceptive orchids promise to provide food to their progeny. There are obvious grey areas between these two strategies — as suggested in the previous story about *Dracula*. It could also be argued that the flowers' principal strategy is food-deception, especially given that many insects do not go as far as to copulate and lay their eggs on the flowers, and the adults may in fact also attempt to feed at those deceptive brood-sites. The line between brood-site deception and food deception is a fine one, and an interesting discussion. But let's get back to that irresistible slipper orchid action.

In China we find a captivating diversity of deceptive strategies involving slipper orchids. Scientists recently discovered that among the *Paphiopedilum* there are species that deceive syrphid flies in a very different way. The larvae of most hoverflies are predatory and the pregnant females lay their eggs near prey. However, many adults feed exclusively on nectar and pollen, and they use colour as a primary cue for locating these food sources. The urine-like smell of the flowers of *P. villosum* may invite the female hoverflies that pollinate its flowers, but they don't elicit egg-laying. Instead, it seems that the staminode mimics droplets of honeydew or moisture as part of a food-deception strategy. Then, as the pollinator attempts to hold on to a perch-like wart on the staminode, it slips into the lip. *Paphiopedilum spicerianum* is also pollinated by syrphid flies, but in this case, the flower is similar to a nectar-rewarding flower of a plant species in the Polygonaceae family — commonly known as buckwheat or knotweed — that shares the same habitat. Researchers suggests that buckwheat plays a very important role in attracting and feeding the syrphid flies. Its presence ensures that the deceptive pollination strategy of *P. spicerianum* is maintained. The study also shows that successful efforts to reintroduce this critically endangered slipper orchid depend on the high density of the rewarding *Polygonum* species to ensure the presence of the orchid's pollinator.

Experiments using syrphid flies show that they have an innate preference for wavelengths of visible light between 460 and 600 nm, which correspond to the common colours of flower anthers and pollen grains. A revealing study carried out on *P. barbigerum* in the Maolan National Nature Reserve, in the Guizhou Province of southeast China, shows that the orchid has tapped into an innate colour preference of the hoverflies to ensure fertilisation. The large, shield-like staminode of this species reflects wavelengths of 500 to 560 nm, which appear green or yellow to human eyes. Rather than mimicking aphids as in other *Paphiopedilum* species, in this case, the staminode provides a colour stimulus for the adult hoverfly. Incredibly, this trick seems to work even outside the natural distribution of these *Paphiopedilum* species. *Paphiopedilum insigne* — a sister species of *P. barbigerum* — has a very similar large, yellow staminode, and in Costa Rica is visited by the local syrphid flies. Despite being very far from its native China, the local hoverflies are still duped into visiting the flowers and get caught in them.

Since Atwood's groundbreaking discoveries about *Paphiopedilum*, syrphid eggs and larvae have been found on the flowers of two other genera in the slipper orchid subfamily, *Cypripedium* and *Phragmipedium*. There is also much more diversity in these strategies than was initially thought. Botanists had long believed that *Cypripedium* species were mainly pollinated by bees through food deception. But we now know that there are several fly-pollinated species that use other strategies. There have been reports that syrphid flies with 'entomophagous' larvae — which feed on living insects — and 'saprophagous' larvae — which feed on decomposing plant or animal matter — are tricked into using *Cypripedium* flowers as brood sites. *Cypripedium fargesii*, a critically endangered species endemic to south-west China, is pollinated by flies of the genus *Cheilosia*. These hoverflies feed on fungal spores and lay their eggs in fungus-infected foliage. The short plant bears a few broad leaves that are covered in dark blotches and spots, and its solitary, dark-red to dull-yellow flower produces a faint, but unpleasant smell of rotten leaves. The whole orchid — plant and flower — seems to be mimicking fungus-infected foliage, with the dark hairy spots on the orchid's leaves serving as visual lures and a floral scent that contains fungal odour molecules. Another slipper orchid, *C. lichiangense*, is pollinated by a fly of the genus *Ferdinandea*. This particular hoverfly looks for humus-rich places to lay its eggs, including diseased trees and the wet fungal decay of infected plant roots. The leaves and flowers of *C. lichiangense* are covered in hairy, maroon spots and blotches, mimicking the hoverfly's

Fig. 2.10.2. The deceptive slipper orchid genus *Cypripedium*. **A.** *C. fargesii*, Jiuzhaigou, Sichuan province, showing the naturally spotted leaves. **B.** *C. lichiangense* pollinated by the syrphid fly *Ferdinandea cuprea*. **C.** The unpleasantly scented *C. sichuanense* is pollinated by dung flies. **D.** *C. fasciculatum* is pollinated by parasitic diapriid wasps.
© Ron Parsons (A, C-D), Chen-Chen Zheng (B).

preferred brood sites. This, together with the flower's strong smell of decaying vegetation, successfully lures the female *Ferdinandea* into laying her eggs on the orchid's flower (see Figure 2.10.2).

*Cypripedium* species are not restricted to pollination by syrphid flies. In China, fruit flies are attracted to the rotting-fruit smell and brownish flowers of *C. micranthum*, while dung flies are lured by the dull flowers and unpleasant scent of *C. sichuanense*. The extraordinary extent of mimicry strategies in *Cypripedium* does not end there. Studies carried out on *C. fasciculatum* in south-west Oregon, in the United States, show that this species is pollinated by wasps in the Diapriidae family. Diapriid wasps — as they are commonly known — are small, sting-less parasitic wasps that use the larvae or pupae of fungus gnats (families Mycetophilidae and Sciaridae) as hosts for their own eggs and larvae. The authors found that the flowering of the orchid was synchronised with the peak activity of both the diapriid wasp and the fungus gnat. They showed that when diapriid wasp activity was low, fruit was less likely to set on the orchids. As with other slipper orchids, the small parasitic wasps become trapped inside the lip of *C. fasciculatum*, removing the pollinaria as they exit. What makes *C. fasciculatum* unique is that it attracts its parasitic wasp pollinator by mimicking the smell of the fungus gnats in which the wasp lays its eggs.

Another peculiar twist to this story comes from a rare Chinese species. *Cypripedium subtropicum* is pollinated by the same syrphid flies that are fooled using the aphid colony mimicry that is reported in *Paphiopedilum*. However, while conducting field studies in Malipo County in south-east Yunnan, a group of scientists made an extraordinary discovery. As the hoverflies visited *C. subtropicum* they landed on the lip and began eating the tufts of white hairs that are found on the sides of the lip (see Figure 2.10.4). No other *Cypripedium* species is known to have edible hairs. In fact, no other slipper orchid is known to be rewarding. It turns out that these food hairs are quite nutritious, containing sugars and amino acids, and they may actually be mimicking the honeydew secreted by aphids, on which many hoverflies feed. Interestingly, this slipper orchid is both deceiving and rewarding its pollinator at the same time.

We have now discussed the pollination of two slipper orchid genera, *Cypripedium* and *Paphiopedilum*, but what about the stunning *Phragmipedium* species found in tropical America? The first account of the pollination of *Phragmipedium* species is found in a paper published by Dodson in 1966. He recounts how he observed a bee and a syrphid fly pollinating the flowers of *P. longifolium* in Ecuador. He noted that the insects slipped into the broad trap formed by the lip while they were inspecting the brown spots and green warts on its folded margins. Years later, the American entomologist Robert Pemberton observed syrphids visiting the flowers of *P. longifolium* and *P. pearcei*. Again, the flies were seen to fly towards the spotted area on the lip entrance, and after a few minutes, they emerged from the lateral windows at the base of the lip. The flies were females of the genus *Ocyptamus*, the larvae of which — as with other hoverflies — feed on aphid colonies. Syrphid flies of the genus *Syrphus* were recorded visiting the flowers and removing pollinaria of the extraordinary *Phragmipedium caudatum*.

Even though some *Phragmipedium* species can self-fertilise without the help of insects, most are dependent on pollinators. An extensive field study carried out on *P. longifolium* in Costa Rica by Melissa Díaz-Morales suggests that while this species is 'self-compatible' — it can produce fruits after being pollinated by hand with pollen from the same plant — fruits did not set if the researchers did not allow the insects access to the flower. Even under natural conditions, there are few visits to the flowers by pollinators. Only 12% of the hundreds of flowers that the researchers observed produced fruits naturally. Such a low fertilisation success is typical of deceptive orchids, which provide nothing for the visiting insect. It seems likely that the plants have adapted to become self-compatible to increase the chances of successfully reproducing under difficult circumstances, isolated conditions and when pollinator services are unavailable or unpredictable.

Díaz-Morales' unpublished studies show that *P. longifolium* is pollinated by syrphid flies which lay eggs on the flowers, especially the lip. The petals, and specifically the glandular, finger-like trichomes found on them, play a decisive role in the chemical attraction of the female hoverflies. These structures are likely to be the source of the fragrances that deceive the insects. At Lankester Botanical Garden in Costa Rica, I have photographed both syrphid eggs and larvae on the flowers of *P. longifolium* (see Figure 2.10.3). Most are indeed found on the lip, close to the area where the green and brown spots are. They may also be found on the petals. The larvae are found crawling aimlessly on the floral segments after hatching from the eggs. They have been sentenced to death by being born on barren land without any type of food.

**Fig. 2.10.3.** The deceptive slipper orchid genus *Phragmipedium*. **A.** *P. longifolium*, a common species in Central and South America. **B.** A close-up of the lip showing the aphid-mimicking spots and syrphid fly eggs. **C.** A recently laid egg on the lip surface. **D.** An empty eggshell on the petal surface. **E.** A syrphid fly larva blindly looking for food on the deceptive flower.
© APK.

Upon initially inspecting the flowers of slipper orchids, Darwin suspected that the glandular hairs inside the lip contained nectar and that pollination occurred as insects reached for the bottom of the lip. In the second edition of *Fertilisation of Orchids*, he amended his initial diagnosis, but, still unwilling to accept the notion that orchids deceived their pollinators, he kept looking for a nonexistent food source — not unlike the hoverfly larvae. Today we know for a fact that slipper orchids fool their pollinators, and have multiple different deceitful ways to do so. From the point of view of the pregnant female syrphid fly looking for a good place to lay her eggs, the exotic flowers of slipper orchids are a cruel artifice. Not only does the orchid require her to slip and fall into the trap-like lip (twice!), but any of the larvae that hatch from the eggs she laid on the floral parts will certainly starve.

**IT'S A LONG WAY TO THE TOP**
Slipper orchids and brood-site mimicry

**SCAN ME**
https://youtu.be/l86zOfX3paw

Slipper orchids around the world share a standard floral morphology and the bare essentials of their pollination mechanism are quite similar. A flying insect is tricked into falling into the sack-like lip. Unable to fly out, it crawls to the top where it is offered only two exits either side of the column. As the insect squeezes out, it transfers the orchid's pollinarium. However, there are several different strategies used by slipper orchids to attract their pollinators, and each one is unique. The flower of *Phragmipedium longifolium* mimics the aphid colonies that the pregnant syrphid fly looks for to lay her eggs. Syrphid larvae are blind and feed on the aphids. As the female lays her eggs on *P. longifolium*, she slips into the lip, removing the pollinarium as she escapes through one of the two lip-column cavities. The larvae will eventually hatch from the eggs and starve to death looking for a non-existent food source.

The videos and photographs shown here are owned by Adam Karremans and Ron Parsons and have been reproduced and edited by the author with permission.

**Fig. 2.10.4.** The Chinese *Cypripedium subtropicum* is unique in offering nutrient-rich edible hairs to its pollinator as a reward.
© Hong Jiang.

Male *Euglossa* visiting the flowers of *Macradenia brassavolae* in Peru.
© Luis Enrique Yupanki.

CHAPTER 3

# **REWARD**

Orchids are famous for being deceitful. Titles such as 'The masters of lying, cheating and stealing', 'Love and lies' and 'Deception and trickery' are commonly used when discussing orchid reproduction. The always delightfully sarcastic Jim Ackerman began his article on food-deceptive pollination in orchids by calling them a 'disreputable family' of 'infamous frauds'. Nonetheless, and this may come as a surprise to most readers, many orchids do not deceive their pollinators. They reward them. It is generally estimated that about one third of all orchids employ food deception and that one in ten may use sexual deception. This means that even though thousands of orchid species deceive their pollinators, a large portion of them are offering some kind of reward.

A reward can be defined as something given in exchange for a useful service. When we imagine flowers rewarding their pollinators, we commonly think of a gift in the form of food. Food is indeed the main handout offered by flowering plants, and this is no different in orchids. A flower's best bargaining chip is pollen. However, in orchids, pollen is typically scarce, so it is tightly packed and well hidden within the anthers. As pollen is needed for reproduction, giving it away is a luxury that most orchids cannot afford. Only a few so-called primitive orchids, which have loose pollen grains scattered over the anther, are still able to expend part of it as a reward for pollinators. More advanced orchids can't squander their pollen, but instead reward their pollinators in other ways. Nectar is by far the number one reward offered by orchid flowers. Diluted sucrose, glucose and fructose are the most common components of nectar and are used to feed a diversity of animals, typically bees, butterflies, hummingbirds and flies. Orchid flowers also offer lesser known collectable rewards, including oils, 'pseudopollen' (fake pollen) and wax.

Many orchids are believed to be deceitful merely because no food source is immediately evident, but an important point to make here is that the rewards on offer are not always obvious. Proteins, which are fed to insects by means of secretory glands, are good examples of non-obvious floral rewards. These also include rewards that aren't meant to be eaten. Contrary to what most people believe, not all rewards are in fact food. The most important is probably fragrance. Hundreds of orchid species are visited by bees looking to collect the fragrances that they require in order to make perfumes to attract a female. Rewards are also not limited to substances that are picked up and carried away by the pollinator. A well-known example of a non-obvious, non-edible, floral reward that can't be carried away is shelter.

It is important to mention that the distinction between rewarding and deceiving orchids is not always clearcut. There are many grey areas. Some orchids typically considered deceitful may actually be providing some sort of non-obvious benefit. At the same time, rewards provided by some flowers may in fact come in such low quantities or qualities that they are not truly beneficial to the pollinator. It is also possible that orchid flowers deceive pollinators in one way and reward them in another. We won't worry about those details just yet. Even though one cannot say for sure that most orchids are deceiving, it is true that the frequency of deception within the orchid family is exceptionally high compared to that within other plant families. It is also true that some of the most mind-blowing pollination syndromes in the family are deceptive systems. However, reward mechanisms may well be as diverse and spectacular as the deceptive ones. A selection of the rewarding pollination strategies used by orchids — and the myths that surround them — is presented in this chapter.

## 3.1 EVERYBODY KNOWS

*"I must stop. I have just received such a box full from Mr Bateman with the astounding Angræcum sesquipedalia* [sic] *with a nectary a foot long — Good Heavens what insect can suck it."*
Letter from CHARLES DARWIN to JOSEPH HOOKER (26 January 1862)

If you have ever ventured into the marvelous realm that is orchid pollination, you've surely come across the tale of Darwin's prediction about the pollination of *Angraecum sesquipedale*\*, popularly known as the Madagascar star or Darwin's orchid. Upon careful examination of the flower's outrageously long nectariferous spur (see Figure 3.1.1), Darwin predicted that there would be an insect capable of reaching the nectar at the very end of that spur, using an equally long tongue. No such insect was known at the time.

Darwin had received the Madagascar star orchid from James Bateman, and was flabbergasted. He mentions the orchid several times in his correspondence with colleagues. Another letter to Joseph Hooker written on the 30 January 1862, reads: "What a proboscis the moth that sucks it, must have! It is a very pretty case." This discovery impressed Darwin so much that he included it in the first edition of his *Fertilisation of Orchids*, published only three months after he first received the *Angraecum* plant. In the famous orchid book, he points out that the large, snow-white, star-like flowers of *A. sesquipedale* bear a whip-like nectary of about eleven and a half inches (around 30 cm) long, causing great admiration among travellers in Madagascar. He also noted that the nectar was only found in the lower inch and a half of that length. Darwin was sure that, even though not known at the time, "in Madagascar there must be moths with probosces capable of extension to a length of between ten and eleven inches." He firmly believed that the presence of nectar at the end of such a long spur had come about through the forces of natural selection; the flower's nectary becoming deeper at the same time that the moth's proboscis became longer.

---

\* The name *Angraecum sesquipedale* comes from *angurek* or *anggrek*, the Malay word for orchid, and the Latin words *sesqui*, one and a half, and *pedalis*, foot; a direct reference to the exaggerated length of its flower.

**Fig. 3.1.1.** The exquisite, long-spurred Madagascar star orchid *Angraecum sesquipedale*, famously known as Darwin's orchid.
© Gustavo Rojas Alvarado.

It is common knowledge that Darwin's thoughts on natural selection were met with scepticism by many of his contemporaries. The detailed observations presented in the orchid book, published after the famous *Origin of Species*, supported many of his views on natural selection and speciation through interspecific interactions. Nevertheless, he was still far from convincing all his fellow scholars. In *The Reign of Law*, George Douglas Campbell, Duke of Argyll and an avid opponent of Darwinism, criticised the naturalist for the language he used to describe the various contrivances by which orchids are fertilised. The Duke, a knowledgeable and respected creationist, found Darwin's explanations "nothing but the vaguest and most unsatisfactory conjectures". With erudition and finesse, rather than ridicule, Campbell argued in favour of what we today refer to as 'intelligent design' instead of natural selection. As an example, he cited Darwin's 'curious and characteristic' passage on the spur length of the Madagascar star orchid.

Darwin's views found support among colleagues, however. "I think your anticipation by analogy of a Madagascar moth with a proboscis ten inches long equals Adams & Le Verrier* — What a triumph it will be to find him", reads a letter from Edward Cresy to Darwin dated 19 May 1862. The moth was indeed eventually found. In 1903, 21 years after Darwin's death, experts Walter Rothschild and Karl Jordan described a hawkmoth from Madagascar with an exceptionally long proboscis, based on two specimens: a male and a female. Their eight-inch-tongued hawkmoth was named *Xanthopan morganii praedicta* and was said by the authors to "do for *Angraecum* what is necessary", referring to pollen transfer, from one flower to the other, while attempting to reach the sugary reward at the very end of the flower's spur. The moth's epithet *praedicta*, which comes from the Latin *praedictus*, is a direct reference to its predicted existence.

However, you may be surprised to learn that Rothschild and Jordan were not crediting Darwin when they named the moth. Even though some authors have suggested that the general prediction, rather than a specific person, was being honoured, the fact remains that Rothschild

---

\* Cresy was referring to the fact that in 1845 the British astronomer John Couch Adams and the French astronomer Urbain Jean Joseph Le Verrier independently postulated the existence and position of the planet Neptune using calculations based on its gravitational effect on the orbit of Uranus.

and Jordan allude to the work of the famous naturalist Alfred Russel Wallace and make no mention of Darwin at all. Wallace had not been the first to speculate about the flower, but he took special interest in this particular case and expanded on the initial prediction. He carefully studied and measured the tongues of diverse moths. He even studied specimens of *Xanthopan morganii* from mainland Africa, without realising that the pollinator he was looking for was in fact a Madagascan subspecies of that very species.

In a research paper published in 1867, Wallace supports Darwin's original prediction and appends it with his own findings. He writes "that such a moth exists in Madagascar may be safely predicted; and naturalists who visit that island should search for it with as much confidence as astronomers searched for the planet Neptune, and they will be equally successful." Accompanying the text was the now famous illustration of the imagined hawkmoth visiting the flowers of the orchid. In a letter addressed to Darwin on 1 October 1867, Wallace wrote: "My article on *Creation by Law* in reply to the Duke of Argyle and the North British Reviewer, is in the present month's Number of the Quarterly Journal of Science … There is a nice illustration of the predicted Madagascar Moth and *Angræcum sesquipedale*. I shall be glad to know whether I have done it satisfactorily to you, and hope you will not be so very sparing of criticism as you usually are." Shortly after, Darwin replied: "You have just touched on the points which I particularly wished to see noticed. I am glad you had the courage to take up *Angræcum* after the Duke's attack; for I believe the principle in this case may be widely applied. I like the Figure but I wish the artist had drawn a better sphynx*."

More than a century after the famous prophecies, pollination studies in Madagascar have definitively shown the close relationship between the long-spurred orchids, including *Angraecum* and its relatives, and the long-tongued hawkmoths (see Figure 3.1.2). Visits are nocturnal and quite rare, which explains why it had been difficult to observe these orchids being pollinated. Night-vision video techniques were required to 'catch' the insects in the act. The white *Angraecum* flowers produce a sweet

---

\*   Sphynx or sphinx is a common name for moths belonging to the Sphingidae family that are also known as hawkmoths.

nocturnal odour that contains more than 200 compounds, dominated by aromatics and terpenoids. From a distance, the hawkmoth is guided towards the flower by its scent, and when closer, the contrast of the white flowers against a dark background visually guides the insect. The moth hovers in front of the flower and inserts its proboscis in the nectary with the help of the large, funnel-shaped lip. The pollen gets stuck to the base of the insect's proboscis while it attempts to reach for the nectar.

Contrary to what Darwin believed, it is unlikely that the relationship between the orchids and moths is truly a case of coevolution. He thought that as the insect's tongue grew longer, so did the orchid's spur. It is more likely that the orchids adapted to the moths. In other words, the moths developed long tongues first — probably allowing them to probe nectar-rich flowers from a distance to avoid predatory animals. The orchids only later adapted to the long-tongued moths by adjusting their spurs. This would explain why, even though there are many different

**Fig. 3.1.2.** *Panogena lingens*, a long-tongued moth pollinating *Angraecum arachnites*, another long-spurred orchid.
© Lutz Thilo Wasserthal.

Fig. 3.1.3. The moth *Autographa gamma* pollinating *Platanthera bifolia*.
© Jean Claessens.

groups of orchids pollinated by hawkmoths around the world, it is Madagascar — where the diversity of long-tongued hawkmoths is very high — that is the world hotspot for long-spurred orchids. In a nutshell, the differences in the length of the spur allow the orchids to shift from one species of pollinator to another. A precise fit is highly important to ensure reproductive success for the plant, and thus it is under high selective pressure. If the tongue is shorter than the spur, the insect will not reach the nectar and will learn to avoid the flowers. If the tongue is much longer than the spur, the insect will reach the nectar from a distance and thus may not come into close contact with the pollen.

Studies on a species of genus *Platanthera*, another hawkmoth-pollinated orchid (see Figure 3.1.3), support the idea of adaptation to the moths rather than coevolution. In September 1862, discussing pollination of *Platanthera*, Darwin wrote to the renowned Harvard botanist Asa Gray: "I suspect its structure may have been arrived at by a process somewhat analogous to that which apparently has produced the wondrous nectary of *Angræcum sesquipedale*." More recent studies found that different populations of the widely distributed *P. bifolia* show large variations in the spur's length. These variations result in pollination by a different species of hawkmoth in each population. The average spur length in each population corresponds to the average tongue length of each of their different species of pollinators. These findings support the hypothesis that pollinator shifts in *Platanthera* have driven population-level differentiation of the spur length. Such differences may eventually isolate each of the populations and cause speciation, as has probably happened with the *Angraecum* in Madagascar. *"That's how it goes, everybody knows."*

## 3.2 LINGER

Some animals — such as owls and jaguars — rarely socialise with other members of their own species, except for mating. Others — including ants, wolves and humans — live in groups of many individuals that are continuously interacting with each other. We generally refer to animals with a high degree of interaction as social animals. Communication is critical for animals that continuously interact and a large part of their organisation is mediated through pheromones. The term pheromone was originally proposed in 1959 and is used to define a substance secreted by an organism outside of its body that causes a specific reaction in a receiving organism of the same species. We can think of it as a chemical substance released by one animal that influences the behaviour of another animal of its kind.

But just how much can a chemical secretion really influence the behaviour of a receiving animal? More than you'd think. There is a wonderful source of evidence in our own kitchens for the power of chemical pheromones. The speed with which a fleet of ants storms onto a drop of honey, or a swarm of flies lands on the bananas in our fruit basket is striking, making us wonder if the insects have perhaps been there all along. But that's not the case. It is well known that both ants and fruit flies are highly social animals that actively communicate with each other through chemical signals. When an individual finds a source of food or an adequate breeding site, it releases pheromones to inform the others. These particular chemical substances are called aggregation pheromones, and they differ from sex pheromones — mimicked by orchids that employ sexual deception — in that they are produced by both males and females and their function is to bring individuals of both sexes to a certain location, rather than to stimulate copulation.

Orchids, resourceful as we have come to expect them to be, have 'picked up' on the social behaviour of fruit flies and use it to their own benefit. Among the species of *Specklinia*, a neotropical genus of tiny epiphytic plants, the six or so members of the *S. endotrachys* group are easily distinguished by their showy, relatively large, bright orange to reddish flowers, which are conspicuously covered in warts. The long inflorescence can produce more than a dozen flowers, but they open one at a time in very slow succession. A single plant can therefore continuously bloom for several months, always with a single open flower. In the field, the exquisite orange flowers may be found completely covered in fruit flies. Contrary to many other orchids that are visited only briefly by pollinators, *Specklinia* are visited by fruit flies that may spend hours running around and socialising on the flowers (see Figure 3.2.1).

The earliest observations on the pollination of *Specklinia* were made by Auguste Endrés in the 19th century. In his correspondence with Heinrich Gustav Reichenbach — the leading orchid authority at the time — Endrés noted that flowers of *S. endotrachys* were viscous inside and much visited by a small fly\*. No additional observations were made until the mid-1980s when the British botanist and author Mark Chase prepared the first detailed account of the pollination of *S. endotrachys*. In his meticulous description, Chase noted that flies of the genus *Drosophila*, commonly known as fruit flies, persistently investigated the rough surfaces of the sepals with their mouthparts. The flies would eventually become trapped between the movable lip and the column, and remove the pollinarium while struggling mightily to free themselves. Despite these early observations, the means by which the flowers of these orchids attract and arrest the flies remained unknown until very recently.

At Lankester Botanical Garden in Costa Rica, my colleague Franco Pupulin and I carefully studied the fruit flies visiting *Specklinia* flowers. The flies were active throughout the day, especially in the early mornings and late afternoons when the temperatures are lower. One to several individual flies could be found on a single flower, lingering for hours at a time. The insects were observed to rest immobilised

---

\* The exact words in Latin were "*Flos intus viscidus a parva mosca quadam diligentissime visitatus*".

**Fig. 3.2.1.** The large, orange-flowered species of *Specklinia* use aggregation pheromones to attract the fruit flies that pollinate them. **A.** A fruit fly regurgitating on *S. pfavii*. **B.** Two fruit flies feed on *S. remotiflora*. **C.** A single fruit fly inspects the sepals of *S. spectabilis*. **D.** A fruit fly feeds after removing the pollinarium of *Specklinia* sp.
© APK.

on a single spot for up to one hour during these prolonged visits. However, most of their time was spent socialising. The most common behaviours included following each other around the flower, fencing with their forelegs, and flapping their wings. Copulation was observed very occasionally. Flies were sometimes seen holding a large drop of liquid with their mouthparts — regurgitated stomach contents that flies offer to their mates as nuptial gifts. The insects also interacted continuously with the floral parts. They repeatedly explored the inner and outer surfaces of the sepals using their mouthparts, walking from one sepal to the other and making contact with the petals, lip and column in the process.

Many questions arose from observing the insects' peculiar behaviour. What makes the *Specklinia* flowers so attractive to these flies? Why do they linger for such long periods of time? Is there a hidden reward system in place? Is each *Specklinia* species visited and pollinated by particular species of fruit fly?

The first step to solve these puzzles was to analyse the floral scent. Franco and I had noticed that during the daytime, the orange-flowered *Specklinia* species emitted a delicate fruity aroma. In order to analyse the components of the floral scent, the flowers were placed for a few minutes in hexane, a commonly used solvent. Then, using a technique called gas chromatography–mass spectrometry (GC-MS), we identified the different substances in the sample. This method does not tell you exactly what substances are present, but it generates a profile of all the components, which can then be compared with existing profiles of well-known compounds. In essence, we could compare the scent profile of *Specklinia* flowers with other scent profiles in a library of previously identified samples. Through the GC-MS method, we were able to identify a series of chemical compounds produced by these flowers, and here is where our story becomes quite remarkable.

Among the diverse components of the *Specklinia* scent profile that we identified were compounds called tiglates. It turned out that the very same tiglates had been previously identified as aggregation pheromones for certain fruit flies. In other words, the *Specklinia* flowers mimic the pheromones that *Drosophila* use to communicate with each other. The flies visit the flowers with high expectations. They are under the distinct impression that a member of their own species has identified a useful resource and is signalling them. Instead, it is the flower emitting a fraudulent message. That certainly explains why the *Specklinia* flowers are constantly being visited by *Drosophila*. However, one would expect the duped insect to eventually lose interest and leave. What makes them stay?

Prior studies had found no evidence of rewards being offered by *Specklinia*. The flower lacks the nectar repository found in many rewarding orchids, and the lip and petals are rather featureless. The large sepals, however, are covered with warts that are sucked on continually by the visiting flies. Despite that, no nectar production had been observed on the sepals either. Then, on a warm morning, we noticed tiny drops of a transparent viscous liquid on the warts of a plant kept in isolation. Eureka! We collected the little drops and slowly but surely the flower restored them. A laboratory test confirmed that the liquid was rich in dissolved sugars. In other words, it was nectar. With the help of micro-photographic techniques, we were able to determine that each sepal wart has a nectar-secreting pore or 'stoma' at its tip. The stoma opens, allowing the liquid to slowly flow to the surface, only accumulating as a visible droplet when it is not consumed by insects (see Figure 3.2.2). *Specklinia* flowers, therefore, have developed a two-step pollination strategy. They release aggregation pheromones, which are very effective at recruiting the flies, and at the same time exude a sweet, viscous nectar, which keeps the insects interested. But how do the *Drosophila* actually pollinate these orchids?

The nectar that the orchid offers comes in relatively low quantities and is mostly presented on the edges of the sepals. It keeps the fly motivated, but flows slowly, forcing the insect to walk around in order to feed. The insect strides from one side of the flower to the other picking up small amounts of nectar at a time, and it is while wandering from edge to edge that the insect eventually steps onto the central lip. Only when positioned precisely above a certain tipping point will the lip tilt and press the fly against the column. The insect becomes trapped between the column and lip for up to half an hour and, after a struggle, escapes with the pollinia attached to its back.

The relationship between these orchids and their pollinators is not species-specific. Any fly that responds to those pheromones, and is the right size to tilt the lip, will do the trick. In fact, more than a dozen different, yet closely related species of *Drosophila* have been observed visiting the flowers of *S. endotrachys* and its relatives.

**Fig. 3.2.2.** The bright-orange flowers of *Specklinia endotrachys* from Costa Rica secrete tiny nectar droplets that accumulate on the apices of warts on the sepals. The fruits flies that pollinate this orchid walk up and down the flower feeding on the droplets.
© APK.

Admittedly, this may not sound like a very efficient system. It can take a long time and several tries before an insect steps on the lip in such a way that it is tilted in the right direction. Simultaneous visits by multiple individuals of different species for several hours are probably needed for this pollination strategy to work at all. It is not surprising, therefore, that in nature not many *Specklinia* flowers actually get pollinated. Fewer than 10% of flowers develop into fruits. The low fruit set is possibly compensated by a continuous production of flowers throughout the year and a large number of seeds per fruit. Oddly enough, some *Specklinia* species form very large populations of thousands of individuals that grow in close proximity, so they can certainly not be described as unsuccessful.

Species of the genus *Masdevallia*, a close relative of *Specklinia*, are among the most popular orchids in cultivation. They are especially diverse in the Andes of Colombia, Ecuador and Peru, where hundreds of beautifully coloured variants can be found. There have been many observations of insects visiting *Masdevallia* flowers, but somewhat inexplicably, there is not a single detailed pollination study for any one species. It isn't clear how *Masdevallia* flowers attract their pollinators, but one can see similarities with *Specklinia* in some species. However, I will refrain from speculating and will stick to what we have actually observed.

Like *Specklinia*, some *Masdevallia* species have a fruity smell and the flowers are commonly found covered in fruit flies. Several species have large sepals with warty surfaces, and flies can be observed wandering about, sucking on them. The large sepals create a tubular flower and the flies are somehow guided to the base. There, they step on the movable lip and are pressed against the column, held in place by the short lateral petals. The hook-shaped pollinarium is removed when the insect is able to exit the cavity, and placed exactly in the same position as in *Specklinia*. How and which flies are attracted to the *Masdevallia* flowers remains to be studied. It is not clear yet what the function of each floral part is in this particular scheme, nor is it known if the flowers are rewarding or deceiving their visitors. But we know they are pollinated by flies that enthusiastically wander the flowers searching for something and eventually get caught in the cavity formed by the column and movable lip.

It is commonly believed that some brightly coloured *Masdevallia* species are pollinated by hummingbirds. However, there is little proof to support this claim. In fact, both direct and indirect evidence suggests quite the contrary. Flies have been observed visiting and removing the pollinaria of flowers belonging to several species of *Masdevallia* (see Figure 3.2.3). This is consistent with the size and function of the lip and petals, which are clearly adapted to holding a small insect. The flowers of *Masdevallia* are very diverse in the size, shape and colour of the sepals, but all the other reproductive structures are virtually identical, which suggests that different species are pollinated by similar agents. Transitioning from the highly precise pollinia placement on a small fly to the removal of pollen by birds would surely have been accompanied by drastic morphological changes, as we have seen in other orchids. *Masdevallia* flowers lack the tubular or cup-shaped nectary typical of bird pollination. The flowers may or may not be secreting small amounts of nectar to feed the flies, but they certainly do not have large nectaries. Finally, some of the brightly coloured *Masdevallia* species, including *M. ignea* and *M. rosea*, cannot be pollinated by a hummingbird simply because the front entrance to the tubular flower is blocked by one of the sepals. A fly, on the other hand, can easily crawl in from the sides.

**Fig. 3.2.3.** Many *Masdevallia* species, such as *M. bicolor*, have large, brightly coloured flowers that are pollinated by tiny flies.
© Rogier van Vugt.

**LINGER**
*Specklinia* pollination by fruit flies

**SCAN ME**
https://youtu.be/LAw54SYZVLg

The bright yellow to orange flowers of species of the *Specklinia endotrachys* group mimic the aggregation pheromones that fruit flies use to communicate the presence of a food source. By producing these volatiles they attract the flies, which belong to genus *Drosophila* and remain on the flowers for hours at a time. On the flowers, the insects interact by chasing each other, fencing, flapping their wings and even copulating. On the three large sepals, the flower secretes nectar from wart-like glands, and the flies eagerly walk from one side of the flower to the other sucking on them for food. Pollen removal occurs when the fly steps onto the lip, which then tilts and presses it against the column.

The videos and photographs shown here are owned by Adam Karremans and have been reproduced and edited by the author with permission. Videos published as Supplementary data in Karremans *et al.* 2015 (https://doi.org/10.1093/aob/mcv086) have been reproduced with the kind permission of *Annals of Botany* and Oxford Academic.

## 3.3 MISUNDERSTOOD

Orchids can also be pollinated by birds, that is an undisputed fact. My own first encounter with bird-pollinated orchids occurred only a few years back. While observing the behaviour of insects on the flowers of orchids at Lankester Botanical Garden in Costa Rica, I noticed something moving very quickly in the corner of my eye. Just a few steps away from where I was working was a large patch of *Arpophyllum giganteum* in full bloom (see Figure 3.3.1). Looking more carefully, I could see a number of bees relentlessly visiting the bright fuchsia-coloured flowers. But every now and then something larger would swiftly swing by. I stopped what I was doing and stared at the *Arpophyllum* flowers. Suddenly, I saw it, clear as day. It was a small hummingbird! And it was quick, very quick. Far quicker than my reaction time — as I would soon find out. After three unsuccessful attempts at pulling out my phone to capture the events with its camera, I gave up.

The next day, determined to 'catch' the bird in action, I fixed a camera to a tripod in front of the flowers and set it to record in video mode nonstop. As the speedy hummingbird visited and revisited the small *Arpophyllum* flowers it briefly but consistently showed up in several frames throughout the film. Scrutinising its behaviour was something quite special. In a matter of seconds — and with utmost precision — the bird systematically pecked each of the dozens of tiny purple flowers* (see Figure 3.3.3). The pecking was also highly effective. When played in slow motion, the footage shows the dull grey pollinaria being picked up on the tip of the bird's beak, only to be lost again after pecking the next flower. In the following weeks, most of the flowers developed into fruits, with the notable exception of those where the birds' flight was somehow physically impeded.

Understandably, proof like this of bird pollination in orchids is relatively scarce. This has caused widespread scepticism. However, ornithophily (bird pollination) probably isn't as rare as one would imagine, considering the small body of evidence that is available. Bird visits are normally quick and intermittent, and may therefore be difficult to record and study. When we think of bird pollination in orchids we immediately associate it with flowers that have bright colours, lack a discernible scent, and have copious amounts of nectar held in long tubes. This is what the floral syndrome hypothesis tells us, and as a matter of fact, several orchid flowers with those features have indeed been proven to be ornithophilous. However, generalisations about pollination based solely on floral morphology can be misleading. Not all bird-pollinated flowers are brightly coloured, unscented and have nectar-rich, long tubes, and not all flowers with these characteristics are ornithophilous.

Besides being swift, birds are also known to casually explore flowers without actually pollinating them. A curious animal hovering in front of an inflorescence may easily trick a naïve observer into thinking it's a pollinator. Not surprisingly, there are several species of orchids that are

---

\* The motion can only be compared with that made by the hand of the most experienced lab technician while pipetting PCR plates.

**Fig. 3.3.1.** The tiny, sturdy, nectar-rich, brightly coloured flowers of *Arpophyllum giganteum* are visited by several animals in search of nectar.
© Grettel Salguero.

popularly believed to be bird-pollinated but are probably not. Common examples include the brightly coloured species belonging to the genera *Epidendrum* and *Masdevallia*. Even though their bright flowers can occasionally be inspected by a curious bird, there is no evidence that they are ornithophilous. Furthermore, every single pollination study involving species belonging to these genera shows that they are pollinated by insects\*, even those that have spectacularly bright-coloured flowers.

There are about a dozen neotropical orchid genera that are said to be bird-pollinated. Unfortunately, for the vast majority no proof has ever been put forward. This makes it especially challenging to produce an overview on the subject of bird pollination in orchids. There are certainly more orchids pollinated by birds than those discussed here. However, because lots of confusion and misunderstanding already exists around the subject, I will not be addressing those for which only anecdotal evidence of ornithophily is available. At the end of this section, I will make an exception for the genus *Comparettia*, which is so tenaciously and persistently portrayed as hummingbird-pollinated that its exclusion from the ornithophilous orchids in this book merits further explanation.

Among the orchid-pollinating birds, hummingbirds are the most common, and also the most charismatic. Today these marvelous creatures are only known from the Americas, where they began to diversify about 22 million years ago. But fossil records show that the ancestors of hummingbirds once had a much broader distribution. They evolved in Eurasia and only later invaded the New World. In neotropical habitats, the presence of these tiny nectar addicts permanently modified many plant communities. Hummingbirds developed a specialised relationship with flowers, putting strong selective pressure on certain favourable floral traits and driving the diversification of several plant lineages.

Hummingbirds are the only birds known to pollinate orchids in the New World. Species of the neotropical genus *Elleanthus* are among those best known for being pollinated by hummingbirds. Many have the obligatory fuchsia, orange or red-coloured bracts or flowers, which are typically odourless, nectar-secreting and narrowly tubular. Detailed

---

\* See sections 2.7 'Never enough' and 3.2 'Linger' for more on the pollination of *Epidendrum* and *Masdevallia* species, respectively.

**Fig. 3.3.2. A.** The orange-flowered *Elleanthus oliganthus*. **B.** A rufous-tailed hummingbird pollinating *Elleanthus* species.
© APK.

studies on the pollination of *Elleanthus* species have shown that their floral nectaries secrete a steady trickle of sucrose. The constant production of sugars is certainly an incentive for the hummingbirds, but it also encourages the presence of many other guests. I have observed several species of bees, beetles and butterflies continually visiting the bright yellow flowers of an *Elleanthus* species in Colombia. The nectar thieves enthusiastically probed several flowers but never removed any pollinaria. Only hummingbirds effectively pollinate the flowers (see Figure 3.3.2).

Producing a small but steady stream of sucrose stimulates the birds to visit flowers of many different plants in the same area. After a while the nectar of the emptied flower is replenished and it may be visited by hummingbirds over and over again. This strategy promotes the outcrossing of pollen between different individuals of the same species and secures high fruit set. The pollinia are removed when the hummingbird inserts its beak into the floral tube or cup while attempting to reach the base. At the base, a pair of nectariferous calli* secrete the precious sugary liquid. Insertion of the beak triggers a lever mechanism that places the pollen just below the point of the bird's beak.

---

* Plural of callus, a term given to any discrete, raised portion of the lip.

The distantly related orchid genera *Arpophyllum*, *Ornithidium*, *Rodriguezia* and *Sacoila* — all belonging to different orchid tribes — are also among those confirmed to be pollinated by hummingbirds. Like *Elleanthus*, they too have the stereotypical brightly coloured, non-fragrant, tubular, nectar-rich flowers. *Ornithidium* literally means 'bird flower'. Many *Ornithidium* species have small, bright purplish, reddish or yellowish, bell-shaped or globular flowers that are produced in large clusters. Careful anatomical and micro-morphological inspection of the flowers of *O. coccineum* and *O. sophronitis* definitively shows that they offer nectar rewards. The sturdy flowers of both species have a small protuberance that produces a slow but constant flow of nectar during the day, as observed in *Elleanthus*. Unsurprisingly, hummingbirds are known to visit the flowers and are assumed to pollinate them. This

**Fig. 3.3.3.** The rufous-tailed hummingbird swiftly visits the nectar-rich flowers of *Arpophyllum giganteum*, transferring pollen as it pecks the flowers.
© APK.

is not a far-fetched assumption given the overall similarity of some *Ornithidium* flowers with those of *Arpophyllum* and *Elleanthus*.

No study has definitively shown that *Ornithidium* species are truly ornithophilous. However, a few observations have been made in Costa Rica. During her undergraduate studies on orchid pollination at the Bosque de Paz private reserve, Noelia Belfort observed hummingbirds visiting and pollinating *O. fulgens* (see Figure 3.3.5C). A recently published photograph — taken in the Monteverde cloud forest — also shows a hummingbird hovering in front of the flowers of *O. fulgens*. The very particular floral features of *Ornithidium* are indeed consistent with ornithophilous pollination, and it is probably safe to say they are exclusively pollinated by hummingbirds.

In Brazil, butterflies, bees and wasps have been seen visiting the bright red, nectar-rich flowers of *Rodriguezia lanceolata*. However, as in the case of *Elleanthus*, only hummingbirds effectively removed pollinaria. Again, the birds visited the flowers only for a few seconds at a time, returning repeatedly throughout the day. They hovered in front of the inflorescence and inserted their bill into the tube to probe for nectar. This action detached the pollinarium. A similar observation was made for the terrestrial orchid *Sacoila lanceolata*. Hummingbirds inserted their bills to reach for the nectar that had accumulated at the base of the flowers. While probing the nectar, they got the pollinarium stuck on the tip of their bills. A single individual bird was seen carrying up to four granular pollinia of *S. lanceolata*. Contrary to those of most orchids, the pollinia of *S. lanceolata* can easily crumble and the fragments may pollinate several different flowers.

Orchid pollinaria are highly variable in shape, size and composition. The colour of the pollinia, nonetheless, is pretty much the same across the whole family. The vast majority of orchids have bright yellow pollinia. However, in a note published in 1971, Bob Dressler emphasised that the pollinia of many hummingbird-pollinated orchids, including *Arpophyllum* and *Elleanthus*, are distinctively purplish or greyish-blue rather than yellow. He found that the dull colours were more similar to those of the beaks of many hummingbirds. Dressler proposed that dull-coloured pollinia would make less contrast with the beak and therefore reduce the chance of the bird attempting to clean them off. There is without a doubt a much higher incidence of dull-coloured pollinia in orchids that are reputed to be pollinated by hummingbirds, so it seems

Fig. 3.3.4. Diverse dull-coloured pollinaria of typical of hummingbird-pollinated orchids — *Arpophyllum spicatum*, *Elleanthus oliganthus*, *E. cynarocephalus*, *Camaridium horichii* and *Ornithidium fulgens* —, a case of evolutionary convergence. © APK.

likely that bird pollination has indeed driven the evolution of greyish pollinia (see Figure 3.3.4). However, as far as I'm aware, Dressler's hypothesis has never been tested. It also raises the question of why dull pollinia aren't found in all hummingbird-pollinated orchids? Part of the answer may lie in the correct determination of which orchids are truly hummingbird-pollinated. But that may not fully resolve the issue. Pollinia of whitish hues are also found among hummingbird-pollinated orchids, including the previously discussed *Ornithidium* and *Sacoila*. Is it possible that those light-coloured pollinia are also far less disturbing to the bird than the bright yellow ones? Experimental studies are needed to answer these questions.

I have underlined the point that not all orchids commonly believed to be ornithophilous are actually bird-pollinated. Typical examples of this confusion are *Epidendrum* and *Masdevallia*. Another is the genus *Comparettia*. Dodson regarded *C. falcata* as a clearcut example of hummingbird pollination, without presenting any evidence. The notion was reinforced by a study from Puerto Rico that claimed that *C. falcata* was pollinated by a bird endemic to the island. Unfortunately, the authors were not able to witness pollen removal or visitation of any sort by the bird. Instead, they relied on the presence of pollinarium remnants

Fig. 3.3.5. **A.** *Comparettia falcata* is commonly believed to be a hummingbird-pollinated orchid. **B.** *Oncidium strictum*, which has been seen to attract hummingbirds. **C.** *Ornithidium fulgens*, perhaps a true hummingbird-pollinated orchid.
© Franco Pupulin (A), Ron Parsons (B), Gustavo Rojas Alvarado (C).

(accessory structures) stuck on the beak of a single hummingbird caught in a net in the vicinity of the study site. Hardly solid proof. However, it did lend support to Dodson's initial proposal, and the belief became widespread. A photograph of a bird inspecting the flowers of *C. falcata* appeared in a paper published in 2011 in the *Orchid Digest*, again strengthening the association between *Comparettia* and hummingbirds. Again, there was a catch, and it is a big one. The orchid featured in the picture is actually *Oncidium strictum*, not *C. falcata* (see Figure 3.3.5).

The only study that definitively demonstrates pollinaria removal by a pollinator in the genus *Comparettia* is presented in a Brazilian paper involving *C. coccinea*. The authors show that the flowers are adapted to pollination by butterflies, beautifully detailing the orchid's floral features and presenting the insects in action. Structurally, *C. coccinea* and *C. falcata* are essentially identical. The large lip, long, bent spur, and spacious entrance to the nectary are consistent with pollination being effected by a butterfly probing the deep nectary, head first.

The myth of *C. falcata* being ornithophilous is a magnificent example of how an initial speculation gains strength by mere repetition, eventually leading to circular reasoning in which authors set out to prove what is already believed to be true, misunderstanding the true nature of what they are so enthusiastically studying.

**MISUNDERSTOOD**
Hummingbird pollination in orchids

**SCAN ME**
https://youtu.be/JnviSNzZW9o

In the Neotropics, certain orchids are adapted to pollination by hummingbirds. These tiny, fast flyers have high energetic requirements and need copious amounts of nectar. Flowers pollinated by hummingbirds are therefore nectar-rich, and typically need to be more or less tubular and sturdy. They commonly have bright red, orange or purple flowers which catch the bird's eye. This has led to the misunderstanding that all brightly coloured orchids are pollinated by hummingbirds, while in fact many are bee- or fly-pollinated. A true hummingbird orchid is

*Arpophyllum giganteum*, which is pollinated by the rufous-tailed and scintillant hummingbirds of Costa Rica. The greyish pollinaria can be seen below the tip of the bird's beak as it systematically pecks the tiny bright fuchsia-coloured flowers. Other typical hummingbird orchids are the members of genus *Elleanthus*. Like *Arpophyllum*, and many other bird-pollinated orchids, *Elleanthus* flowers have converged towards having greyish pollinaria which are said to be less likely to be removed by the bird.

The videos and photographs shown here are owned by Adam Karremans and Noelia Belfort (Bosque de Paz Biological Reserve), and have been reproduced and edited by the author with permission.

## 3.4 MADNESS

In the humid jungles of Southeast Asia live a group of terrestrial orchids that are unlike any other. The members of the Apostasioideae, a rare subfamily that comprises only two genera and some 16 known species, are sisters to all the other orchids known today. They are commonly thought of as living orchid fossils of sorts because their flowers lack many of the classical orchidaceous features. Strictly speaking, however, they are not ancestors but rather an early diverging lineage within the orchid family. This is evidenced by the presence of traits considered to be advanced in the subfamily.

Nevertheless, the Apostasioideae do have certain morphological features that can be considered primitive (see Figure 3.4.1). A striking example is the lack of a column and the presence of multiple anthers. In most orchids, the style and stamens are fused into a single central structure known as the column. The column commonly bears a single anther at the very tip, and a stigma just below that. In *Apostasia* and *Neuwiedia*, the two genera of this odd subfamily, the style and stamens are only partly fused, with either two or three separate anthers loosely surrounding an independent stigma. They are the only orchids that lack a true column, and one of the few with multiple anthers.

*Apostasia* and *Neuwiedia* also lack pollinia, the distinct packets of aggregated pollen that, together with their accessory structures, form the pollinarium in many orchids. The pollen in Apostasioideae is made up of pollen grains loosely placed along the anthers. This is quite an important difference. Many flowering plants have copious amounts of

**Fig. 3.4.1.** *Neuwiedia veratrifolia*, growing in its native habitat in Sulawesi. © Ron Parsons.

pollen. So much so that pollen is not only used for reproduction but also as a reward offered to visiting animals.

There is a strong and puzzling tendency towards reduced numbers of pollen grains in orchids. This leads to a condensation of pollen into discrete units that enter the stigma and move as a mass into the ovary below the flower where they pollinate the ovules, rather than moving individually. That may sound like a disadvantage, but it is not necessarily so. The evolutionary path that orchids have taken reduces the squandering of pollen and favours a more precise placing of each individual unit into the stigma. There is a high risk that no pollen makes it to the stigma at all, but when it does, the large amount of pollen ensures that many ovules are fertilised by the multiple compacted grains. Economically speaking, that means little investment in pollen but a high production of tiny seeds. By contrast, because the flowers of Apostasioideae lack nectar and have a lot of granular pollen, it has been speculated that they offer pollen as a reward to pollinators. This is otherwise virtually unheard of in orchids and, if confirmed, sets them apart from their sisters even more. But following the discovery of these peculiar orchids, it would take almost 200 years to finally unravel how they are pollinated.

The first botanist to recognise and describe any member of Apostasioideae was the German-Dutch botanist Carl Ludwig Blume. Born in Germany in 1796, Blume soon moved to the Netherlands where he studied medicine at the University of Leiden. Working for the Dutch in their colonies in Southeast Asia, he became the second director of the Buitenzorg Botanical Garden (today the Bogor Botanical Gardens) while stationed in Java (part of present-day Indonesia). In 1829, back in the Netherlands, he was appointed the first director of the Rijksherbarium*. Blume was one of the greatest scientists of his time and received the oldest and highest civilian order of chivalry when he was made a knight of the Order of the Lion of the Netherlands. But he also had a reputation for being a stubborn and antagonistic person, handling the collections under his management suspiciously and monopolistically.

---

\* The national herbarium of the Netherlands, or Rijksherbarium, was originally founded in Brussels, but moved to Leiden when Belgium became independent a year later. It was renamed the Nationaal Herbarium Nederland at the end of the 20th century, and is today still in Leiden where it is known as Naturalis Biodiversity Center.

The doors of the Rijksherbarium remained virtually closed during Blume's directorship, which lasted until his death in 1862. Despite his eccentricity, he was known to be an excellent taxonomist. Famously perseverant and highly active, Blume extensively studied the flora of Southeast Asia, particularly on the island of Java. In 1825, while stationed in Java, he proposed the genus *Apostasia,* immediately recognising it as an orchid, despite its anomalous morphological features. A few years later he also described its sister genus *Neuwiedia*, identifying it as a close relative of *Apostasia*. Blume's hypotheses would be challenged by several subsequent authors. Some considered *Apostasia* and *Neuwiedia* to be unrelated to each other, while others thought they were not orchids at all. However, modern DNA sequence data proves he was right all along. *Apostasia* and *Neuwiedia* are indeed close relatives, and sisters to the rest of the orchids.

Several other authors have built on our knowledge of *Apostasia* and *Neuwiedia* over the years, including a few other Dutch botanists associated with Leiden\*. Noteworthy are the works of Johannes Jacobus Smith, who described several species of *Apostasia* and *Neuwiedia* about a century after Blume. Another Dutchman, J. J. Smith, was likewise stationed in Java, where he initially had a position as inspector of a coffee plantation. But his interest in the local flora secured him a job at the Buitenzorg Botanical Garden shortly after. Smith was initially appointed assistant curator of the herbarium (Herbarium Bogoriense) and afterwards headed the institute until his retirement. He returned to the Netherlands, settling first in Utrecht and finally in Oegstgeest, near Leiden. True to form, the most comprehensive modern monograph of the Apostasioideae was later published in the 1960s by another Leiden botanist, Eduard Ferdinand de Vogel. Ed thoroughly revised both genera and described half the species of *Neuwiedia* known today.

Despite having been known to scientists for about two centuries, the first pollination reports on any species of Apostasioideae were

---

\* The close relationship between Leiden and the flora of Southeast Asia is a longstanding one. This tradition became a *modus operandi* of sorts when in 1955, through a gentleman's agreement, the Rijksherbarium in Leiden, the University Herbarium in Utrecht and the Herbarium Vadense in Wageningen, resolved to concentrate their activities on the floras of Asia, tropical America and Africa, respectively. This policy of geographical separation between the three main Dutch herbaria was still in place until quite recently, when both the Utrecht and Wageningen herbaria were moved to Leiden.

not published until the 1990s. Just like many of its floral features, the pollination strategy of these orchids is unlike any other. Their flowers are devoid of nectar but nonetheless reward their pollinators. Japanese scientists have shown that the fresh *Neuwiedia* flowers are foraged for pollen grains by stingless bees in the early mornings. When the flowers open, the loosely placed pollen grains fall from the anthers easily onto the floral segments. Bees belonging to the genus *Tetragonula* not only collect the fallen pollen grains but also scrape them out from the anthers (see Figure 3.4.2). They are said to hold on to one of the three stamens with their legs and collect the pollen grains using their mandibles. In the process, part of the pollen falls straight onto the stigma, or is transferred to it by other visiting bees. The stingless bees store the gathered grains in their corbicula or pollen basket, a smooth area on their hind legs surrounded by hairs that is used to transport pollen. The bees move from flower to flower mainly within the same inflorescence. This means that the flowers are mostly fertilised by autogamy (fertilisation of a flower with its own pollen) or geitonogamy (fertilisation with pollen from another flower on the same plant). The high fruit set indicates that these plants are self-compatible and efficient selfers. Cross-pollination — which is important for genetic diversity — does occur thanks to the action of the bees, which occasionally move pollen from one plant to another.

Fig. 3.4.2. A. *Neuwiedia zollingeri* being pollinated by the *Tetragonula* (*Trigona*) *melina*. B. *Tetragonula* (*Trigona*) *laeviceps* pollinates *N. veratrifolia* through pollen collection.
© Alexander Kocyan.

Unlike those of *Neuwiedia*, the flowers of *Apostasia* hang downwards, with the sepals and petals bent backwards and two anthers forming a cone-shaped structure. No pollination records have been published for any of the species belonging to this genus, but it has been speculated that they too offer pollen as a reward to buzzing bees.

Apart from the Apostasioideae subfamily, the only other orchids in which direct pollen collection from anthers has been reported are *Epistephium* and *Psilochilus*. Both are terrestrial orchids that have granular pollen and belong to relatively primitive orchid groups. In *Epistephium* and *Psilochilus*, the bees appear to actively collect the pollen using their front legs. As in *Neuwiedia*, they gather the collected pollen in their corbiculae, and during this process also transfer some of the pollen to the stigma. Contrary to the situation in the vast majority of orchids, in these genera, the granular pollen is not aggregated into discrete pollinia and transported as a unit by pollinators. This allows for part of the pollen to be used as a reward. Do other orchids offer pollen as a reward? Perhaps. It is certainly possible that other orchids with granular pollen may dispense part of it as a reward. However, in most orchids pollen is a limited resource that is presented as a single unit, so it is unlikely that this is a generalised mechanism in the family.

## 3.5 TWO OUT OF THREE AIN'T BAD

We have learned that even though many flowering plants offer pollen as a reward, this occurs very rarely in orchids. Pollen rewards in the Apostasioideae and a few other orchids are the exception and the reason is straightforward. In most orchid species, pollen is a scarce resource and the grains are clumped into a single unit that is kept stowed away within the anther — unaccessible to many pollen-foraging animals. Apart from some more primitive groups, orchid pollen grains are typically tightly packed into larger packages that work as a single functional unit, the pollinarium. In advanced orchids, the entirety of the pollen of a single flower goes into forming the pollinarium and therefore offering pollen as a reward to the pollinator would be a problem. However — as we have come to expect — orchid flowers have figured out a way to circumvent this limitation and to profit from the pollen-foraging behaviour of certain insects. They still offer their pollinators food in the form of pollen, but without compromising their own reproductive fitness. Rather than wasting their precious pollen grains, these orchids offer a pollen-like substance with a similar appearance and nutritional value.

The idea of pollen mimicry in orchid flowers was first put forward by J. M. Janse in the 19th century. Just like the Dutch scientists mentioned earlier, Janse was also a botanist based in Leiden and he too was associated with both the Bogor Botanical Gardens and the Rijksherbarium. But unlike the contributions of his colleagues — which were centred around Asian plants — Janse made his discoveries on a group of neotropical orchids. While studying the flowers of species in the genus *Maxillaria*, he noticed a pollen-like reward. This counterfeit pollen, or 'pseudopollen', is a pollen lookalike that does not originate in the anther and cannot fertilise any ovules. Certain *Maxillaria* species provide pollinators with this mealy pseudopollen, which appears as a whitish, flour-like powder on the lip. The pollen-mimicking material, which can be easily picked up by foraging animals, is formed by the fragmentation of multicellular trichomes — hair-like growths that form short chains of individual cells.

Pseudopollen is sometimes referred to in the literature as food-hairs. The term refers to edible trichomes found mostly on the lip of orchid flowers. The difference between the terms pseudopollen and food-hair is not clearcut and researchers may use them interchangeably. Some authors argue that true pseudopollen is formed by the fragmentation of trichomes into rounded, individual component cells and that these are therefore distinguished from other types of food-hairs that consist of relatively few cells that become detached as a unit from the lip. Here, we refer to pseudopollen as those food-hairs that detach naturally from the floral segments and therefore do not require an animal to disturb them in order to become severed. Just like real pollen, pseudopollen is loosely aggregated and easily detached.

Even though insects evidently collect or consume pseudopollen, is it actually a reward? Does it contain nutrients? It has been established that the main food material found in food-hairs is protein. However, Vogel suggested that even when being offered to and collected by visiting insects, pseudopollen may in some instances lack nutrients. Kevin Davies, a botanist at Cardiff University in the United Kingdom, calls this a 'dual deceit strategy' in which pseudopollen not only mimicks food-laden powdery pollen, but may sometimes lack a food reward completely. A variety of substances have been reported to be contained within these edible trichomes, including oil droplets, protein bodies and starch. The main macronutrient found in the pseudopollen offered by orchid species in the genera *Maxillaria* and *Polystachya* is protein, whereas lipids tends to be absent. Starch has been reported in *Dendrobium* and is generally present in *Maxillaria*. However, it is uncommon in *Polystachya*. According to a thorough study on a series of closely related *Maxillaria* species, whether pseudopollen contained any nutrients was highly species-dependent. Some species contained abundant protein bodies and starch, others contained only one of the two, and yet others contained insignificant amounts of either. Many *Maxillaria* species thus offer either protein or starch, or both. Lipids, on the other hand, occurred in negligible quantities and are therefore not considered important as a food reward. I guess two out of three ain't bad, and pseudopollen can be safely regarded as a reward.

Stingless bees belonging to the genus *Trigona*, a member of the Meliponini tribe, systematically visit and pollinate the flowers of *Heterotaxis* (a *Maxillaria* relative) while gathering the lip's trichomes. The insects continue to do so throughout the entire flowering period of the orchid, which suggests that the trichomes provide them with useful nutrients. In Brazil, bees have also been proven to collect the nutrient-rich pseudopollen offered by *Polystachya* species. The bees were seen collecting the fake pollen using their forelegs and later transferring it to their hind legs to be stored. I have personally observed halictid bees, of genus *Lasioglossum*, collecting pseudopollen in this exact manner from the lip of *P. masayensis* at the Monte Alto Reserve in Costa Rica. The bees used their forelegs to collect the pseudopollen and then carried it on their hind legs. They worked until they could carry no more. Then they left, only to return after a few minutes to load up again. To date, however, there is no unequivocal evidence that pseudopollen is directly ingested by these insects.

*Trigona* species are common in many tropical forests, where they visit and pollinate a wide array of flowers. They are social bees that live in large colonies of several thousand individuals. Just like honeybees, the main resources of stingless bees are pollen and nectar, and they may visit flowers solely for the purpose of collecting these resources.

**Fig. 3.5.1.** Pseudopollen is offered as a reward by diverse orchids. **A.** *Heterotaxis (Maxillaria) maleolens*. **B.** *H. sessilis (Maxillaria crassifolia)*. **C.** *Polystachya foliosa*. The granular substance can be observed on the lip.
© APK.

But *Trigona* species also collect materials such as resin, wax, salt, and protein as a food source or to build their nests. Carbohydrates in nectar are the main energy source, while pollen typically serves as a source of proteins for both adults and larvae. Interestingly, individual stingless bees appear to specialise in one type of foraging. This may guarantee a more efficient exploitation of the pollen or nectar resource.

It isn't clear how widespread the use of false pollen in the orchid family might be. Only a handful of genera have been shown to deploy it. Nevertheless, these are geographically and genetically notoriously distant genera. This supports the hypothesis that pseudopollen evolved independently several times in the Orchidaceae. Pseudopollen is known to occur in the genera *Dendrobium* and *Mycaranthes* (also known as *Eria*) from Southeast Asia, in the genera *Heterotaxis* and *Maxillaria* endemic to the Neotropics, and in the widespread, pantropical genus *Polystachya*, of which most species are found in tropical Africa and Madagascar (see Figure 3.5.1). It seems likely that pseudopollen will eventually be found in other orchids as well.

### TWO OUT OF THREE AIN'T BAD
Pseudopollen and *Polystachya*

**SCAN ME**

https://youtu.be/Knz0Qs0Jv7I

Instead of providing real pollen as a reward, certain orchids have instead resorted to offering a pollen-like substance to profit from foraging bees. This pseudopollen can be found as a granular substance on the lip of species belonging to the genera *Maxillaria* and *Polystachya*, among others. Pseudopollen has been found to have nutritional value and is collected intensively by certain bees, such as the *Lasioglossum* (Halictidae) shown here, which in turn pollinate the orchids in the foraging process.

The videos and photographs shown here are owned by Adam Karremans and Miguel Méndez, and have been reproduced and edited by the author with permission.

## 3.6 DARK NECESSITIES

Spend an evening in the warm tropical rainforest without proper insect repellent and you will stare into the very eyes of evil. There are some places where you will be feasted upon by insects, no matter how many layers of clothing or mosquito nets you deploy. As you patiently sit in front of an exotic tropical orchid waiting for pollinators to perform, you may feel an annoying, hurtful sting, and then another, and another: a seemingly unending attack by invisible pests that can penetrate your clothes. In the height of desperation you may think you are going mad. The problem is that your mind is set on looking for mosquitoes, when actually the devilish creatures that are getting the best of you are something far worse. They are popularly known as no-see-ums — and rightfully so! These tiny blood-sucking, biting midges can easily slip through closed doors, screens and clothing. They are so small you may not even realise they are crawling on you.

Biting midges belong to the fly family Ceratopogonidae and include several thousand species distributed worldwide. To produce eggs, female no-see-ums need protein-rich meals, which they get by sucking blood from host animals. As if they weren't dreadful enough, some Ceratopogonidae are 'kleptoparasitic'. Kleptoparasitism is a macabre form of parasitism in which an insect feeds on the haemolymph — a blood equivalent found in invertebrates — of dead insects, which it steals from other animals that have captured and killed them. Bizarre as this may appear, it is even more remarkable that plants have found a way to take advantage of these tiny insects' dark necessities. The term kleptomyiophily has been used to describe the strategy employed by fly-pollinated flowers that mimic the compounds released from freshly killed insects. Kleptomyiophilous pollination, in which flowers take advantage of the kleptoparasitic insects that prey on corpses stolen from other animals, has now been reported in a few unrelated plants, including *Aristolochia*, or Dutchman's pipe, in the Aristolochiaceae family, *Ceropegia* in the Apocynaceae family, and, as always, orchids.

In the previous section, we learned about orchids providing highly nutritious food hairs as a reward to visiting pollinators. The stingless bees that pollinate the orchids forage for protein-rich pollen on other flowers. Presenting them with pollen-like substances therefore becomes an effective means for the orchids to secure pollination. There

Fig. 3.6.1. A–C. *Trichosalpinx blaisdellii* is pollinated by kleptoparasitic female biting midges, belonging to the family Ceratopogonidae, in search of proteins secreted by the flower.
© APK.

are, however, other ways to provide protein rewards. Species of the neotropical orchid genus *Trichosalpinx* — a close relative of the sexually deceptive genus *Lepanthes* and the nectar-rewarding genus *Specklinia* — are pollinated through kleptomyiophily by the tiny biting midges in the Ceratopogonidae family (see Figure 3.6.1).

How exactly do the flowers of *Trichosalpinx* benefit from klepto-parasitic biting midges? A team of researchers led by my colleague Diego Bogarín uncovered the remarkable mechanism. Species of *Trichosalpinx* have dark purple flowers in which the hairy lip is so delicately hinged to the base that it trembles in the wind.\* The colour, texture and movement of floral parts, together with the production of particular fragrances, are believed to be part of a complex system for mimicking dying insects. *Trichosalpinx* are visited exclusively by female biting midges that land on the sepals and then head straight towards the lip. The biting midges have well-developed mouthparts that they need to draw the protein-rich haemolymph from their prey. The flies seek and suck on — rather than pierce — the secretory glands present on the margins of the lip. Through the use of colour dyes, it has been established that both *T. blaisdellii* and *T. reflexa* secrete the precious proteins that the midges feed upon.

Whether the proteins are abundant enough to satisfy the insects' needs is still unclear; they may be nothing more than a nutrient tease. Nevertheless, they are sufficient to elicit the protein collection behaviour of voracious female Ceratopogonidae flies, and that is the only thing that matters to the orchid. In order for pollinia removal to occur, the fly needs to crawl far enough into the lip for it to tilt, briefly pressing the insect against the column. After struggling to release itself, the insect ultimately exits the flower with the pollinarium attached to its scutellum, the plate between its wings.

Even though both *T. blaisdellii* and *T. reflexa* are regularly visited by biting midges in Costa Rica, it is extremely difficult to spot the minute flies. Other morphologically similar *Trichosalpinx* species throughout the Neotropics are almost certainly pollinated in the same manner, but this remains to be proven. Small, dark flowers with narrow segments, and hairy vibrating lips are also found in a few other orchid genera besides

---

\*   You can read more on the role of vibrating lips in *Bulbophyllum* pollination in section 5.7, 'Blowin' in the wind'.

*Trichosalpinx*. Species of the closely related genera *Lankesteriana* and *Pendusalpinx*, as well as some belonging to the genus *Anathallis*, and the much more distantly related *Bulbophyllum*, can have almost identical flowers. Based on their floral resemblance, it is possible that they are using a similar pollination strategy. If confirmed, they represent yet another wonderful example of how evolution drives the flowers of unrelated species to look alike when they are destined to perform a similar function.

Biting midges are also known to pollinate many other flowering plants. The most prominent is the highly esteemed cacao or cocoa tree — *Theobroma cacao* — from which the raw material used to make chocolate is harvested. Perhaps unsurprisingly, one can observe a number of superficial similarities between the flowers of cacao and those of these orchids. Without pollination by biting midges, cocoa trees would not produce the highly valued cocoa beans. Annoying as they may be to us, these devilish creatures are responsible for producing one of the world's most heavenly delights: chocolate!

**DARK NECESSITIES**
*Trichosalpinx* pollination by bitting midges

**SCAN ME**

https://youtu.be/8zZDgFu1XQk

*Trichosalpinx* species are pollinated by tiny female biting midges. These flies require proteins for egg production and that is exactly what the orchid offers. Glandular trichomes on the floral organs secrete the proteins that the female Ceratopogonidae flies are after. It is while feeding that the biting midges move from the apex to the base of the lip. The weight of the insect causes the lip to tilt, pressing the female fly against the column. After a few minutes the midge is able to free itself, removing the orchid's pollinarium in the process. Several flowers that are also pollinated by biting midges have a similar floral morphology to *Trichosalpinx*, and are likely using a similar pollination mechanism. These include other orchid genera such as *Anathallis*, *Bulbophyllum* and *Lankesteriana*, and also *Theobroma cacao* — the chocolate tree.

The videos and photographs shown here are owned by Adam Karremans, and have been reproduced and edited by the author with permission.

## 3.7 YOU'RE SO VAIN

Have you ever wondered why you prefer a certain perfume over another? Perfumes are typically made up of blends of essential oils and aromatic compounds. Every perfume has a particular composition of diverse compounds, each one in a specific concentration. Only a true connoisseur may be skilful enough to identify each fragrance component. Their exact composition and concentration, however, is impossible to figure out and is kept top secret by perfumeries. It is the presence and prevalence of each compound in the blend, be it a major or minor contributor, that makes a scent more or less attractive. Plants are by far the largest source of fragrant compounds used in perfumery, and fragrance production to attract pollinators is in fact one of the key innovations of flowering plants. Many pollination syndromes in Orchidaceae employ scent as part of the strategy, and the intricate relationship between orchid flowers and fragrance-collecting bees of the Euglossini tribe is one of the most fascinating interactions involving plants and their pollinators.

Euglossine bees, which are commonly referred to simply as orchid bees, are well known for their unique scent-collecting behaviour. These bees are not only attracted to particular fragrances but they actively select certain compounds, pick them up and use them as part of their own reproduction strategy. Thanks to this very special behavioural trait, they play an essential role in a series of elaborate pollination syndromes that involve many fragrant neotropical orchids. Male orchid bees use perfumes to attract a mate. Unable to produce their own chemical attractants, these bees collect volatile aromatic compounds from diverse natural sources, such as resins, fungi, rotting vegetation and, of course, flowers. The bees search for and collect the volatiles to concoct highly specific perfume blends that they store in pouch-like pockets on their hind legs.

Once the male bees have accumulated sufficient quantities of each of the compounds needed for their perfume, they perform an elaborate courtship display in which the perfumes are dispersed. The females locate these and recognise the males based on the composition and frequency of the release of the fragrances. The scent is believed to play a critical role in their selection of mates. The vain males spend a good portion of their day looking for the compounds they need in order to complete their bouquet.

There are more than two hundred Euglossini species known today. They are easily recognised by their extremely long tongues and the broad variety of colours they display, from the classical black hues seen in other bees to some of the most spectacular bright metallic blue, green, red and bronze. Orchid bees are found only in the tropical forests of the New World, from Mexico to southern Brazil, where they are the exclusive pollinators of hundreds of different orchid species. In fact, these flying jewels are responsible for the reproduction of about 10% of all neotropical orchids. To ensure pollination, the plants use an incredibly diverse range of ways to take advantage of the bees' persistent need for fragrances.

The first observations of euglossine bees visiting and pollinating orchids were made by the German botanist Hermann Crüger in 1864. As director of the Botanical Garden in Trinidad, Crüger was in a unique position to witness the interaction between bees and orchids first hand. He prepared detailed descriptions of what he believed were 'humble-bees' (bumblebees) visiting flowers of the orchid genera *Catasetum*, *Coryanthes*, *Gongora* and *Stanhopea*. "The insects are attracted at first by the smell of the flower," the observant director noted. "But the smell probably only gives notice to the insects; the substance they really come for … is the interior lining of the labellum, which they gnaw off with great industry." Scent collection was completely unknown at the time, and the perfume collecting behaviour of euglossine bees was evidently not obvious to Crüger.

The true nature of the interaction didn't become apparent until the 1960s. Contrary to what Crüger, Darwin, and others believed, Orchidaceae that are pollinated by male euglossine bees do reward their pollinators. The floral rewards, however, are not in the form of nectar or pollen. Rather, these particular orchids produce large quantities of floral scents, which act as both an attractant and a reward. Instead of producing food, diverse fragrant orchids produce the volatile aromatic compounds that orchid bees are looking for. Male bees collect the compounds and in the process pollinate the flowers. Upon visiting a flower, the bee brushes the surface of the floral segments using a patch of hairs on its forefeet. This brushing behaviour is typically brief. The bee then hovers close to the flower while scrubbing its legs together, transferring the collected substances to pouches on its hind legs, before brushing the flower once again.

Fig. 3.7.1. Different neotropical orchids depend on the fragrance-collecting behaviour of male euglossine bees for pollination. **A.** *Lycaste bradeorum*. **B.** *L. candida*. **C.** *L. xytriophora*. **D.** *Macradenia brassavolae*. **E.** *Kefersteinia orbicularis*. **F.** *Trichopilia fragrans*. All are visited by different *Euglossa* species. © APK (A–C, F), Luis Enrique Yupanki (D), Karen Gil (E).

This procedure may be repeated by the bee for long periods of time. Some authors have suggested that the bees become somewhat inebriated during their visits and thus have reduced mobility and coordination. What is known is that during this process, mostly guided by the orchid's intricate floral morphology, the bee eventually gets stuck in an awkward position where it removes the pollinia.

Pollination by male orchid bees has evolved at least three times independently within the tribe Cymbidieae, resulting in separate radiations in the orchid subtribes Catasetinae, Coeliopsidinae, Maxillariinae, Oncidiinae, Stanhopeinae and Zygopetalinae. Among them we find the genera originally observed by Crüger — *Catasetum*, *Coryanthes*, *Gongora* and *Stanhopea* — but also many of the most charming orchid genera found in cultivation — such as *Acineta*, *Cochleanthes*, *Cycnoches*, *Dressleria*, *Lycaste*, *Mormodes* and *Trichopilia* — and even some of the tiny neotropical gems — including *Dichaea*, *Macradenia*, *Macroclinium* and *Notylia* (see Figure 3.7.1). Orchid bees also pollinate the world famous *Vanilla planifolia* and several of its close relatives. So euglossine bees are not only required for the survival of some of the most charismatic orchids in the world, but are also responsible for producing the fruits from which we extract the queen of all flavours: vanilla. But why have so many different orchid groups resorted to being pollinated by these particular bees?

Pollination by male euglossine bees may have one crucial advantage for the orchids. Male orchid bees leave the nest upon emerging from the pupa and do not return, travelling like vagabonds from one place to another in search of food and fragrances. In fact, research suggests that they can fly up to 100 km in a few days. Considering that orchid pollinia remain viable for a relatively long period, and are well attached to the insect's body, orchid bees are likely to serve as long-distance pollinators of orchids and thus as agents of gene dispersal over long distances.

The relationship between orchids and male Euglossini bees is relatively specific, but in a one-sided way. They are not mutually dependent. This 'asymmetric dependency' suggests that the pre-existing behaviour of male euglossine bees drove the diversification of euglossine bee-pollinated orchids. The orchids are highly dependent on the bees to reproduce, whereas the bees are likely to have multiple sources for their precious perfumes. There are many different

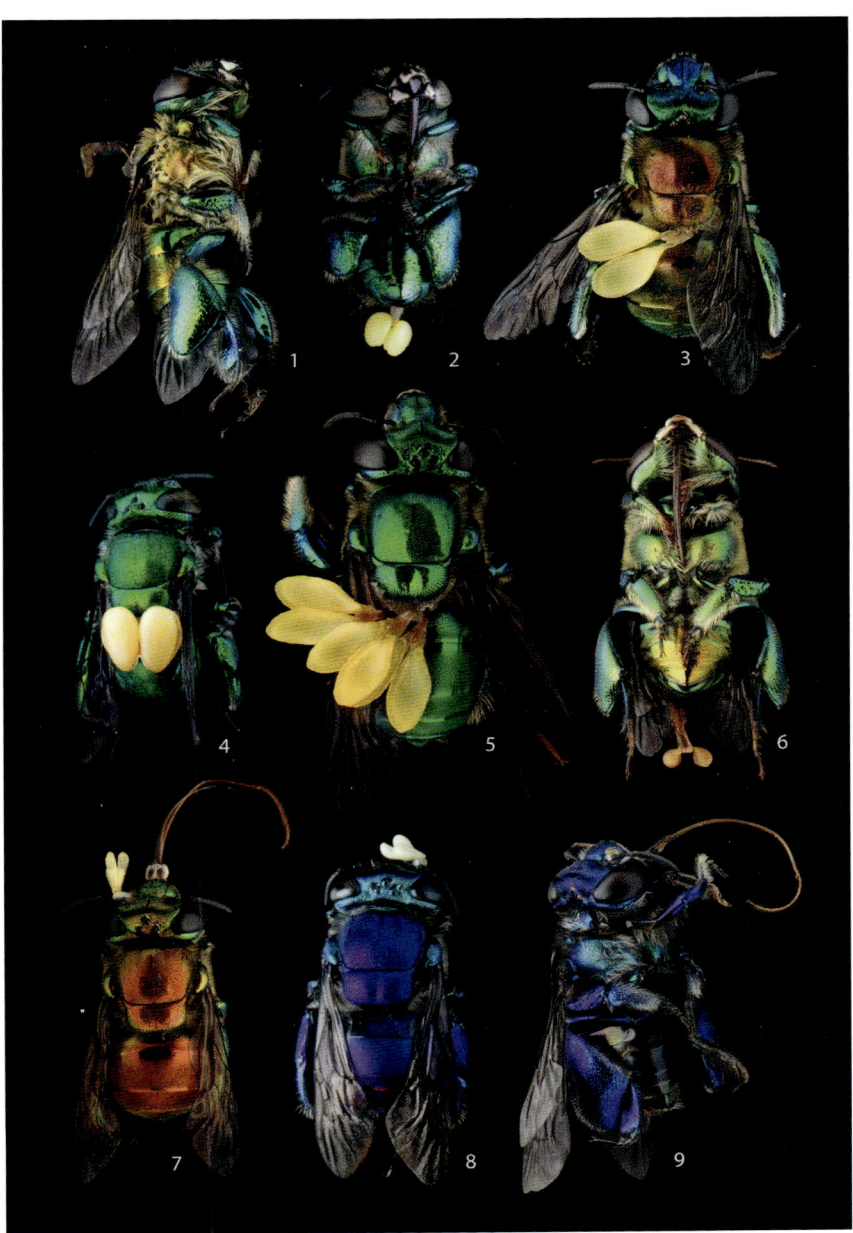

**Fig. 3.7.2.** Different *Euglossa* (Euglossini) species, collected around Costa Rica, carrying orchid pollinaria on different body parts. *Coryanthes* (4), *Mormodes* (1) and *Stanhopea* (3, 5) pollinaria are placed in the middle of the bees' back; the pollinaria of *Dichaea* (9) and *Kefersteinia* are placed on the antennae (7, 8); whereas the pollinaria of *Cycnoches* (6) and *Lycaste* (2) stick out on the very tip of the bee's abdomen.
© APK.

euglossine bees and each has a certain preference for fragrances. Every orchid flower, by contrast, produces a unique perfume that may attract only a subset of orchid bee species. A single orchid bee species may visit and pollinate several different orchid species. To avoid hybridisation (a cross between two different species) many of these orchids place their pollen on very specific places on the bee's body (see Figure 3.7.2). This allows for an individual euglossine male to carry the pollinaria of several different orchid species at the same time. To accomplish this precise positioning of the pollinaria, each orchid has developed intricate, yet extremely accurate, pollination mechanisms that rely on the naïve insect's fragrance-collecting behaviour. Some of the most outrageous mechanisms are described in more detail in the following stories.

**YOU'RE SO VAIN**
Pollination by male Euglossini bees

**SCAN ME**

https://youtu.be/rRVr7qmmJ7U

Several different neotropical orchids have converged on being pollinated by male Euglossini bees. These insects, commonly known as orchid-bees, have a unique behaviour that the orchid flowers exploit for their own purposes. Male orchid-bees collect aromatic compounds from many natural sources in order to concoct highly specific perfume blends, which they store in little pockets on their hind legs as part of a complex courtship behaviour that involves liberating samples of the scent. Orchids pollinated by male euglossine bees produce the fragrances, which attract the male bees. This is known as male euglossine bee-pollination and it has evolved independently in orchids of the subtribes Catasetinae, Coeliopsidinae, Maxillariinae, Oncidiinae, Stanhopeinae, and Zygopetalinae, in the subfamily Epidendroideae, and also in the genus *Vanilla* in the Vanilloideae subfamily.

The videos and photographs shown here are owned by Adam Karremans, Karen Gil and Luis Enrique Yupanki, and have been reproduced and edited by the author with their kind permission.

## 3.8 THAT SMELL

*"To those fortunate mortals, whose eye for structure has not been forever dimmed by the rapt contemplation of unending ranks of hybrid Cattleyas, nature offers many marvels."*
American botanist PAUL H. ALLEN (1954)

No phrase seems more appropriate to introduce this story than the opening words of Paul Allen's treatise on *Gongora* pollination. The genus *Gongora*, which our 'botanist-for-all-seasons', Robert Dressler, nominated as the most confused group of orchids, has always been a headache for botanists. The flowers are structurally very unusual. They are carried on a long, pendent raceme, with the fleshy lip uppermost, the lateral sepals twisted backwards, the dorsal sepal pointing straight downwards, and two tiny petals flanking the long, slide-like column. The thick lip is more or less conical, with a pair of horns at the base, another pair near the middle and a pointy, triangular apex. *Gongora* are fascinating, but they are also the plant taxonomist's worst nightmare.

*Gongora* flowers are unlike those of any other orchid. However, the many different species in the genus show only slight variations of the same standard floral form. This makes *Gongora* species really hard to tell apart on the basis of the shape of their flowers alone. However, they are extremely variable in colour. Some have a solid hue but most are beautifully adorned with elaborate colour patterns that may resemble intricate animal prints. Particular floral patterns result from a blend of blotches, spots, streaks and stripes of a diversity of colours, tones and intensities. These patterns may differ significantly within the same *Gongora* species; the density of the spots and the intensity of the colours are variable even among individuals within a single population. To make matters worse, researchers have shown that having similar patterns may not be indicative of a close relationship at all. Different species may look superficially more or less the same.

At this point, I am going to take the liberty of challenging you. Look very carefully at the black and white plate (see Figure 3.8.1). It shows flowers of different *Gongora* individuals from all around Costa Rica. Try your best to identify similar shape patterns among the flowers. Without looking at the colour plate, choose five or six silhouettes which you believe are close. Remember the ones you chose. Now try finding them in the corresponding full-colour plate on the next page (see Figure 3.8.2).

**Fig. 3.8.1.** Silhouettes of *Gongora* flowers from different individuals belonging to multiple species from around Costa Rica. Try searching for morphological patterns that may group the different species.
© Franco Pupulin & Diego Bogarín.

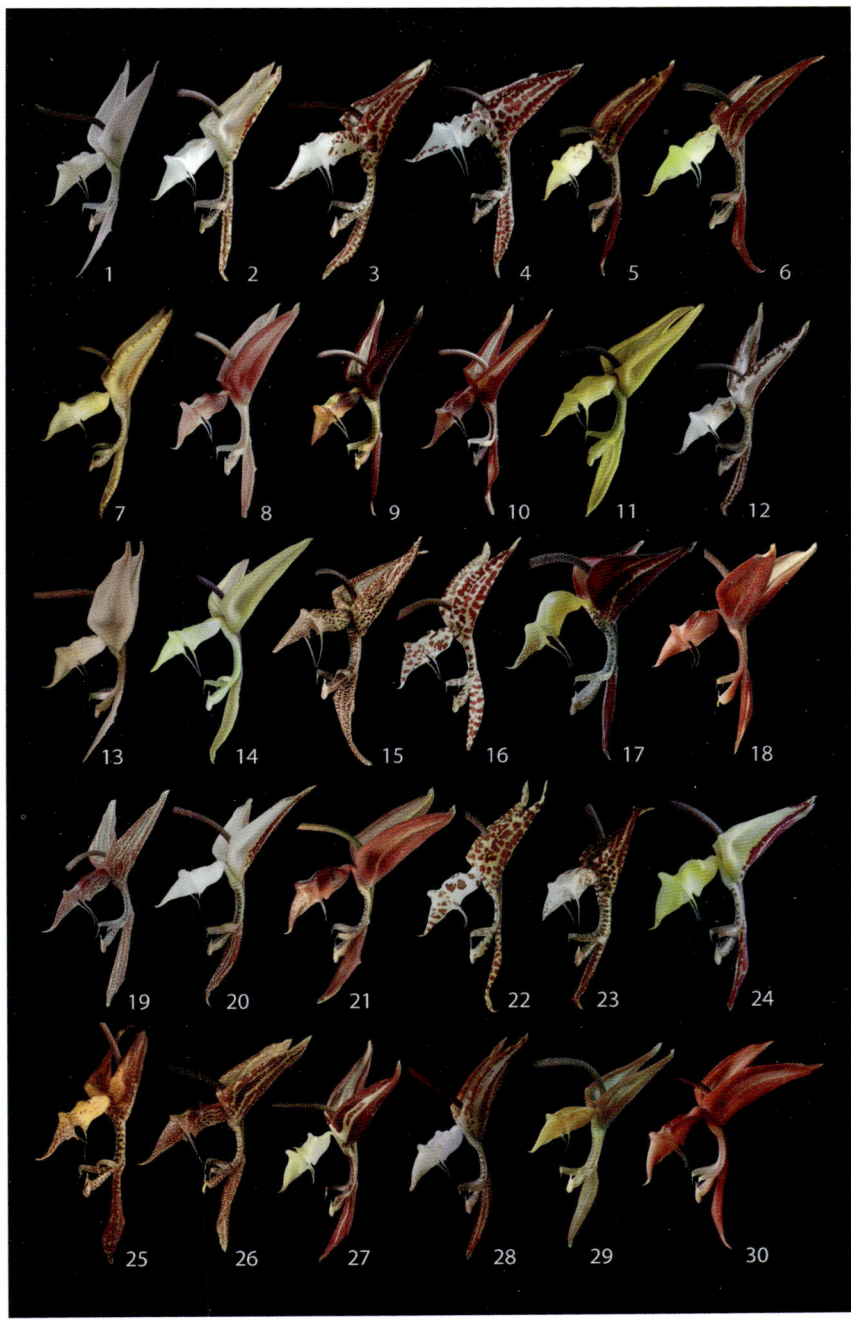

**Fig. 3.8.2.** The intricate colour patterns of *Gongora* flowers from different individuals belonging to multiple species from around Costa Rica. Try grouping the different flowers by their colour patterns.
© Franco Pupulin & Diego Bogarín.

Did the ones you choose have similar colour patterns? Now try doing it the other way around. Select a few flowers with similar colour patterns and then look up their corresponding silhouettes. Can they be set apart from the rest by shape alone? Did you get the result you expected? Most probably, the shape and colour of the flowers you selected do not agree very well with each other. Even though there are a few patterns to be found among these flowers, you are probably unable to say with confidence how many different *Gongora* species are shown. Don't worry, neither are we!

If flower shape and colour patterns are not consistent enough to distinguish between these individuals, how can we be sure we are actually dealing with different species at all? Well, we have to leave that to orchid bees. The male euglossine bees that pollinate *Gongora* are frequently of a gorgeous metallic green, red or blue, and may visit the flowers in large numbers. They hover around the intensively fragrant flower and land on the lip. The bee then attempts to collect the aromatic compounds found on the underside of the lip (see Figure 3.8.3). As the bee frantically clings upside-down to the waxy lip trying to reach the aromatics, it slips and drops straight onto the curved surface of the column. As it slides on its back, kept in place by the two lateral petals, the bee removes the pollinarium while passing the apical portion of the column. Despite the complex mechanism, the rate of pollinaria removal in *Gongora* is apparently high. Fruit-set, on the other hand, is rarely observed. This is in part due to the fact that *Gongora* flowers only last a few days before they wither. But probably also because the pollinia need a few hours to dehydrate and bend before they can fit into stigma, a strategy that may have developed to prevent selfing (as further discussed in chapter 5, section 5.4). Once the bee has removed a pollinarium, it has to visit another flowering *Gongora* plant to deposit and complete the cycle. Once again, the bee collects the fragrances on the lip and while reaching for the underside slips and falls onto the column. It again slides on its back, but this time places the pollinarium in the stigmatic slit, successfully pollinating the flower.

Pollinator attraction is often dictated by several signalling mechanisms, including floral morphology, colour and scent. Authors agree that floral scent can play a central role in mediating pollinator attraction and specificity. Dressler noted that by placing two similar *Gongora* species on each side of a trail in Panama, he could easily sort out

**Fig. 3.8.3.** Different Euglossini bees collecting fragrances on the flowers of *Gongora*. **A.** *G. cruciformis* visited by a *Euglossa decorata* male carrying the pollinarium of genus *Paphinia*. **B.** *G. unicolor* visited by *E. alleni*. **C.** *G. latisepala* visited by *E. aureiventris*. **D.** A spotted *Gongora* species is visited by a male *Euglossa*. **E.** A *Euglossa* species carries the pollinarium of *G. quinquenervis*. **F.** The same orchid is visited by a far too big *Eulaema* male.
© Günter Gerlach (A–C), Luis Enrique Yupanki (D), APK (E–F).

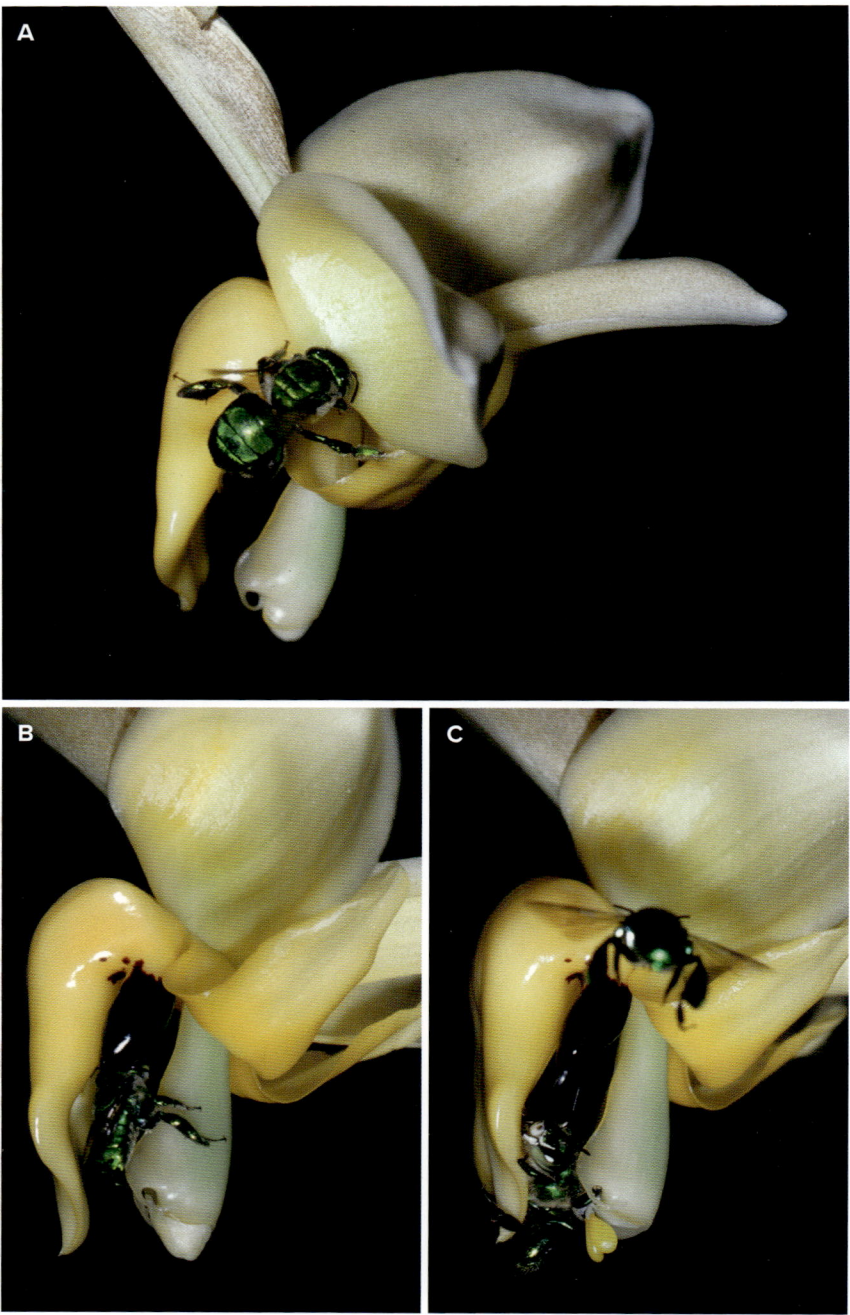

**Fig. 3.8.4. A–C.** A male *Euglossa imperialis* visits the perfumed flowers of *Stanhopea cirrhata*, in Quepos, Costa Rica. This bee fits perfectly between the column and lip, removing the pollinarium as it slides down the slippery lip.
© Günter Gerlach.

the local bee species. A bee that visited one orchid species never visited the other. This suggests that even though their floral shape and colour are similar to our eyes, the unique scent profiles of closely related orchid species may be enough to keep them reproductively isolated. More than 80 different combinations of floral volatiles have been described from *Gongora* species, each with a unique combination and proportion of those compounds. Every species of orchid bee looks for the particular compounds it needs, obtaining their precious perfumes from a diversity of natural sources. Some of the aromatic substances are easily found in other flowers, but there are others that are apparently only found in orchid-bee-pollinated orchids. A few compounds provided by the orchids have also been found in the pouches on the hind legs of male euglossine bees, proving that they do collect those particular volatiles. Accordingly, it has now been shown that visually indistinguishable, co-occurring species of *Gongora* can be reproductively isolated by having their own unique set of bee pollinators regulated through specific floral scents. They may look the same, but... can't you smell that smell?

The pollination strategy of the genus *Stanhopea* — a close relative of *Gongora* — is similar (see Figure 3.8.4). The flowers are quite different, but they too rely on male euglossine bees that, during fragrance collection, slip and fall onto the column, unknowingly taking the pollinaria with them. Like *Gongora*, the structurally complex *Stanhopea* flowers fuel the imagination. The lip of these 'little bulls' — as they are popularly called in Costa Rica — is divided into three distinct sections, each with a specific function. The basal part, named the hypochile, is where the fragrances are produced. It is this section that the fragrance-crazed bees try to hang on to. The smooth, waxy surface isn't fair to the unfortunate bee, and it tumbles down towards the apex of the lip, which is held in place by a pair of upright lateral horns (from which these orchids get their nickname). It is by passing through the apical cavity of the lip, known as the epichile, that the bee comes into contact with the pollinarium. The male euglossine bee seems not to be disturbed by the process and it has been reported to revisit a single flower multiple times for a few minutes each time. As with *Gongora*, the pollinaria need to dehydrate before they can fit into the stigma of the next flower, thus avoiding self-pollination during these repeated visits. This process will take at least half an hour, which is plenty of time for the bee to locate a different *Stanhopea* plant flowering in the vicinity.

It is important to note that even though several different species and genera of euglossine bees may visit a *Stanhopea* flower in search of fragrances, only some of them act as effective pollinators. This is because the bee needs to be of a particular size to fit between the lip horns and through the apical cavity sufficiently tightly to remove the pollinarium. If it is too big, it simply will not pass through. If it is too small, it will not remove the pollen. The late Pedro Ivo Soares Braga, a renowned Brazilian orchidologist, referred to those bees that are unable to remove pollen as *ladrão de odor*, which is Portuguese for fragrance thief. Dressler pointed out that flower size may be an important isolating factor in sympatric *Stanhopea* species that attract the same bee species. Size alone can prevent the two species from hybridising.

The pollination of *Gongora* and *Stanhopea* provide a couple of important takeaway messages. First, the wizardry of chemical perfumes may well be selective enough to ensure reproductive isolation among similar flower species that share the same habitat. Floral perfumes can therefore be a powerful tool in sorting pollinators and, sometimes, even more so than intricate floral shapes and colour patterns. Perfumes may also attract multiple visiting insects. In such cases, a snug fit between the floral parts and the pollinator makes the difference between an effective pollinator and a fragrance thief.

**THAT SMELL**
Pollination of the highly variable *Gongora*

**SCAN ME**
https://youtu.be/KDShkZxp-Sc

Genus *Gongora* (Stanhopeinae) is one of the most famous orchids pollinated by male euglossine bees. The bees visit the colourful and fragrant flowers to collect fragrances. As the bees brush the slippery sepals they fall onto the column and remove the orchid's pollinaria. *Gongora* flowers are known to be extremely diverse and variable in their colouration patterns and so present a headache for orchid taxonomists. However, the bees seem to have no problem telling the different species apart chemically. For the pollen transfer of *Gongora* to be successful, the bees need to be of a specific size: the right bee will fit the flower perfectly.

The videos and photographs shown here are owned by Adam Karremans and have been reproduced and edited by the author with permission.

## 3.9 WHAT HAVE I DONE TO DESERVE THIS?

*"Whether for ultimate good or ill, in Coryanthes we see an evolutionary masterpiece, which, surely, should excite our most profound wonder and admiration."*
P. H. ALLEN (1950)

The flowers of *Coryanthes* look nothing like what we usually imagine a flower should look like, and the evolutionary explanation is nothing short of extraordinary. The odd flowers, however, are just the tip of the iceberg in the unusual biology of this peculiar orchid. Every detail regarding its ecological interactions is a delicate piece in the complex biological puzzle that is this orchid. It is nothing short of an evolutionary masterpiece!

Specimens of *Coryanthes* are rare. By this I mean they are not commonly seen. They are virtually impossible to find in cultivation, though this is certainly not because they are unappealing. They are also uncommon in nature, and not because of over-collecting. So, what makes these orchids intrinsically rare? The generalised scarcity of *Coryanthes* is mostly a consequence of their strict ecological needs. These orchids are restricted to the very warm and highly humid lowland forests of Central and South America — the thought of its habitat alone makes me break into a sweat. In the wild, *Coryanthes* plants are found growing exclusively in strict association with ants. And, unless copiously nourished, in cultivation they are likely to perish if devoid of their faithful companions. The ants that associate with *Coryanthes* are not ground dwellers as we mostly imagine these insects to be. Instead, they build their nests on the twigs, branches, and trunks of large tropical trees, and *Coryanthes* plants are normally found growing right on top of them (see Figure 3.9.1).

**Fig. 3.9.1.** *Coryanthes kaiseriana*, the ecologically complex bucket orchid. **A.** Its bizarre flowers hang down from arboreous ant nests. **B.** The ants patrol the buds and flowers, chasing away intruders. **C.** Euglossine bees collect fragrances from the highly perfumed flower.
© APK.

Being associated with ant nests is not exclusive to the *Coryanthes* orchid. In fact, they are part of a whole community of diverse organisms that have adapted to live among arboreal ants. Several flowering plant species, belonging to different families, are adapted to growing on epiphytic ant nests. Besides orchids, it is commonplace to find multiple species of aroids, bromeliads, cacti, gesneriads, and peperomias cohabiting on a single nest. Together, they form what are known as 'ant gardens'. The gardens consist of masses of soil and plant remains assembled on tree branches and forming large clumps. The ants maintain a strict control over their gardens — they take care of the plants that are useful to them and prune out the unwanted ones. To keep the ants interested, the plants of *Coryanthes* feed them with extrafloral nectar, which is secreted in large quantities from various delicate plant parts. The ants return the favour by jealously protecting their source of sugar against hungry herbivores and daring, or perhaps naïve, orchid collectors.

How the orchid's seeds actually get to the ant nest has been something of a mystery. Observations of *Dendrobium insigne* in Papua New Guinea, and *Epidendrum* in Costa Rica, suggest that the ants themselves collect and carry the seeds to the inside of the nest. To feed their larvae, ants are known to collect seeds that have oily lipid- and protein-rich appendages called 'elaiosomes'. However, orchid seeds, such as those of *Coryanthes*, *Dendrobium* and *Epidendrum*, are small in size and have very little nutritional value. They lack the lipid- and protein-rich elaiosomes. Why would ants carry the seemingly rewardless orchid seeds into their nests?

Some authors have suggested that it is not the ants but the wind that carries the minute orchid seeds onto the ant nests. But if this is the case, why don't more epiphytic orchids grow there? Why is it that only specific orchid species do this? A recent study finally came up with an answer. It turns out that the ants can in fact distinguish between diverse seeds, actively selecting and dispersing those of the species that we find in ant gardens, and ignoring those that do not regularly live on the nests. However, this doesn't explain why the orchid seeds are selected despite lacking the precious elaiosomes that the seeds of other common ant garden plants produce. One possible explanation is that the orchid seeds may be chemically deceiving the ants, mimicking the volatiles produced by the rewarding plant seeds. There is some indirect evidence that the ants collect the seeds and carry them into their nests. When sown in flasks, *Coryanthes* seedlings initially develop long, vine-like, slender growths with scale-like leaves. This elongate growth habit is inconsistent with that of other orchids and with adult *Coryanthes* plants. However, it is consistent with the hypothesis that the seeds, collected by the ants themselves, germinate deep within the nests. These seeds subsequently develop into a clump of orchid pseudobulbs, which is typical of the adult stage, but only once they reach the exterior surface of the nest.

*Coryanthes* is commonly called the bucket orchid, and there is a good reason for that. The inflorescence is long and strictly pendent, and bears two to five, large and heavy\* flowers. *Coryanthes* flowers arguably have the most complex structure of any orchid flower. The lip is composed of three parts: a hood-shaped base (hypochile), a tubular middle (mesochile), and a bucket-shaped apex (epichile). The column extends downward into the apex of the lip, and is bent outwards at the apex, leaving a space between it and the bucket. At the base of the column is a pair of horn-like glands that secrete a watery substance even before the flower bud has opened, and that continue to drip the liquid periodically afterwards. The fluid is not nectar. It does contain a few sugars (such as glucose, fructose, saccharose and mannose), as well as ions, mucilage, and other substances, but all of them in very low concentrations. The liquid drips into the bucket formed by the lip, and partially fills it. The hood-shaped base of the lip emits the fragrances that male euglossine bees find highly attractive.

On a warm morning in the Caribbean plains of Costa Rica, I witnessed a group of metallic-green bees hovering around the freshly opened flowers of a large *Coryanthes*. The male bees attempted to find their footing on the underside of the cap-like base of the lip, competing for a place on the edge. Occasionally, the competition, combined with the slippery lip surface and the lure of bee-attracting fragrances, causes a bee to tumble into the soapy liquid that fills the bucket. The bee, unable to fly out of its involuntary bath, tries to climb out. But the slippery bucket walls impede the insect's ascent, forcing it to search for an alternative. The bee instinctively fights not to drown. It holds on to anything that will allow it to stay afloat. Eventually, the exhausted male

---

\* The German botanist Günther Gerlach tells me that he believes them to be the heaviest flowers in Orchidaceae.

finds the opening between the outward, bent column apex and the lip's bucket. This narrow passage is the only escape route. It crawls through in a desperate attempt to break out, and at last succeeds. The first bee to exit through the tight space will remove the pollinarium, optimally positioned so that it is inevitably taken by the escapee.

To pollinate a *Coryanthes* flower, a pollen-carrying bee needs to repeat this traumatic experience (see Figure 3.9.2). As if it had forgotten all about its agonising first experience, the overpowering drive to obtain the precious fragrances stimulates the insect to continue visiting *Coryanthes* flowers. Eventually it finds itself falling into the bucket again. With what must be a cruel case of déjà vu, the bee once more takes a while to recover and figure out how to avoid drowning. It locates the previous escape route once more. This time, while tightly passing the stigma on the column's apex, it drops off its precious cargo: the orchid's pollinarium.

The pollen of *Coryanthes*, like that of most other orchids, is tightly packed into this single unit. Each flower has one pollinarium, and it is removed and subsequently deposited as a unit by a single bee. *Coryanthes* plants bloom only once a year, producing one or a couple of inflorescences per plant, each bearing an average of three flowers. This means that each *Coryanthes* plant has just a handful of shots at reproducing each year. To improve the odds, this orchid has not only developed a precision-oriented pollination system, in which a specific species of bee is attracted through the release of particular volatile compounds and the pollinarium is meticulously placed on its back, but has also evolved a much higher pollen density than other orchids. A greater pollen density means that there is a larger pollen load per pollinarium. More pollen deposited on the stigma translates into more fertilised ovules, and consequently a larger number of viable seeds in each capsule. We can think of this orchid as having placed a very risky evolutionary bet. It has invested all its resources in a few, very large, attractive flowers, that are able to effectively place large loads of pollen, in the form of very dense backpacks, onto a small number of very devoted and precise carriers. An evolutionary masterpiece!

> '*Coryanthes* probably demonstrates the apex of complex pollination mechanisms to be found in the plant kingdom.'
> CALAWAY DODSON (1965)

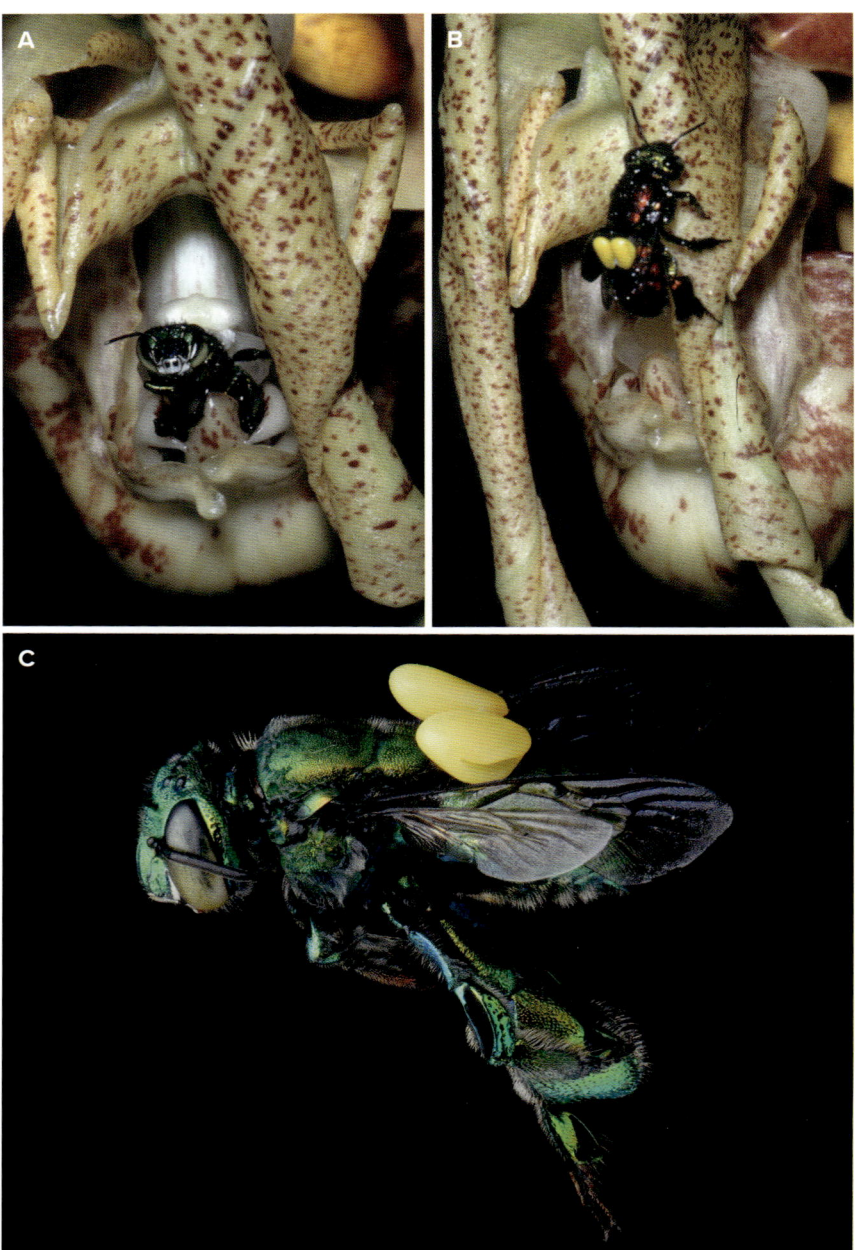

Fig. 3.9.2. As the male *Euglossa* bees collect fragrances from *Coryanthes* flowers, they drop into the liquid-filled bucket, escaping through a passage through the column. **A.** A male *Euglossa* struggles to exit the floral trap. **B.** It finally succeeds, taking away the orchid's pollinarium. **C.** A *Euglossa* carrying a *Coryanthes* pollinarium.
© Günter Gerlach (A–B), APK (C).

## WHAT HAVE I DONE TO DESERVE THIS?
The unusual ecology of the bucket orchid *Coryanthes*

SCAN ME

https://youtu.be/qn6CYYt5P7g

Species of the genus *Coryanthes*, commonly known as bucket orchids, are rare neotropical orchids that grow on arboreous ant nests. The exotic flowers of *Coryanthes* dangle under the ant nests and produce an enticing fragrance that attracts male orchid bees of the Euglossini tribe. In Costa Rica, *Coryanthes kaiseriana* is visited by metallic-green bees belonging to the genus *Euglossa*. These are exclusively males and they scrape the flowers to collect the fragrances, which are said to be used for courtship. During this process, the bee slips into the bucket formed by the lip of the orchid flower. Escaping from the bucket is not easy. It takes several minutes of struggling and crawling. When the bee finally does escape the bucket, it will have the orchid's pollinarium attached to its back. After a quick rest, the bee once again resumes fragrance collection on the flowers, eventually falling into another bucket. This time, it deposits the pollinarium backpack that it picked up earlier.

The videos and photographs shown here are owned by Adam Karremans and Karen Gil, and have been reproduced and edited by the author with their kind permission.

## 3.10 HURT

*"The seeming absurdity of Catasetum violated not only scripture-based notions of the 'special creation' of each immutable species as specifically generative of its own continuing kind or stock. It also contravened the Linnaean taxonomy — it had classified Catasetum, Monacanthus [sic], and Myanthus — in which species generate within and only as the selfsame species."*
WHITNEY DAVIS (2005)

We have a tendency to associate flowers with femininity. But the vast majority of everyday flowers are actually 'hermaphroditic'. The word is derived from Hermaphroditos, the son of Hermes and Aphrodite, who according to Greek mythology possessed both male and female physical traits. Hermaphroditic flowers have both a male and a female function. A single flower is capable of producing pollen that can fertilise other flowers, and of receiving pollen, becoming itself fertilised. There is a strong evolutionary advantage in being hermaphroditic. It means that every flower produced by the plant can develop into a fruit, but also that a single flower is enough for reproduction to occur. For single-sex flowers to reproduce, there must be at least one male flower and one female flower. Only the male will provide pollen and only the female will develop fruit. In short, being hermaphroditic is in general terms more efficient.

However, not all plants are hermaphrodites. Some are 'dioecious'. Dioecious plants have individual male and female flowers. A flower with the male function will produce pollen while a flower with female function will receive pollen, its ovules becoming fertilised and consequently forming fruits. About one in ten flowering plants produces separate male and female flowers. Even though dioecious plants are much less common than hermaphroditic ones, you may unknowingly be consuming some of them on a regular basis. Among the most commonplace single-sex plants are asparagus, dates, papaya and spinach. But the most famous is certainly cannabis, also known as hemp or marijuana*.

Orchid flowers are generally hermaphroditic — having both male and female reproductive parts in a single flower — but there are a few exceptions. Species belonging to the neotropical genera *Catasetum*, *Cycnoches* and *Mormodes*, which are members of the Catasetinae subtribe, are exceptional among orchids because they are dioecious. They can produce separate male and female flowers (see Figures 3.10.1 and 3.10.3). Contrary to the crops previously mentioned, in which male plants bear male flowers, and female plants bear female flowers, in the

---

\* Marijuana growers are well aware that each seed they sow will have an equal chance of developing into either a male or female individual. Male *Cannabis sativa* plants are unwanted because they are unable to produce the much esteemed floral buds of the female. The male buds are not only smaller but are of a lesser quality too. The presence of male flowers, and therefore of pollen, increases the number of seeds produced by the females — another undesired side effect. Getting rid of the unwanted male plants at an early stage is in fact a major concern, but telling them apart before flowering is virtually impossible and usually proves to be quite a hassle. Experienced growers will sow additional seeds to account for the male plants that will need to be weeded out later when the flowers reveal their sex.

**Fig. 3.10.1.** Fragrance collection on *Catasetum*. **A.** The extraordinary male flower of *C. maculatum*. **B–C.** A male *Eulaema* collects fragrances from the lip of a male *C. saccatum* while carrying its pollinaria.
© APK (A), Luis Enrique Yupanki (B–C).

Catasetinae, the sex of each individual plant is interchangeable. Rather than having a fixed sex — as in cannabis — individual plants can produce either male or female flowers, and they may develop one sex or the other at different stages during their lifetime. This phenomenon was not always well understood. John Lindley, an English botanist and the leading orchid authority of his time, was probably the first, back in 1826, to note a single plant of *Catasetum* that bore two flowers with dissimilar forms. It was a "curious monster", he wrote. The striking differences between male and female flowers led Lindley to propose two different genera for such plants, *Myanthus* and *Monachanthus*. He hinted at their relatedness but believed they were two completely different species, clearly not properly grasping their reproductive mechanism. Referring to an 'abnormal' specimen of *Catasetum* he later stated "one of the greatest curiosities that our gardens ever produced… It is that of a plant of *Myanthus cristatus* changing into a *Monachanthus* … combining in its own proper person no fewer than three supposed genera, *Myanthus*, *Monachanthus*, and *Catasetum*." He appeared to realise that the three might actually be the same. In his book, *The Vegetable Kingdom*, Lindley referred to this peculiar case again, cautioning: "such cases shake to the foundation all our ideas of the stability of genera and species" — evidently still somewhat boggled.

The suspicion that these floral forms represent different sexes of a single species — rather than different species — was first raised by the German-born naturalist Robert Schomburgk, based on his 1836 observations in Guyana in South America. He noted that hundreds of *Catasetum* and *Monachanthus* plants grew together in the same area and only the latter produced fruits. A *Catasetum* was produced from the seeds of a *Monachanthus*, he pointed out, before concluding that they form a single genus. Charles Darwin elaborated on the subject, publishing a paper specifically on this case in the early 1860s. His paper, which begins "having kindly permitted me to examine the remarkable specimen … of an Orchid bearing flowers of two supposed genera, and known sometimes to bear the flowers of a third genus", meticulously explains how each floral form has a definite function in the reproduction of a single species. Finally settling the matter, in *The Origin of Species* Darwin writes: "As soon as the three Orchidean forms, *Monachanthus*, *Myanthus* and *Catasetum*, which had previously been ranked as three distinct genera, were known to be sometimes produced on the same plant, they were immediately considered as varieties; and now I have been able to show that they are the male, female and hermaphrodite forms of the same species."

It is now well established that male or female flowers in the Catasetinae are produced in response to particular environmental conditions. In the case of the genus *Catasetum*, plants that are exposed to full sunshine will produce female flowers, while plants growing in the shade will produce males. If the conditions under which the plants are grown change, so will the sex of the flowers. An intermediate may be produced on rare occasions. I once found a plant of *Cycnoches* bearing both male and female flowers on a single inflorescence. The plant had most likely been growing in full sun on top of a tree and fallen in the middle of developing the flowers, triggering a late change in the sex of some emerging flowers. This phenomenon is known as environmental sex determination and it is not exclusive to orchids. In fact, it is commonly found in animals, including fish, crocodiles and turtles*. However, environmental sex determination is very rare in plants and has only been reported in three families. In these particular orchids, the amount of exposure to sunlight and the resulting level of ethylene synthesis in the developing inflorescence determines the sex of the flowers. The general rule of thumb is that more sunlight results in the production of higher ethylene concentrations by the plant so that only female flowers are made. Plants in shady conditions will have lower ethylene concentrations and therefore will develop male flowers instead. We now know how the dioecious Catasetinae members produce single sex flowers. But what about pollination?

In species of *Catasetum*, the pollinarium is placed in a position that is rather difficult for pollinators to access. In fact, the insect doesn't come into close contact with the pollinarium as it enters the flower. But when it touches a pair of highly sensitive horns,** the pollinarium is literally sling-shotted towards the insect (see Figure 3.10.1). In the words of Darwin, "when certain definite parts of the flower are touched by an insect, the pollinia are shot forth like an arrow, not barbed however, but having a blunt and excessively adhesive point." The insect is struck with such force that it desists from visiting the offending flowers again. Why would a *Catasetum* appear to hurt its pollinator? You may have already

---

\* Remarkably, the temperature of the sand in which turtles lay their eggs will determine their sex. Higher temperatures tend to result in more females.

\*\* The so-called horns are extremely sensitive extensions of the rostellum, which is a lobe of the stigma.

**Fig. 3.10.2.** Fragrance collection on *Mormodes*. **A.** A male *Euglossa* carrying the accessory structures of a *M. colossus* pollinarium after depositing the pollinia onto the stigma. **B–C.** Male *Eufriesea* vist a *Mormodes andicola* in Peru. **D–E.** *M. horichii* is visited by several *Euglossa* species in Costa Rica.
© APK (A, D–E), Luis Enrique Yupanki (B–C).

guessed what the evolutionary advantage of such a strategy is. By placing pollinaria with a blow, the bee is discouraged from visiting other male flowers and sent on its way to visit the harmless, female flowers instead. Male and female flowers not only differ morphologically, but are also chemically distinct. The insects are attracted to both, but are able to tell the difference between them. Even though male flowers are much more common in nature, they are, as one would expect, visited less frequently than females. The frantic interaction between a swarm of large bees and *Catasetum* flowers on a warm morning is quite a sight to see.

Pollination in *Mormodes* is similar to that in *Catasetum*, with some noteworthy differences. Not all species of *Mormodes* are dioecious, but most of them have another unusual reproductive syndrome. Flowers of most *Mormodes* species are hermaphroditic, with both male and female parts, but the two sexes are temporally isolated. The male part of the flower is functional first and the female part only becomes functional later on. As far as I know, the male part is always functional first. This strategy of separating the two sexes temporally is known as protandry\*, and is relatively common in flowering plants. How does it work? Well, in the case of *Mormodes*, the column is conspicuously bent to one side at its base, in such a manner that its apex rests on the upturned lip blade (see Figure 5.4.3). The column is always twisted to the side that is opposite to the flower stalk. Like those of *Catasetum*, *Mormodes* flowers have a long filament that is extremely touch-sensitive, and the pollinarium is flung onto the insect's back as soon as it brushes the sensitive filament. At this stage, the flowers of *Mormodes* have female organs, but these are still not operational. About half an hour after the pollinarium is removed, the column twists, becoming fully stretched, consequently exposing the stigma. The flower has now become functionally female. Again, the insect is discouraged from remaining on the same plant after the pollinarium has been flung at it. If it were to remain, however, the temporal unavailability of the female organs would prevent fertilisation of the flower with its own pollen. Together these features conspire to encourage the insect to look for a different flower in which to place the pollinarium. This favours out-crossing, and therefore genetic diversity.

Considering their sexual plasticity, polygamous nature and sadistic means, the Catasetinae are surely the kinkiest species in the Orchidaceae

---

\* Read more about protandry in orchids in chapter 5, section 5.4.

**Fig. 3.10.3.** Fragrance collection on *Cycnoches*. **A.** Female *C. warszewiczii* flowers are visited by numerous male *Eulaema*. **B–C.** The male flowers of *C. egertonianum* are visited by a *Exaerete* species
© Jose Martín Murillo Murillo (A), Luis Enrique Yupanki (B-C).

family. No wonder Darwin considered them the most remarkable of all orchids. In his orchid book, he spent several pages carefully describing the pollination mechanics of these particular orchids. About the reproduction of *Catasetum*, he wrote: "Who would have been bold enough to surmise that the propagation of a species should have depended on so complex, so apparently artificial, and yet so admirable an arrangement?" It is a shame that Darwin was not able to observe the pollination of these orchids himself. He inferred the mechanism by studying the flower's structure, but knew nothing about the fragrance-collecting behaviour of the male euglossine bees that pollinate these orchids. He would have been flabbergasted.

**HURT**
Sexual plasticity and sadomasochism in the Catasetinae

SCAN ME

https://youtu.be/j_RGeiKBEec

Orchid flowers are regularly hermaphroditic; that is, they have both male and female organs in a single flower. However, there are a few exceptions. Species in the genera *Catasetum*, *Cycnoches* and *Mormodes*, members of the Catasetinae subtribe, are exceptional among orchids because they tend to have separate male and female functions. They are also pollinated by orchid bees that collect fragrances, but it is the mechanism that makes them unique. The shape and structure of male and female flowers of both *Catasetum* and *Cycnoches* may be so different that they appear to be completely unrelated. In male *Catasetum* flowers, pollinarium is placed in a position that is rather difficult to access by the pollinators. While collecting fragrances in the lip, the bee contacts a pair of highly sensitive horns that cause the pollinia to be literally sling-shotted towards the insect. *Mormodes* pollination is similar to *Catasetum* and *Cycnoches*, but rather than having male and female flowers, they have flowers with a twisted column that are first functionally male and only later, after the column straightens up, functionally female.

The videos and photographs shown here are owned by Adam Karremans, Luis Enrique Yupanki, Marco Cedeño and Jose Martín Murillo Murillo, and have been reproduced and edited by the author with their kind permission.

## 3.11 WHOLE LOTTA LOVE

*Bulbophyllum* is one of the most species-rich genera in the orchid family. There are an estimated 2,000 species and many remain undiscovered. They are mainly fly-pollinated and as such were widely assumed to be pollinated through sapromyiophily (see chapter 2, section 2.9, 'Fear of the dark'). But just like the other major fly-pollinated orchid groups, *Bulbophyllum* species are pollinated through a multitude of strategies and by many different fly families.

In 1890, upon publishing the results of his studies on the pollination of *Bulbophyllum macranthum*, English naturalist Henry Nicholas Ridley noted that the flowers were visited by a small diurnal fly that "delights in bright sunshine". The flies would spend a long time on the flowers, licking the sepals up and down, and leaving only to start the whole process again in another flower. "I examined the sepals carefully with the microscope to see what it was the flies obtained," Ridley stated. The large flowers of *B. macranthum* appeared to be nectarless, but somehow kept the flies entertained long enough for them to eventually step onto the lip and remove the pollinarium (see Figure 3.11.1). "I was unable to detect any sweetness of taste on the sepals, and they always appear to be quite dry," Ridley wrote after having a go at them with his tongue. Finally, he concluded: "We have in this *Bulbophyllum* a flower with no visible nectar regularly visited by a species of Dipteron … and furthermore this insect is evidently not disappointed in its search, for it spends hours licking this flower, and if driven away speedily returns." He was obviously puzzled. What Ridley didn't realise was that he had stumbled upon one of the most highly specialised and peculiar pollination strategies employed by orchids. It would take more than a century for botanists to finally figure it out. Be patient, we will get there soon enough, but let's first talk about geckos.

Geckos, the kind you find at home if you live in a warm climate, are nocturnal animals that are particularly well adapted to hunting invertebrates in urban habitats. They are voracious predators that feed on a multitude of insects, from butterflies and moths, to bees and flies, and even cockroaches and termites. That is why my wife loves to keep geckos around the house. Their eating habits also make them perfect to test the capacity of invertebrates to avoid predation. In an experiment using *Bactrocera papayae*, a member of the fruit fly family

**Fig. 3.11.1.** The nectarless *Bulbophyllum macranthum* offers a unique reward to its pollinator.
© Ron Parsons.

Tephritidae, researchers tested how the insects' feeding habits affected their palatability to geckos. Male *B. papayae* are known to seek out food that contains several naturally found compounds, including methyl eugenol (ME). The function of ME had remained unknown until a team of researchers from Malaysia and Japan discovered that fruit flies that had eaten ME became repulsive to common house geckos. The geckos either avoided them completely or vomited after eating them, presumably because the compounds made the flies toxic or distasteful.

But consuming ME had another effect on the flies. In *Bactrocera* flies, chemicals produced in the insects' rectum play an important role as sex pheromones during courtship. The ME consumed by male flies is converted into various compounds that are stored in their rectal glands,

**Fig. 3.11.2.** The fragrance-rewarding *Bulbophyllum patens*. © Ron Parsons.

which secrete sex-attractant pheromones. Mating regularly occurs just before dusk, when the anxious males gather on leaves waiting for the females. As they perch, the males produce an audible, high-pitched buzz as they vigorously fan their wings to disperse the pheromones released from their rectal glands. Males jump on any female that enters their mating territory. The female either escapes or engages, in which case their lovemaking lasts through the night and the couple remains paired until sunrise. Several compounds have been identified in the rectal pheromone gland secretions, but relatively little is known about them. But the experiments showed that male flies that had fed on the ME competed significantly better for mates and were preferred by virgin females. It seems that male *Bactrocera* flies seek out and feed on naturally occurring compounds, especially ME and raspberry ketone (RK), that lead to the production of a more potent sex pheromone. More than 500 fruit flies are known to occur in Southeast Asia and Northern Oceania, where many are serious agricultural pests. Each *Bactrocera* species appears to be attracted to one or the other compound. Perhaps you are wondering what all of this has to do with orchid pollination? Bear with me.

Those seemingly rewardless, fruit fly-pollinated orchid species in the genus *Bulbophyllum* are actually producing ME and RK, which the flies so eagerly search for. *Bulbophyllum cheiri* produces ME and lures only the ME-sensitive *Bactrocera* species, whereas *B. apertum* produces RK and attracts the RK-sensitive *Bactrocera*. Another extraordinary discovery was made when studying the chemical composition of *Bulbophyllum* fragrances. In the lowlands of Borneo, Peninsular Malaysia and Sumatra, *B. patens* is visited and pollinated by male fruit flies of several different *Bactrocera* species — in a similar manner to its close relatives (see Figure 3.11.2). Researchers have now found that *B. patens* produces a compound named zingerone (ZN). This compound is structurally similar to ME and RK, but is attractive to both ME- and RK-sensitive *Bactrocera* fruit flies. Producing ZN, rather than either ME or RK, therefore secures *B. patens* the interest of a broader range of flies. It has now been established that ZN accumulates in the rectal glands of male fruit flies, where it plays an important role in sexual selection. Males that fed on ZN were more attractive to females and had increased mating success. All in all, ZN is a powerful aphrodisiac.

Curiously, in other species of *Bulbophyllum* that are pollinated by these fruit flies (see Figure 3.11.3), there seems to be variation in the production of ZN, ME and RK. In *B. macranthum* the major floral component in plants from Malaysia and Thailand was zingerone, while in plants from the Philippines, it was ME. In *B. praetervisum*, the major constituent seems to be RK, but ZN is also found in certain individuals, and ME has been found together with both RK and ZN in one individual. These differences in fruit fly-attracting floral scents are evidence of versatility in the floral biosynthetic processes as variation may occur within a single species.

Like many other insects, the *Bactrocera* female practises polyandry — meaning she will regularly mate with many males. In an extraordinary twist, it seems that the gland products of male fruit flies can influence the propensity of their former mates to copulate later with other males. In other words, compounds secreted by male *Bactrocera* have an important role in deterring their prior female partner from 'sleeping around'. The interaction between *Bulbophyllum* and their fly pollinators may represent a true case of mutualism: the orchid flower gets pollinated, and the fruit fly — by feeding on the compounds offered as a reward — not only boosts its defence system against predators but also its sexual competitiveness.

**Fig. 3.11.3.** *Bulbophyllum praetervisum* visited and pollinated by *Bactrocera* fruit flies. © Ong Poh Teck.

## 3.12 GIMME SHELTER

*"The fertilisation of orchids does not necessarily always occur through the visits of insects in search of honey, but may also be due to their seeking a safe hiding place for a night's lodging."*
MASTERS JOHN GODFREY (1920)

When we talk about floral rewards, we mostly think of discrete substances that are produced and emitted by the flowers. These substances are then collected and either consumed directly or taken away by a visiting animal. In previous sections, we have learned about the variety of riches orchid flowers that offer their guests. Food rewards can include edibles such as sugars or proteins, which are directly fed upon for their nutritional value. Other rewards may come in the form of collectible fragrance molecules, which bees and flies desperately require for their courtship displays. Floral incentives, however, are not always takeaways. The flowers of some European terrestrial orchids are said to take advantage of the pollinator's need for a place to rest. They offer the visitors a place to overnight as a reward. Among these shelter-providing orchids are many of the 20 or so species belonging to the genus *Serapias*, popularly known as tongue orchids. Indeed, the flowers of tongue orchids form a well-protected and comfortable chamber. But are *Serapias* species truly becoming fertilised by giving shelter to their pollinators?

The thesis that the flowers of tongue orchids offer refuge to their pollinators is not new. The earliest mention of this unusual form of pollination was probably by Oskar Kirchner. In 1900, he wrote "after the absence of nectar, I think it is probable that the flowers are used by some insects as temporary shelter and are pollinated in this way."* Colonel M. J. Godfrey** elaborated the idea further in a series of observations published between 1920 and 1931. The Colonel provides detailed descriptions, supported by personal insights, on the interaction between tongue orchids and their pollinators. He concludes: "The flowers of both [*Serapias*] species are specially adapted for the concealment of any insect from birds or other enemies, and also afford protection from rain. To enable it to render this service to the insects necessary for cross-pollination, the flower has been modified in architecture, shape, and coloration …"

Shelter mimicry in plants is not only little known, it appears to be quite rare. Curiously, it is currently restricted to the Mediterranean region and neighbouring areas, where it has been identified in only two different plant families: orchids and irises. Altogether, maybe two dozen plants are reported to have shelter mimicry. Both the orchids and the irises grow as terrestrials in dry Mediterranean-type climates, and share the dark reddish colour of their flowers. They attract primarily solitary male eucerine bees as pollinators. It has been suggested that these plant groups have converged evolutionarily in their form and function. The pollinators are not only similar, their behaviour towards the flowers is also very much alike. The bees mostly ignore the flowers during the daytime. They generally search for rock crevices or hollow wood stems during harsh weather conditions or late in the afternoon. It is during that process that they also seek out flowers for refuge.

Flowers of *Serapias* species serve as night shelters to diverse species of bees (see Figure 3.12.1). The dark-red, nectar-less, tubular flowers exclusively attract solitary male bees. The pollinating insects are seen

---

\* His exact words in the original German text are: "*nach dem Fehlen von Nektar in den Blüten halte ich es für wahrscheinlich, dass die Blüten von manchen Insekten als zeitweises Obdach benützt und hierbei bestäubt werden*".

\*\* Alternatively spelled Godfery.

entering the flowers in the afternoon and leaving in the early morning hours. They generally explore several flowers on the same plant before choosing the right one to spend the night in. During this exploratory phase, the bees may spend from a few minutes to half an hour in a single flower. They have been observed moving from one to the other loaded with pollen. After their initial explorations, the bees finally settle and can be found immobilised within the floral tubes. Bees that sleep in *Serapias* flowers enjoy protection from harsh weather conditions and predators. Studies show that the temperature within the floral tunnels of shelter-mimicking species is a few degrees higher than that of the ambient air around them. The microclimate formed within the *Serapias* flower is in fact another reward provided to the pollinator in addition to protection. The bees pollinate the plants while inspecting multiple flowers in search of a refuge. Similarly, the shelter-mimicking irises, found primarily in the semi-desert areas of the Middle-East, Turkey and the Caucasus, have large, tunnel-like flowers that provide no reward other than a place to overnight.

A flower offering shelter to a pollinator may seem a bit odd, but haven't we already come to expect the unexpected from orchids? Ironically, the shelter-providing orchid may be badly in need of protection itself. The underground tubers of *Serapias* species are used to make salep powder, the raw material for a traditional "Turkish beverage — that bears the same name —, ice-cream and even some drugs. The tradition of using ground-up orchid tubers as an aphrodisiac beverage dates back to Ancient Rome. Salep use has been reported in Turkey from the 8th century, being well established during the Ottoman Empire era. The name *Serapias* may in fact honour the Egyptian god Serapis, in whose temple pilgrims lacked legal and moral sexual restraints. Unfortunately, the use of orchid tubers in traditional food and medicine, together with habitat loss, has resulted in a decline in orchid populations throughout Europe. Sustainable solutions are needed to maintain tradition without extinguishing the orchid by over-harvesting. A few years back, while visiting my great grandmother in Liguria, Italy, I had the privilege to see the unique tongue orchid in full bloom. The orchids seemed to be doing well in that particular area, and it can only be hoped that these marvelous plants are allowed to continue thriving.

*Oh, a storm is threat'ning my very life today, if I don't get some shelter, oh yeah, I'm gonna fade away…*

**Fig. 3.12.1.** The Mediterranean shelter-rewarding orchid. **A.** *Serapias lingua*, Mallorca, Spain. **B.** *S. cordigera*, Lesbos, Greece. **C.** *S. cordigera*, Sicily, Italy. **D.** A bee carrying pollinaria caught sleeping in the flower of *S. cordigera*.
© Ron Parsons (A–C), Jean Claessens (D).

*Thelymitra jacksonii* visited by beetles.
© Jeremy Storey.

CHAPTER 4

# MISFITS

The previous two chapters covered a broad assortment of deceptive and rewarding pollination systems found in orchids. We have learned that the variation in orchid flowers is an evolutionary consequence of the incredibly diverse means by which they are fertilised. The chapter 'Deceit' explored pollination strategies that involve sexual deception, generalised food deception, Batesian mimicry and brood-site mimicry; while in the 'Reward' chapter, we turned to flowers that provide nectar, pollen, pseudopollen, proteins, fragrances or shelter to their pollinators. However — despite the seemingly unending diversity of pollination contrivances — a very large portion of orchid flowers are pollinated exclusively by bees, flies or wasps, with smaller portions being pollinated by butterflies, moths or birds. These six groups represent only a fraction of the animal kingdom, and yet they are responsible for the pollination of more than 95% of all orchids. Are there no other animal groups that pollinate orchids? In this chapter, we will discuss these misfits.

Let's first talk about insects, given that they play a major role in the pollination of orchids. Insects have existed for at least 400 million years. There are about one million insect species currently known to science — that is more than all the species of other animals, plants and fungi combined. Despite the extraordinary number of insects already known, it has been estimated that between 2.5 and 10 million insect species remain to be discovered. In terms of their interactions with other organisms, insects are the most important of all terrestrial animals. They comprise 24 large groups known as orders. However, more than 80% of all insects belong to the four dominant orders: *Coleoptera* or beetles, *Lepidoptera* or butterflies and moths, *Diptera* or flies, and *Hymenoptera* or bees, wasps and ants.

Regardless of where you are in the world, these four insect groups will be present and the chances are you are familiar with them. Hymenopterans, specifically bees and wasps, are considered to be the main animal group responsible for the pollination of orchids*. Not only do bees and wasps pollinate a large proportion of orchid species, they are also responsible for pollinating the greatest diversity of orchid groups. Ants are another well-known member of the Hymenoptera, and while it is unusual, they too are known to pollinate certain orchids. Bees and wasps are only rivaled by flies in their importance as orchid pollinators. Although pollination by dipterans is not as widespread in orchids as pollination by hymenopterans, dipterans do pollinate the most species-rich groups within the family. Lepidopterans, specifically butterflies and moths, are responsible for the pollination of a few thousand orchid species as well. Coleopterans, or beetles, despite their predominance in the insect world, are relatively minor contributors to the pollination of orchids. Finally, out of the 20 or more other orders of insects, only three have been known to pollinate orchids and only on exceedingly rare occasions. Cited as occasional orchid pollinators, we find the order Orthoptera, specifically crickets, the order Thysanoptera, to which thrips belong, and the order Blattodea, of which only termites have been observed to pollinate orchids.

Orchid pollination strategies are incredibly varied and undoubtedly many remain to be discovered. Curiously, the pollinating agents or vectors themselves are not as diverse as the mechanisms. Every single strategy discussed in previous chapters revolves mainly around the more common orchid pollinator groups: bees, wasps, flies, butterflies, moths and birds. But several other groups of animals have been cited as pollinators of orchid flowers, and even though they are mostly minor contributors, they may be important to certain groups of orchid species or in particular geographical areas. Beetles are the most commonly cited minor pollinator group in orchids, but ants, crickets, termites, bats, lizards and mice have also been proposed and — in some cases — proven to be pollinators. The history behind these notions and the current evidence to support or disprove each are the subject of this chapter.

---

\* This is further discussed in chapter 6, section 6.5.

## 4.1 YOU CAN'T ALWAYS GET WHAT YOU WANT

We are bombarded with information about the crucial role that bees play in maintaining ecosystems. But what about ants? A relatively common strategy among orchids to avoid being eaten by herbivoers is to offer sugar-rich nectar to ants. The ants thoroughly patrol the plant parts that produce the nectar, and fiercely defend their food source against herbivores. Nectar is therefore frequently offered to the ants in the vicinity of the most delicate and important plant organs: the flowers. So-called extrafloral nectaries are exactly that — a source of sugar outside the flower that has evolved to feed ants in exchange for protection. They secrete a watery liquid that mainly contains sucrose, glucose and fructose. This appears to us as persistent drops, viscous to the touch and extremely sweet-tasting. In orchids, extrafloral nectaries can be observed in species belonging to genera as diverse as *Brassia, Catasetum, Cattleya, Chelonistele, Coelogyne, Coryanthes, Cycnoches, Cymbidium, Dendrobium, Encyclia, Epidendrum, Gongora, Laelia, Myrmecophila, Notylia, Oncidium* (see Figure 4.1.1), *Otochilus, Pholidota, Prosthechea, Vanda, Vanilla*, and many others.

**Fig. 4.1.1.** An *Oncidium* species showing extrafloral nectar drops being secreted on the inflorescence.
© APK.

They are typically found on the developing inflorescence and flower buds, on the ovary (see Figure 4.1.2) or floral bracts (modified leaves that protect the floral bud), and on branches of inflorescence, and it is thanks to them that the flowers of orchids are often covered in ants.

Ants have a close relationship with many flowers. Research has shown that these insects play a crucial role in protecting the flowers from herbivores. This protection of floral organs is especially important for orchids given how infrequently they are visited by pollinators and overall low fruit production. Orchids don't seem to be too fussy about which ant defends them. An orchid species may be associated with many different species of ants. However, different ant species may be more or less effective in their defense. Studies on the orchid genus *Myrmecophila* in Mexico have shown that the absence of ants harms the plant's reproductive success. But the presence of smaller ant species also resulted in poor defences, and in a dramatic reduction in fitness as a result. Larger ants defended the plant much better, resulting in a larger floral display, increased pollinator visitation and higher fruit-set. A study on the importance of ants for a Brazilian *Epidendrum* species shows that they reduced the presence of herbivores without

**Fig. 4.1.2.** Ants feeding from the extrafloral nectaries on the apex of the ovary of *Cattleya dowiana*.
© APK.

affecting the activities of the butterflies that pollinate the flowers. A study on a species of *Coelogyne* orchids in Nepal got similar results. Extrafloral nectar was exuded by modified stomata. The presence of the sugary reward had a significant impact as ants played a key role in deterring leaf-eating beetles.

But ants may also be a serious problem for orchids. In the African genus *Aerangis*, arboreal ants have been shown to rob the flowers of their nectar. In this particular study, the ants were seen robbing the whole nectar content of over 60% of the flowers. This forces the flowers to produce sufficient nectar to not only reward the pollinator but also feed the thieving ants. Incredibly, the *Aerangis* was found to reabsorb unused nectar in order to avoid its loss through ant depredation. In Mauritius, significant robbing by invasive ant species was also reported in *Aeranthes* flowers, with nectar loss ranging from 64.3% to 100%. Not only were robbers shown to exact an enormous resource cost, but they also negatively affected the orchid's fruiting success. A study in Japan found that the flowers of *Goodyera foliosa* were also harmed by nectar-robbing ants. In this case, the ants frequently got stuck to a sticky pad on the orchid's pollinarium called the 'viscidium'. This may sound unimportant but it prevented pollinating bees from removing a significant number of pollinaria. The flowers of some neotropical orchids, such as the fly-pollinated genus *Pleurothallis*, are occasionally found covered in ants — presumably robbing nectar. They are likely to harm the orchid because they not only do not pollinate the flowers, but I have observed them intimidating the actual pollinators (see Figure 4.1.3). In a surprising twist, however, nectar-robbing ants have been shown to have a positive effect on one species of *Aerides* in India. Ants pierced the nectary and caused a decrease in its volume, which in turn forced the pollinators to visit more flowers to become satiated. This resulted in higher flower visitation and therefore pollen flow between different individuals. By contrast, when ants were absent, this resulted in more herbivory (being eaten by herbivores), which prevented pollinators from visiting altogether.

Ants are therefore intimately interacting with orchid flowers. They can be both antagonistic — robbing nectar and discouraging pollinators — and mutualistic — deterring herbivores and promoting outcrossing. On the other hand, pollination by ants is very rare. So rare in fact that you can almost count the cases of ants pollinating orchids on your

**Fig. 4.1.3.** Ants may rob pollen or nectar intended for actual pollinators. **A.** The bee-pollinated *Spiranthes tuberosa* photographed in Massachusetts. **B.** An ant visits the fly-pollinated *Masdevallia guttulata*. **C.** Ants can deter flies attempting to pollinate *Pleurothallis homalantha* in Costa Rica.
© John Gange (A), Andreas Kay (B) and APK (C).

fingers. This extreme rarity of adaptations to ant pollination is all the more remarkable since their close relatives — bees and wasps — are among the most important pollinators of orchids worldwide. As we have learned, many plants actively recruit ants for their protection, so ants are frequently found marauding around flowering plants. Yet plants rarely use them as consistent pollinators and they may have good reasons for that. Ants are mostly flightless and may cover significantly less distance than some regular flyers. They also tend to be aggressive and even predatory towards other potential pollinators, and — as if the previous were not reason enough — their bodies are coated with antibiotics. These prevent the growth of fungi and bacteria on the ants but can also reduce the viability of pollen.

From any perspective, ants appear to be terrible pollinators, and many plants have developed strategies to avoid them altogether. Yet there are a few orchids that are pollinated by these insects. Regular pollination by ants has been documented in two orchid species belonging to the genera *Chamorchis* and *Dactylorhiza* in the European Alps. These alpine orchids are compact terrestrials that grow well hidden among the grasses. They are exposed to low temperatures and strong

**Fig. 4.1.4.** In the European Alps, ants aid in the pollination of the terrestrial orchid *Chamorchis alpina*.
© Jean Cleassens.

Fig. 4.1.5. *Epipactis palustris* from Japan turns to pollination by ants when conditions are unfavorable.
© Motohiro Sunouchi.

winds, and these harsh climate conditions deter visits by flying insects. Ants are better suited for these extremes because they can walk from flower to flower — even between different plants — if the distances are not too great. *Chamorchis alpina* is also pollinated by beetles and wasps, but they are regarded as occasional, accessory pollinators, given that ants are predominant and the flowers have been shown to produce a scent that is attractive to certain ant species (see Figure 4.1.4). This suggests that *C. alpina* has adapted to pollination by ants. Another European orchid that is pollinated by ants is *Neotinea maculata*, which is a widespread species with small flowers. In a study conducted in wild populations in Valladolid, Spain, researchers found that minor worker ants removed pollinaria from one third of the flowers, carrying up to four pollinaria each. *Neotinea maculata* is mainly self-pollinated, which promotes effective fruit production but lowers the level of genetic diversity in a population. Although limited, pollination by ants allows for some gene-flow between individuals to promote outcrossing.

Similar explanations have been given in other pollination systems involving ants. Flowers of the monotypic (having a single species) genus *Chenorchis*\*— endemic to subtropical China — have turned to self-

---

\* The genus has alternatively been treated as part of *Holcoglossum*, but it represents an interesting divergent lineage however classified.

pollination to cope with the particular environmental conditions of a transitional habitat between temperate and tropical climates. The orchid is only known from the Gaoligong Mountains in Yunnan Province, and has formed mechanisms that promote selfing and inbreeding. Ants are required for short-distance pollen transfer. They transport pollinia from one flower to another within the same inflorescence, in what can be defined as an ant-mediated self-pollination strategy. In Japan, the terrestrial orchid *Epipactis palustris* — which is usually pollinated by hoverflies — turns to pollination by ants when weather conditions are unfavorable (see Figure 4.1.5). In Europe, another species of *Epipactis* is known to be pollinated by several different insects — including ants — depending on pollinator availability. Pollination of a species of the largely Australian genus *Microtis* is not very different from the previously described; foraging ants visit the flowers repeatedly in search of nectar and during these visits they transfer pollen, mostly between nearby flowers, resulting in a high degree of ant-mediated self-pollination.

Pollination by wingless ants has proven to be highly effective, but as expected, it consistently increases the probability of selfing. Self-pollination ensures reproduction but reduces genetic variability, and is therefore avoided by the orchids whenever possible. Indeed, you can't always get what you want, and orchids under harsh conditions will prefer to be pollinated by ants rather than risk not being pollinated at all. Sometimes, you really do just get what you need instead.

There is one more orchid species that has been proven to be ant-pollinated. It belongs to the Australian genus *Leporella*, which is typically dormant during the summer in the form of underground tubers. Only a small portion of the plants in a colony actually bloom in autumn every year. It seems that the flowering of these orchids may respond to fire stimulation, blooming more vigorously after the vegetation above has been burnt. Pollination of *Leporella* is, however, quite different from any other involving ants. It was first observed in 1979, when the Australian botanist Robert Bates noted: "I observed several insects darting rapidly from flower to flower in a colony of *Leporella*, they exhibited the same frenetic behaviour I had observed in a swarm of male red inch ants attempting to copulate with a virgin queen earlier that day." Detailed observations demonstrated that in fact the orchid is pollinated by sexually deceiving winged male ants of the genus *Myrmecia*, which attempt to copulate with the flower (see Figure 4.1.6).

The male typically approaches a target flower in flight after circling the colony, subsequently landing on the inflorescence and crawling towards the flower's lip. After aligning itself sideways on the lip, the male probes the lip with its exposed and pulsating genitalia. During this attempted copulation — which may be quite aggressive — the ant effectively removes and deposits the pollinaria. This is the only known pollination strategy involving pseudocopulation by ants and the only one that involves winged ants as vectors. Ants with wings are functionally more or less equivalent to wasps or bees. In this case, the pollinating ants are winged males that are sexually deceived by the orchids. Floral scent has been shown to be one of the most important cues for ants, eliciting a positive response in pollinating ant species, which is certainly the case in sexually deceptive orchids. For an exploration of other orchids that are pollinated through pseudocopulation, see Chapter 2, 'Deceit'.

Despite being less popular or famous than their cousins — the bees and wasps — ants have been shown to have intricate, mutualistic relationships with plants. These interactions may include seed dispersal

**Fig. 4.1.6.** In Australia, ants are occasionally observed transferring pollinaria of *Microtis media* (**A**), whereas in the case of *Leporella fimbriata* (**B**), the large *Myrmecia urens* acts as the primary pollinator through sexual deceit.
© Jeremy Storey.

and plant protection, but ants also play a role in the pollination of several different plant groups. To date, ant pollination has been recorded in 36 plant species belonging to 18 families. Generally, plants that benefit from ant pollination live in habitats where these insects are very common. The plants tend to be short in stature with small, sessile (stalkless) flowers that offer nectar rewards. As new information accumulates worldwide, so does the number of herbs, trees, shrubs, epiphytes and parasitic plants that are reported to be ant-pollinated. Among these — of course — are a handful of orchids. Ants are sometimes only casual or accessory pollinators, but there are orchids that are adapted to ant-mediated selfing or exclusive pollination by ants. Given the diversity and ubiquity of ant species, more ant-pollinated orchids are sure to turn up. However, the numbers are not likely to increase dramatically considering the unsuitability of ants as pollen carriers, especially since the vast majority of orchid species are epiphytic (growing on trees) and thus unlikely to adapt to depend on pollination by a predominantly flightless animal.

## YOU CAN'T ALWAYS GET WHAT YOU WANT
Ant pollination in orchids

**SCAN ME**

https://youtu.be/bbn6yPu5d-A

Ants are regularly observed visiting orchid flowers, often in search of the extrafloral nectar that many species offer. These fierce insects will patrol and defend their food source, which makes for a good defence system against herbivores. They are not often employed as pollinators, especially as they cannot travel long distances, unlike their flying insect relatives. On the contrary, they may rob nectar from the flowers and even block or intimidate actual pollinators. Nevertheless, under harsh environmental conditions, ants are known to pollinate a few orchid species, and although they are not as efficient as bees and wasps, they occasionally contribute to orchid pollination.

The videos and photographs here are owned by Adam P. Karremans and Jean Claessens (https://europeanorchids.com/) and have been reproduced and edited by the author with permission.

## 4.2 WITH A LITTLE HELP FROM MY FRIENDS

There are between 350,000 and 400,000 known species of beetle worldwide. This is 40% of all the insect species and about one quarter of the entirety of living beings on the planet. In other words, out of every four known organisms on earth, one is a beetle. Think about that. This makes Coleoptera the most species-rich order known to science, and yet it is estimated that three out of every four beetle species remain undiscovered today. Let that sink in! Recent studies have estimated that in total there may be about 1.5 million beetle species on Earth. Despite their extremely high diversity and ubiquitous nature, beetles are rarely found to pollinate orchid flowers. They are no strangers to orchids, which they will often eat — unless of course the plant is well defended by ants. Relatively few orchids have been shown to profit from beetle pollination. And, even when beetles are found to fertilise orchid flowers, they usually do so in unspecialised pollination systems where multiple insects are effective pollinators. Such is the case of the frog orchid, *Dactylorhiza viridis*, and the man orchid, *Orchis anthropophora*, which are pollinated by several hymenopteran families and diverse species of beetle (see Figure 4.2.1). Beetle pollination also occurs in populations at the edge of a species' distribution range where pollination is limited and the main pollinator is absent. Orchid flowers are therefore generally not specialised for beetle pollination. But there are a few cases in which beetles have been proven to be the exclusive pollinators of orchid species. Let's explore those.

First we travel to the African savanna to talk about one of its most important orchids: the genus *Eulophia*. Despite the broad distribution of the 200-plus species that belong to the genus, *Eulophia* are primarily found in central and southern Africa. Their flowers — although morphologically variable among species — tend to be rather standard for orchids. Even Darwin found more interest in other South African orchids. "These orchids (except *Eulophia*) are so surprisingly different from anything that I have seen that I could hardly make them out for some time," he wrote in a letter to the British-South African naturalist Roland Trimen. As a result, most *Eulophia* species have been poorly studied and were assumed to be pollinated by bees in a more or less monotonous fashion. Pollination studies in *Eulophia* remained limited until Craig Peter's PhD dissertation — under the supervision of Steven

**Fig. 4.2.1.** Beetles are often found removing pollinaria and mating on orchids in Europe. *Orchis anthropophora* (**A**) with copulating *Cantharis rustica* and *Dactylorhiza viridis* (**B**) with beetles pollinating during copulation.
© Jean Claessens.

Johnson at the University of KwaZulu-Natal in South Africa — revealed that their flowers showed a varied array of fertilisation strategies and intricate mechanisms, beetle specialisation being one of them.

Specialised beetle pollination had previously been described in the deceptive South African orchid *Ceratandra*. South African botanist Kim Steiner found that beetles looked for mates on the yellow flowers of *Ceratandra grandiflora*. When pollination is ensured or enhanced by the presence of mating males and females on a flower or inflorescence, this is known as 'rendezvous pollination'. The beetles that visit *Ceratandra* have been observed to pollinate flowers while feeding and mating. And, delightfully, they sometimes do both at the same time. Researchers noted that beetles remained on the flowers for long periods of time despite the absence of a food source. Perhaps the prospect of finding a mate is compelling enough for them to stick around. Rendezvous pollination is usually included as one of the categories of deceptive pollination in orchids. I personally fail to see why it should be classified as a form of deception if the insects are successfully mating. Perhaps, the rendezvous behaviour in this case is best described as part of a food-deceptive strategy, in which the beetles are duped into looking for a food source. But that is a discussion for another time. The fact remains that these beetles pollinate the orchid flowers while feeding and mating. Isn't that the dream!

Let's get back to *Eulophia*. The first observations of beetles pollinating this particular group of orchids were made on the Argentinian *E. ruwenzoriensis*. Beetles of the tribe Cetoniinae — otherwise known as flower chafers — were commonly seen to visit flowers to feed on pollen and nectar, but also on sap, fruits, leaves and other green tissues. Beetles belonging to the genus *Euphoria* entered the flowers of *Eulophia* in the afternoon hours, crawling from the older to the younger ones. The entrance to the flowers is narrow and the beetles have to force their way in by pushing the sepals and petals. Only half of their body actually enters the flower, where they may remain for up to 20 minutes. During this time, the insect seems to feed on a jelly-like exudation produced at the base of the flower. It removes the pollinarium while reversing out of the flower. The pollinaria — which are placed on the back of the beetle's head — are easily removed by the broad, sticky stigma of the next flower that the beetle visits.

In their studies, Peter and Johnson later found that several South African *Eulophia* were also adapted for beetle pollination. They discovered that *E. foliosa*, a species that forms large populations in grasslands throughout the eastern parts of South Africa, is pollinated exclusively by beetles. Although the flowers host many visitors, only beetles, especially a click beetle species in the genus *Cardiophorus* (Elateridae), removed their pollinaria. The beetles removed the pollinarium and anther cap as a single unit as they probed the flowers. Observations showed that the anther cap drops off after just about eight minutes on average, which suggests that this particular orchid species uses 'anther retention' as a strategy to prevent self-fertilisation. Anther retention is a phenomenon whereby the anther cap is removed with the pollinarium, and is retained for several minutes on the insect to avoid the possibility of placing the pollinarium onto the stigma immediately after removal. The anther cap eventually drops after it dehydrates, which ensures that the beetle will already have left the inflorescence from where in was picked up. This increases the likelihood of cross-pollination.

*Eulophia* species are non-rewarding, and the basis of their attractiveness to beetles seems to be a resemblance between their inflorescence and that of Asteraceae — the daisy family — which chafer beetles use as food or rendezvous sites. Pollination by beetles that aggregate on the deceptive inflorescences for mating purposes is also known to occur in the orchid *Ceratandra grandiflora*. Further studies on *Eulophia* have now also found chafer beetles to be the main pollinators of *E. ensata* (see Figure 4.2.2) and *E. welwitschii* — two South African species that, like *C. grandiflora*, have crowded inflorescences that bear yellowish-cream flowers and have compact heads. The cetoniid beetles that pollinate these orchids spend quite some time climbing around the inflorescence without systematically probing the flowers for food, which suggests a similar rendezvous behaviour. However, beetle rendezvous for mating typically takes place on rewarding flowers.

Not all *Eulophia* species are beetle-pollinated — many rely on bees for pollination. Interestingly, studies on the floral variation of *E. parviflora* have provided evidence that the diversification

**Fig. 4.2.2.** In South Africa, *Eulophia ensata* is pollinated by beetles. The brown *Atrichelaphinis tigrina* is too big to remove pollinaria, but the smaller *Leucocelis haemorrhoidalis* effectively pollinates the orchid.
© Craig Peter.

of *Eulophia* may have been driven by their pollinators. *Eulophia parviflora* is a morphologically variable species, with multiple flower forms identified. What authors found is rather remarkable. The two forms — which have slight differences in their flower shape, scent and flowering time — are adapted for pollination by different insects. Despite sometimes growing in the same area, there is a long-spurred form that is always pollinated by bees, and a short-spurred form that is exclusively pollinated by beetles. The two forms are still compatible with each other, which suggests that they are perhaps in the early stages of diversification. But, the accumulation of enough differences over time may eventually lead to speciation.

Beetles are not usually the exclusive pollinators of orchid species. The rare *Satyrium microrrhynchum* — another South African orchid — is specialised for pollination by both cetoniid beetles and pompilid wasps. This orchid secretes nectar that both types of insect sweep from the 'lollipop hairs' on the orchid's lip. A shared pollinator role of beetles has also been found in the sexually deceptive South African orchids *Disa atricapilla* and *D. bivalvata* (see Figure 4.2.3). The two *Disa* species are primarily regarded as being pollinated by male wasps looking for a mate, while the beetles are considered less efficient, secondary pollinators. However, a recent study shows that at least one species of *Disa* is primarily pollinated through sexual deception by beetles. *Disa forficaria*, a rare species that is endemic to the Cape Floristic Region of South Africa, is pollinated by a longhorn beetle that exhibits sexual behaviour when visiting the flowers. In this case, the beetles become so stimulated that they ejaculate on the flowers. Sperm deposition on orchid flowers is unusual but has also been registered in the Australian sexually deceptive *Cryptostylis* — which is pollinated by male wasps (see section 2.2). The pollination of *D. forficaria* seems to be the first unambiguous case of sexual deception of beetles by orchids.

However, beetles have recently also been confirmed to be involved in pollination through sexual deception in Asian orchids. In Japan, the floral scent of the epiphytic orchid *Luisia teres* sexually attracts male Cetoniinae beetles (see Figure 4.2.4). Even though in one area, the orchid is pollinated by offering nectar rewards to both males and females of two cetoniid beetles species, in another locality a few hundred miles away, it has specialised in the sexual deceit of males of *Protaetia* scarab beetles. Apparently this species has adopted divergent pollination

Fig. 4.2.3. The sexually deceptive South African orchids *Disa atricapilla* (**A**) and *D. bivalvata* (**B**) are primarily regarded as being pollinated by male wasps. Beetles are known to pollinate them secondarily.
© Craig Peter.

systems through geographical isolation — which may also eventually lead to speciation. In this orchid, male beetles were seen to lick the nectar at the base of the column, extruding the genitals from their extended abdomens as they fed. In Thailand, the closely related *L. curtisii* is also known to be pollinated exclusively by beetles. In this particular case, food or brood-site deception has been suggested as the strategy, based on the dull colour of the flowers and their unpleasant scent.

Be it as primary or secondary pollinators, Coleoptera have certainly gained recognition as pollinators in the Orchidaceae. They are now known to be involved in both rewarding and deceptive strategies, including rendezvous, food and sexual deception. Beetle pollination is now known to occur in several unrelated orchid groups, from geographically distant regions, such as Argentina, South Africa and Thailand. However, beetle pollination — for all its heterogeneity — was unknown for neotropical orchids until a very odd beetle-mediated pollination strategy was recently identified in the Atlantic forests of Brazil.

This particular beetle is a weevil belonging to the genus *Montella* which pollinates the flowers of an orchid in the genus *Dichaea*. The female beetle lays her eggs within the orchid's ovary — which later

Fig. 4.2.4. In Japan, the epiphytic orchid *Luisia teres* is pollinated by cetoniid beetles.
© Motohiro Sunouchi.

becomes a fruit. The eggs eventually hatch and the larvae of the weevil feed on the fruit capsule of the orchid. This type of herbivory is of course generally detrimental to the orchid as it leads to seed loss. But a team of Brazilian researchers has discovered something truly amazing. The weevils need the orchid's fruit capsules to develop in order to lay their eggs and provide food for their larvae. *Dichaea* are generally pollinated by the fragrance-collecting male Euglossini bees. But rather than depend on these unreliable pollinators, the beetles pollinate the orchid flowers themselves. Pollination of flowers by parasites of the seeds or ovules is known as 'nursery pollination'. A classic example of nursery pollination is what happens in figs. Figs are always parasitised, but only a part of the fig's seeds are eaten by their parasite. The remaining seeds develop and both the parasite and the fig benefit from this interaction.

The fig–parasite relationship is known as 'mutualism', because the interaction has positive effects for both species. By contrast, when the effects of the parasite are detrimental to the survival of the host, we talk about an 'antagonistic' relationship. The larvae of the weevil that pollinates the flowers of *Dichaea* consume the whole fruit. When larvae are present, the net result is zero seeds, and therefore the relationship should strictly speaking be considered antagonistic. However, the pollinating weevil increases the number of fruits produced by the plant by almost 50%, and given it only parasitises an average of one fifth of all the fruits, the net effect may actually be beneficial for the plant. But one more key element in this complex interaction definitively shifts the balance towards mutualism. The weevil's larvae also have natural enemies. After the beetle has pollinated the orchid and laid its eggs in the fruits, a portion of the fruits are 'rescued' by parasitic wasps that kill the weevil larvae before they consume everything. The parasite gets parasites, and the orchid gets by with a little help from its friends!

**WITH A LITTLE HELP FROM MY FRIENDS**
Beetle pollination in orchids

SCAN ME
https://youtu.be/Ez7Xxx1WZzA

Beetles are the largest group of organisms on Earth. They are often plant pests, but it has now been well established that some species may also be important pollinators of certain orchids. One orchid genus with several beetle-pollinated species is *Eulophia*, many of which can be found growing in South Africa. Both male and female beetles are often found visiting a single inflorescence, mating and feeding while they remove pollinaria. This particular strategy is known as rendezvous pollination. Perhaps the most curious pollination strategy involving beetles is that of the *Montella* weevil, which pollinates a *Dichaea* orchid in Brazil. In this case, the female weevil wants to lay her eggs in the orchid's fruits. But rather than waiting for them to become pollinated by themselves, the weevil actively reaches for the pollinia and places them onto the stigma. This causes the *Dichaea* to self-pollinate and the fruit to grow. The beetle then lays her eggs in the ovary and the larvae will feed on the developing fruit. The orchid is still able to reproduce because some of the beetle larvae are themselves parasitised.

The photographs reproduced here are owned by Craig Peter and Jean Claessens, while the video is owned by Carlos E. P. Nunes (https://doi.org/10.1007/s00442-016-3703-5). They have been reproduced and edited by the author with permission.

## 4.3  WHAT YOU'RE PROPOSING

Among orchid pollinators, thrips are surely the smallest. Thrips are tiny insects of the order Thysanoptera that have very weak, feathery wings that are not well-suited for conventional flight. Their favourite hobbies include puncturing and sucking the contents of plant and flower tissues, and they are unwelcome pests of commercially important crops. Thrips are also common vectors for viruses that cause plant diseases. They are rarely thought of as pollinators. Yet — since the time of Darwin — they have been proposed to be involved in the transport pollen, and therefore in pollination. On discussing the pollination of the Bird's-nest orchid, *Neottia nidus-avis*, Darwin noted: "The spreading of the pollen seems to be in part caused by the presence of Thrips, many of which minute insects were crawling about the flowers, dusted all over with pollen." But the role that thrips play in orchid pollination has remained largely unexplored.

In Japan, researchers have noted that thrips often live on the terrestrial orchid *Habenaria radiata* — famous for its extraordinary white flowers that resemble a bird in flight (see Figure 4.3.1). The white egret orchid, as it is commonly called, has a long nectary spur and is known to be pollinated by hawkmoths. Nevertheless, researchers observed that thrips were able to transfer pollen grains and act as a secondary pollinator. They hypothesise that this is a mutualistic relationship in which the orchid offers a brood site while the thrips assist in self-pollinating the flowers. Given the thrips' minute size, the authors were not able to observe this first hand. However, using paper bags, they established that inflorescences that were accessed by thrips had a much higher fruit set than inflorescences from which thrips were excluded. In another study on Japanese orchids, thrips were observed to move pollen from the anthers to the stigmas. However, in both *Epipactis thunbergii* and *Pogonia japonica* (see Figure 4.3.2), the thrips consumed and damaged the pollinia, and the seed mass was significantly lower than that under open and manual pollination. In the case of *E. thunbergii*, the thrips were found to be inferior to hoverflies as pollinators. The authors therefore warn that, "while thrips contribute to pollination, they have the potential to inflict heavy costs in terms of plant fitness". In *P. japonica*, the researchers suggest: "Thrip's self-pollination may supplement low fruit set as a result of limited visits by major pollinators."

**Fig. 4.3.1.** In Japan, thrips have been observed acting as secondary pollinators of the beautiful white egret orchid, *Habenaria radiata*.
© Motohiro Sunouchi.

In 1986 — in a paper that has largely passed unnoticed — the American missionary and orchid specialist Donald Dod reported that aphids and thrips were pollinators of different orchid species belonging to the Pleurothallidinae subtribe in the Dominican Republic. Dod observed aphids attempting to reach the nectar at the base of the lip in the tiny flowers of the genera *Lepanthes* and *Lepanthopsis*. In the process, the aphids disrupted the anther and the pollinaria became attached to their minute bodies. According to the author, a very similar process occurred in the genus *Stelis*, but in this case, courtesy of thrips.

**Fig. 4.3.2.** A secondary role of thrips in pollination has also been suggested to occur in both *Epipactis thunbergii* (**A–C**) and *Pogonia japonica* (**D–E**).
© Motohiro Sunouchi.

**Fig. 4.3.3.** An aphid infestation is not uncommon in orchids. Small flowers such as those of *Stelis kefersteiniana* (subtribe Pleurothallidinae), may occasionally have their self-pollination assisted by these tiny animals.
© Henry Oakeley.

The flowers of these pleurothallids are well-known to be adapted to pollination by flies. However, it is possible, given the exceptionally close proximity of the anther and stigma in these extremely reduced flowers, that the tiny animals are essentially helping the flowers of *Lepanthes*, *Lepanthopsis* and *Stelis* to self-pollinate by dislodging and slightly displacing pollinaria onto the stigma (see Figure 4.3.3). These genera of Pleurothallidinae are mostly absent from the Antilles, which is very far from the Andes, where they have their highest diversity. Selfing, and perhaps assisted selfing, could help to compensate the reduced availability of pollinators in these areas.

Thrips and aphids — normally considered plant antagonists — may contribute to orchid reproduction. However, they are certainly not very efficient pollinators and although they may have positive effects, they may also have negative ones. Orchid flowers are not adapted specifically for pollen removal and deposition by these insects. Nevertheless, under certain conditions in which pollinators are scarce and the flower morphology permits it, these tiny insects may assist in self-pollination.

## 4.4 SECRET GARDEN

*"The idea of a subterranean orchid is like finding life on Mars"*
MARK CLEMENTS (2020)

In 1928, a farmer by the name of John Trott was ploughing virgin soil in preparation for cultivation near the town of Corrigin in the wheatbelt of Western Australia. As he removed the top layers of soil, he discovered what seemed to be an unusual underground flower. The keen-sighted Mr Trott would find additional specimens at two nearby locations that same year. In total, he uncovered 36 plants, always at the base of the mallee scrub, *Melaleuca*. It became clear that what he had found was a chlorophyll-deprived plant that fully grew and flowered underground. Material was sent for study to Richard Sanders Rogers, the expert on Australian orchids at the time. Almost immediately, Rogers prepared and published a detailed description of the remarkable plant, which was completely new to science. He called it *Rhizanthella gardneri*. In 1931, another underground orchid was discovered, this time in Eastern Australia at Bulahdelah in New South Wales, by an orchid hunter who was digging up a hyacinth orchid and found an unusual plant tangled in its roots. Unlike *R. gardneri*, *R. slateri* was found in eucalyptus forest and did not appear to be associated with a *Melaleuca* species.

We don't normally think of underground plants, certainly not underground flowers! The discoveries caused great excitement, both in Australia and across the rest of the world. Orchids are known to require the help of fungi — known as mycorrhiza — to germinate and develop. The mycorrhiza supplies the orchid with the food that it requires for its initial growth, but as the plant develops green leaves, it eventually becomes independent. Some orchids stay dependent on their fungal partner and never grow green leaves, however, and such is the case for the underground orchid. Orchids that lack green leaves and continue to be dependent on fungi are known as mycoheterotrophs — a form of parasitism that is both elusive and poorly understood. Mycoheterotrophic plants obtain carbon from other photosynthetic plants through a shared mycorrhizal fungal network, rather than by photosynthesis. Some are very rare, growing in deep shade under leaf litter, and are easily overlooked. *Rhizanthella* is unusual in being the only flowering plant that remains underground through germination, growth, flowering and seed development. For much of its life, the underground orchid exists in the soil as a small white rhizome — a thickened underground stem. Its pinkish flowers remain hidden under leaf litter and soil close to the surface (see Figure 4.4.1). Surveys are difficult as they entail excavation, which involves careful removal of the top centimetre of soil to reveal the tips of the bracts. Before 1979, *R. gardneri* discoveries had only been made accidentally from clearing or farming activities. Searches between 1980 and 1984 involving thousands of volunteer hours identified the orchid in fewer than 4% of its likely habitats.

*Rhizanthella gardneri* completely lacks chloroplasts\*. The absence of chlorophyll gives the plant a whitish-purple colour and prevents it from photosynthesising. It also has limited access to nutrients because it lacks a developed root system. So how does this species feed? The answer presumably lies in parasitism. When the plants of *R. gardneri* were discovered, they were always found growing among the roots of

---

\* Recent DNA studies have proven that this orchid has lost most of the genes associated with photosynthesis and has the smallest 'organelle genome' yet described in any land plant.

**Fig. 4.4.1.** *Rhizanthella gardneri*, the western underground orchid growing *in situ* in Western Australia. Note that the organic matter and top soil layer have been removed to expose the orchid.
© Jeremy Storey.

*Melaleuca* shrubs. Most researchers believed that the orchid was either feeding on the shrub or on the decaying matter around it. A study using carbon and nitrogen isotopes was finally able to determine that in fact *R. gardneri* obtains its nutrients via complex fungal connections. Isotopes that were applied to the shrub's leaves and fed to the substrate's fungi were later found in the orchid. This demonstrated the existence of an underground ménage à trois connecting the roots of the *Melaleuca* shrub, the fungi living in the decaying organic matter, and the orchid. *Rhizanthella* thrives in a secret garden where it parasitises fungi and depends on them to survive.

What does this subterranean lifestyle mean for the reproduction of this orchid? How does an underground flower become pollinated? Contrary to what you might expect from a ground-dwelling flowering plant, *Rhizanthella* is insect-pollinated. During the winter months, this orchid produces a solitary inflorescence that emerges to just below the leaf litter. It looks like a single large flower but is actually a cup-like inflorescence composed of many densely clustered small flowers. The flowers are completely covered with the top layer of organic matter, but tiny insects are able to reach them by crawling between the spaces in the litter. Observations of insects removing pollen from the plant are obviously rare and extremely difficult. But the pollinators are known to include species of the fly genus *Megaselia* (family Phoridae), fungus gnats, and a species of wasp. The principal pollinator seems to be the phorid fly, which feeds on decaying fungal matter and presumably encounters the underground orchid while foraging through litter in search of the fungi. Self-pollination has also been suggested. However, the most exceptional of the alleged pollinating agents is unquestionably termites. Termites, which we normally think of only as those horrible wood-devouring pests, have been observed visiting flowers. For now, it is unclear whether they are lured to the scented inflorescence, or whether they simply stumble into it. What is certain is that the termites systematically remove and deposit the pollen of flowers within a single inflorescence. This astonishing case remains the only known termite-mediated pollination in the orchid family — and the only extant case in flowering plants.

After pollination — of course — comes fruiting, and again you must be wondering how an underground orchid disperses its seeds. The fruits of *Rhizanthella* are indehiscent, that is they do not open naturally like those of most orchids. They are fleshy and filled with hundreds of tiny, dark, spherical seeds. The seeds are much less numerous and heavier than those of most other orchids and are thought to be dispersed by marsupials that may be attracted by the smell of the fruits. The seeds of *Rhizanthella* are produced in fruits that take several months to mature, and are known to emerge above the soil line. Seed dispersal has never been observed, and even though wind-dispersal has been suggested, the most likely dispersal agent seem to be a native 'fossorial' or underground-dwelling marsupial. However, these animals are now extinct in part of the region where the plant occurs. Unlike other orchid seeds that are wind-dispersed, the seeds of *Rhizanthella* possess a thickened and somewhat resistant testa or seed coat. The robustness of the seeds suggest that they are indeed probably ingested by small fossorial marsupials. Laboratory studies have confirmed this.

Despite being discovered in the early 20th century, few specimens were found after that, and by the early 1970s, *Rhizanthella* were generally considered extinct. Systematic searching did result in their rediscovery and came with additional surprises. Until very recently two species of *Rhizanthella* were known, but three more have been discovered within the past two decades. Today, we know five species of underground orchids. Two are endemic to south-west Western Australia — *R. gardneri* and *R. johnstonii* (see Figure 4.4.2) — while another three are only known from Eastern Australia — *R. omissa* from south-eastern Queensland, *R. slateri* from the central coast of New South Wales, and *R. speciosa* from north-eastern New South Wales. The latter is the most recently discovered *Rhizanthella* species. It was only described in 2020, after Maree Elliot had found the plants growing under deep leaf litter a few years before while looking for native truffles for her PhD studies in the forests of the Barrington Tops National Park. Who knows if other secret orchids are awaiting discovery underground in Australia and elsewhere.

We have learned that *Rhizanthella* is unique for many reasons, the main one being that the plant goes through its whole life cycle

**Fig. 4.4.2.** *Rhizanthella johnstonii*, known as the south coast underground orchid. Photographed in Munglinup, Western Australia.
© Mark Brundrett.

underground. This makes the orchid — often growing in impoverished soils — dependent on fungal associations to survive. In those relationships, *Rhizanthella* is also very specialised — being intimately linked to the root systems of specific host trees and shrubs. Finding underground orchids is extremely difficult and unpredictable, which makes understanding their diversity, distribution and population size a challenge. Flowering seems to be enhanced by good rainfall, but it is unknown what the ratio between flowering and non-flowering individuals is in a population. Nor is it known how much genetic variation there is between individuals within a single population. *Rhizanthella* plants are capable of vegetative division, and close-by rhizomes may therefore be identical clones of each other. Further research is needed on both the pollination and seed dispersal of these extraordinary and cryptic orchids, all of which remain critically endangered.

*'The author and naturalist, W. H. Hudson, wrote of an atheist who went to an orchid show and left believing in the devil'*
ANDREW YOUNG in *A Prospect of Flowers* (1947), cited in THOROGOOD et al. (2019)

## 4.5 IMAGINE

In Mexico, crickets and grasshoppers are culinary delicacies, best served with lemon juice and spices. But to the majority of us, these insects are mostly known as plant pests. Crickets and grasshoppers belong to the order Orthoptera and plants typically go to great lengths to deter them, for example through the accumulation of toxins in their leaves. A relatively common strategy to avoid these and other herbivores is to offer sugar-rich nectar to attract ants (as discussed above in section 4.1).

Orthopterans not only feed on leaves, but also flowers. Florivory is the type of herbivory in which the floral parts are consumed. Damaged flowers are generally less attractive to pollinators or may not develop fruits at all. Therefore, florivory is likely to come at a high cost for the plant. Crickets and grasshoppers are also known to feed specifically on pollen. Again, this comes at a high cost for the plant as the loss of pollen can decisively impact reproductive success. Their florivorous nature may explain, at least in part, why crickets and grasshoppers

Fig. 4.5.1. Orthopterans are often florivorous, feeding on orchid flowers rather than serving as pollinators. **A.** The flowers of a *Lycaste desboisiana* in Costa Rica. **B.** The Australian orchid *Cyanicula ixioides*.
© APK (A) and Jeremy Storey (B).

are seldom thought of as regular pollinators, even though they have been found visiting many different flowers and carrying pollen traces. In the Orchidaceae, florivory by these insects is apparently uncommon (maybe due to those overprotective ants), but it is not unheard of (see Figure 4.5.1). In the terrestrial orchid *Habenaria sagittifera*, juveniles of the bush cricket *Ducetia* have been observed nibbling on pollinia and anther caps while visiting the flowers. The cricket occasionally removed pollinia and therefore could potentially act as a pollinator. However, its inefficiency and the removal of only partial pollinaria suggests that there must be a more suitable pollinator for this particular orchid species.

Pollination by Orthoptera is virtually unheard of in flowering plants, though there are a couple of exceptions. In the Galapagos Islands, one species of grasshopper was found to carry pollen of at least five different plant species. Even though pollen transfer remains to be demonstrated, the necessary basis for pollination under those isolated island conditions seem to be met. But it should not come as a surprise to you by now that the only flowering plant proven to be pollinated by a cricket is an orchid. The species in question is a small-flowered cousin of the famous *Angraecum sesquipedale*, or Darwin's orchid, discussed in the *Everybody knows* story in Chapter 3 (section 3.1). *Angraecum cadetii* is found on La Réunion, an island east of Madagascar that is known for being the place where the technique for manually pollinating *Vanilla* was invented. *Angraecum* species are typically adapted to pollination by long-spurred moths, but the fleshy white flowers of *A. cadetii* emit a peculiar nocturnal bouquet that lures its pollinating cricket. The insect — *Glomeremus orchidophilus* — was observed on multiple occasions probing flowers in search of nectar, removing and depositing pollinaria in the process (see Figure 4.5.2). These Gryllacridinae, the family to which the cricket belongs, are endemic to the island Archipelago in which they are found. They are nocturnal and explore their surroundings with a pair of extremely long antennae. Mainland species of Gryllacridinae are normally omnivores, eating diverse plant parts. However, on islands they may have adapted to nectar consumption to compensate for the scarcity of food sources.

It isn't clear whether pollination by crickets in *A. cadetii* represents an isolated case or whether there are other such instances in which orthopterans are pollinating plants. It is easy to imagine that more pollinators are to be found among the dozens of Orthoptera species that have been recorded to visit flowers. Nevertheless, the fact remains that they are very infrequent pollinators. Why? A possible explanation is that the conditions for a transition from florivory to pollination are rare. The night-scented white flowers of *Angraecum* are, in a way, pre-adapted to pollination by nocturnal insects, while the restricted sources of food on islands may be a strong evolutionary driver, stimulating the switch to nectar feeding. Thus, a critical combination of conditions may have prompted the evolution of a rare pollination syndrome.

Intriguingly, the mutualism between orchids and orthopterans is not limited to pollination under strenuous conditions. Crickets have been shown to act as seed dispersers in certain orchids too. Recent studies in Japan have revealed that the crickets eat the fruits and excrete the seeds of primitive orchids belonging to the genera *Apostasia* and *Yoania*. The orthopterans are said to obtain nutrients from the pulp of the greenish or whitish fruits, and to disperse the small, hard-coated seeds after they pass through their digestive system undamaged. In fact, passage through the digestive systems of animals can scarify the hard seed coat and improve germination, as seems to be occurring with these orchids. In Costa Rica, I have personally observed and caught crickets belonging to the genus *Idiarthron* (Tettigoniidae) that had fed on the fleshy fruits of *Vanilla planifolia*, its tiny black seeds passing entirely through the digestive system and showing up in their faeces. But I will talk more about *Vanilla* in the next section.

Fig. 4.5.2. The nectar-rich flowers of the rare Madagascan orchid *Angraecum cadetii* are pollinated by nocturnal crickets.
© Claire Micheneau and Jacques Fournel.

**IMAGINE**
Cricket pollination in *Angraecum*

**SCAN ME**
https://youtu.be/CZ5ziOOqHhA

Grasshoppers and crickets are not typically thought of as flower pollinators. In fact, many of them are plant pests, and in orchids they may occasionally be found munching away on the fleshy flowers. There is a single known case of the pollination of an orchid flower by crickets. On the island of La Réunion, *Angraecum cadetii* is pollinated by the raspy cricket, *Glomeremus orchidophilus*. The whitish flowers are fragrant at night, luring the nocturnal cricket, which feeds on the nectar offered by the flowers. It is during this careful inspection of the flowers that the raspy cricket removes the orchid's pollinarium.

The videos and photographs shown here are owned by Claire Micheneau and Jacques Fournel and have been reproduced and edited by the author with permission. Videos published as Supplementary data in Micheneau *et al.* 2010 (https://doi.org/10.1093/aob/mcp299) have been reproduced with the kind permission of *Annals of Botany* and Oxford Academic Press.

## 4.6 LADY IN BLACK

It was a warm Sunday morning on the Osa peninsula in Southwestern Costa Rica. I was rushing up and down a dirt road deep in the jungle close to Corcovado National Park. The weather in Osa is known to be extreme. It becomes increasingly warm during the morning hours and severe rainfalls follow virtually every afternoon. Most days the heat and humidity are so intense that by 7am you are breaking a sweat merely by sitting down to have breakfast. This was such a day. It was early but the sun was already scorching hot and the air full of moisture. Two Master's students stood patiently waiting for me to show up. Each of us was stationed at a different spot along the dirt road, equipped with a professional camera and a tripod. Every now and then, I would sprint from one location to the other, changing the exhausted batteries for fully charged ones, and swapping each filled memory card for an empty one. The students had to stay still, wiping the sweat from their brows and chasing away the pestering mosquitoes, without making too much of a racket. They waited quietly for something to happen in front of their lenses. That morning, we were monitoring the pollination of wild *Vanilla* flowers.

We are all familiar with vanilla as a food flavouring. It is an essential ingredient in ice-cream and almost every other kind of sweet treat, from cakes and cookies, to custards, puddings, flans and oatmeal. It has also been used to make milk-based drinks more palatable for children. Vanilla essence is found in many cosmetics, too, as well as in aromatherapy oils, where it is said to promote relaxation and improve sleep quality. However, *Vanilla* is also the name of the orchid that produces the fruit from which the world-famous spice is obtained. Vanilla flavouring comes from the organic compound vanillin, which is found exclusively in the fruits of these tropical orchids. There are more than 100 different species of *Vanilla* known worldwide. They prosper in the tropical regions of Africa, America and Asia. But only certain species native to the American continent produce the highly esteemed aromatic compound that we all love. Mexico is of course known as the origin for vanilla culture, but fragrant *Vanilla* plants are found in the wild in most Latin American countries, even today. In Costa Rica, several fragrant *Vanilla* species can still be seen thriving in their natural habitats. Large populations with hundreds of plants can be spotted climbing the tall trees in many coastal areas. A few native *Vanilla* species are common in the vicinity of Piro Station in the Osa Peninsula, where we have been studying them for several years.

Despite being a highly valued crop, little is known about the most elemental biological aspects of *Vanilla*. The story of how a 12-year-old African slave called Edmond Albius discovered how to hand-pollinate cultivated vanilla flowers has been told many times. In 1841, Albius placed the pollen from the anther of a vanilla flower onto its own stigma, ensuring self-fertilisation. As a consequence, he was able to produce the much-valued fruits outside of their natural habitat for the first time. Until then, vanilla plants were taken from Middle America — mostly Mexico — to Europe and its tropical colonies, where the plants grew and flourished, but never produced any pods. The importance of Edmond's contribution to the transformation of vanilla from a Mexican curiosity to a worldwide, multi-million dollar industry is

**Fig. 4.6.1.** *Vanilla dressleri*, a beautiful species with creamish white flowers and a citrusy smell.
© APK.

thoroughly presented in Tim Ecott's highly insightful *Vanilla: Travels in Search of the Ice Cream Orchid*. The book is without a doubt the richest source of information on vanilla culture and history. However, the ecological interactions and ecosystem role of the wild relatives (see Figure 4.6.1) of the commercial *Vanilla* are still largely a mystery. Virtually nothing is known about their natural pollination and the dispersal of their seeds.

*Vanilla* is the only orchid — as far as I can tell — that has been associated with pollination by bats. Bat pollination, or chiropterophily, is occasionally mentioned in *Vanilla* literature, where it is mostly regarded as unlikely. Recent studies on bat-pollinated flowers explicitly exclude *Vanilla*. But where did the belief come from in the first place? I discovered that the vast majority of publications that mention bat pollination in *Vanilla* either don't cite any source, or cite another paper that is not actually the source — most are dead ends. But after persisting for several hours, my diligence finally paid off. The notion that *Vanilla* flowers could be pollinated by bats has a single source: it was proposed by the Brazilian naturalist Augusto Ruschi. In a 1978 publication dealing with the bats in the Espírito Santo state in Brazil, he noted: "It is important to point out in this work, that we give publicity for the first time, to what we observed at the Estação de Biologia Marinha, in Santa Cruz, in 1972 and the years that followed, in relation to the pollination of flowers of the orchid: *Vanilla chamissonis* var. *longifolia* Hoehne, made by the bat *Glossophaga soricina soricina* (Pallas), as it searches for the pollen of this species, and yet another bat that visits it, to eat its flowers, and also the fruit when ripe, is *Artibeus jamaicensis planirostris* (Spix, 1823)."\*

The observations made by Ruschi are certainly fascinating, but bats are not likely to pollinate the flowers of any *Vanilla* species. *Glossophaga* is a small, generalist bat that visits many flowers in search of nectar, while *Artibeus* is a very large frugivore (fruit eater) that mostly thrives on raw fruits and succulent, highly nutritious plant parts. Neither is well suited

---

\* The original Portuguese reads: "*É importante ainda assinalar neste trabalho, que pela primera vez damos publicidade, ao que observamos na Estação de Biologia Marinha, em Santa Cruz, nos anos e 1972 e seguintes, em relação á polinização das flores da Orquidea: Vanilla chamissonis var. longifolia Hoehne, efectuada pelo morcego Glossophaga soricina soricina (Pallas), uma vez que vai em busca do pólem dessa espécie, e ainda outro morcego que a visita, para comer suas flores, e também o fruto quando maduro, é Artibeus jamaicensis planirostris (Spix, 1823).*"

for pollen removal in the diurnal (daytime flowering), nectar-less, fragile flowers of *Vanilla*. Flowers pollinated by bats share certain characteristics described by the so-called 'chiropterophilous syndrome'. The most common features found in flowers that are bat-pollinated include: nocturnal anthesis, meaning that they open at night; dull colours, mostly white or green; a musty, fetid smell; large, tubular shape; and especially the production of copious amounts of nectar, pollen and other highly nutritional edible plant tissues. In essence, floral characteristics associated with bat pollination have evolved to attract relatively big, nocturnal, colour-blind, flying pollinators. *Vanilla* flowers may be relatively large — compared to those of other orchids — and have dull colours, but they definitely do not offer the profuse and highly nutritious rewards that a bat needs. Most importantly, the flowers of most *Vanilla* species, including *V. chamissonis*, only open for a few hours during the warmer morning hours of the day. By nightfall the flowers are shriveled and their delicate tissues have deteriorated. Bats certainly cannot pollinate *Vanilla* flowers, but perhaps Ruschi was on to something and their presence has another explanation. We will return to his observations later on.

Back in the intense jungles of Osa, we were still trying to catch the natural pollinators of *Vanilla* in the act. It is a rare and largely undocumented event. *Vanilla* flowers are extremely short-lived — in most species they have withered by midday. They are at their most active in the morning, which closely correlates with the peak activity of the bees that are often suggested to pollinate them. In the hope of increasing our odds, we had positioned ourselves at three different locations where we had identified mature floral buds the previous day. Despite the large number of individual plants, floral visits are scarce and brief, and thus difficult to document. We came up with the idea of continuously recording flowers for several hours so that we would't miss any of the action. But non-stop recording is complicated and expensive, and Piro Station is a good eight-hour drive from Lankester Botanical Garden. Going there for a few days without knowing whether the *Vanilla* plants would be in bloom or whether the rain would make our efforts futile meant our chances of getting anything were not great. Nonetheless, after several trips to the station and repeating the strategy over three years, we were finally able to record Euglossini bees visiting the flowers of *V. pompona* (three times) and *V. trigonocarpa* (twice). It was clear that the only way to study this further was by collaborating

Fig. 4.6.2. The large, fragrant, yellow flower of *Vanilla pompona* is highly attractive to scent-collecting male euglossine bees.
© Sebastián Vieira Uribe.

with someone locally. That is when PhD student Charlotte Watteyn joined the team. She settled in Osa for her studies, and finally some data on the rare occurrence of *Vanilla* pollination started coming in.

What she discovered during the studies on *Vanilla pompona* flowers in Costa Rica was that they apply a very unusual, dual pollination strategy that involves both fragrance collection and food deception (see Figure 4.6.2). *Vanilla pompona* flowers are pollinated by male euglossine bees that collect fragrances from the floral parts*. Unlike those orchids from which the bee removes and deposits pollen during fragrance collection — as in *Coryanthes* and *Gongora* — in *Vanilla*, fragrances seem to be merely a means for initial, long-distance attraction. Even though a few *Vanilla* species were known to be pollinated by Euglossini bees — and fragrance collection had been observed or suspected to occur — it had never been shown to be related to pollination. Fragrance collection occurs only on the sepals and petals of the flower and is not, on its own,

---

\* See Chapter 3, section 3.7, *You're so vain* and subsequent stories for further detail on this particular pollination strategy.

enough to ensure that the flower becomes fertilised. For pollen removal and deposition to occur, the bee needs to crawl into the long tubular lip, where the fragrances are absent. How does a flower induce a bee to stop collecting fragrances on the sepals and petals and instead crawl into the tube where the pollen is kept? Very simple. The large male bees from the genus *Eulaema* expend a lot of energy scraping the fragrances off the sepals and petals of the *Vanilla* flower. Their weight doesn't always allow them to land on the delicate flowers for support, so they have to hover while doing this. After a while, they are hungry and need to feed. Inside the lip of *V. pompona* is a series of glandular hairs that produce traces of nectar — far too little to actually feed the bees, but enough to have them probe the lip for food. With their tongues extended, the male *Eulaema* bees crawl into the tube, only to exit once again shortly after with nothing but the flower's pollen on their backs (see Figure 4.6.3).

In the Neotropics, many *Vanilla* species are pollinated by Euglossini bees, but not all of them reward their pollinators with floral fragrances. In fact, most are pollinated through food deception. Species such as *V. planifolia* and *V. odorata* have food-deceptive flowers that lack a strong floral aroma. The long column forms a narrow, dry nectary together with the base of the lip. The nectaries are provided with hairs and extremely little — if any — actual nectar. These are of course metabolically active sites, producing volatiles and possibly also sugary traces that entice the bees to probe the flowers. Euglossini have high energetic requirements, and these orchid flowers simply don't have what they need. Visits to these deceptive flowers are therefore sporadic and very brief, but enough to transfer pollinia successfully.

*Vanilla hartii*, another locally abundant species, differs from its close relatives in having a true nectary with measurable quantities of the precious sugary liquid to offer its pollinators. It is visited by multiple Euglossini bees, but the small floral tube only allows for the smaller species to enter the lip and remove pollinia as they feast on the nectar. Nectar robbers include larger Euglossini — which can't enter the floral tube but reach the nectar with their long tongues — and hummingbirds. There are only a handful of other pollinator groups that have been shown to pollinate *Vanilla* species around the world. All of them are species of bees and wasps — mostly belonging to the tribes Centridini and Allodapini — and in all cases pollination seems to involve food deception.

Fig. 4.6.3. A male *Eulaema cingulata* (Euglossini) carrying *Vanilla pompona* pollinia masses on its back.
© APK and Charlotte Watteyn.

Studying ecological relationships in the field is not easy: observing visiting animals and determining the extent of their interaction with an orchid flower can be a real challenge. A very recent paper suggests that *Vanilla palmarum* is pollinated by hummingbirds. In the *Misunderstood* story (Chapter 3, section 3.3), we discussed how multiple orchids have mistakenly been assumed to be pollinated by hummingbirds, following the observation that these fast-flying birds hovering in front of the flowers. Hummingbirds are curious birds that will examine virtually anything that catches their eye — even brightly coloured, inert objects. They have previously been seen inspecting the flowers of both *V. chamissonis* and *V. planifolia*, and our team has recorded them robbing nectar from *V. hartii* flowers too. But they have not been shown to transfer *Vanilla* pollinia. Ornithophily in *Vanilla* is mechanically quite difficult to explain, and unfortunately pollinia removal and deposition by the birds was not shown in the *V. palmarum* study. Unlike the compact, tubular flowers of the hummingbird-pollinated genera *Arpophyllum*, *Elleanthus* or *Ornithidium* — for example — the flowers of *Vanilla* have a long, broad, tubular lip expressly made for accommodating bees. For pollinia to be transferred, a bee needs to enter the tube, contact the column, flip the rostellum (a flap-like extension of the stigma) and get the pollinia smeared onto its back while exiting. It's a very precise mechanism, with a tight fit, and not every bee is suited to transfer pollinia. Some are too small and can't reach the column, whereas others are too big and can't enter the tube. A bird's beak is far too narrow for the broad tube, and even if it contacts the column, it seems unlikely to exert the pressure required to flip the rostellum and remove pollinaria from one flower, and deposit them in the next. Hummingbirds may show an interest in *Vanilla* flowers, but the long column, broad lip, and large rostellar flap suggest that these flowers are not well adapted for ornithophily. We have learned that orchid flowers may indeed shift to new pollinator groups and become generalists under certain conditions. But unless pollen transfer can be shown and the mechanics explained, birds are perhaps best treated as a floral visitor in *Vanilla* rather than a pollinator.

In Mexico, vanilla was known by the Nahuatl name *tlilxochitl* — meaning black flower. We know that vanillas actually have greenish or yellowish-cream flowers and it is the fruit that becomes black when ripe. In his *Vanilla* book, Tim Ecott blames the Aztecs — whom he accused of a lack of familiarity with vanilla plants under natural conditions — for this supposed mistake. Ecott suggests that the Spanish conquerors, who also lacked first-hand knowledge of vanilla, helped to disseminate the blunder. This idea of the Aztec mistake was repeated by Kevin Ashton in his *How to Fly a Horse: The Secret History of Creation, Invention, and Discovery* — another most entertaining book. I personally believe that the name 'black flower' isn't a mistake at all. Alonso de Molina's dictionary of the Spanish (Castellano) and Mexican (Nahuatl) vocabulary — originally published in 1555 — was the first dictionary published in the New World. The 1571 edition of the dictionary included a Nahuatl to Spanish section in which we find the word *tlilxuchitl* defined as "*ciertas vaynicas de olores*" [certain aromatic beans]. The black flower of the Aztecs obviously alludes to the fruits, not to the actual flowers. But did they mistake the fruits for flowers? Probably not, and it is more probable that the name shows they had in intricate knowledge of the plant in its natural habitat. It is a misconception that the mature vanilla beans turn blackish and highly aromatic only after harvesting and processing. They also ferment and become aromatic naturally. Anyone who has spent time in a forest trying to locate *Vanilla* fruits through their enticing perfume can testify to this. The reason for the name 'black flower' most likely comes from the fact that the fruits of many wild aromatic *Vanilla* species naturally dehisce when mature — meaning they split longitudinally — while still hanging from the vines. The two sides of the long, bean-like fruit open up, exposing the black seeds and releasing their delightful, sweet, overpowering aroma. When the aromatic fruits of *Vanilla* split, they give the impression of being a sweetly scented black flower — something the Aztecs surely knew very well. The dehiscent aromatic fruits hanging from the *Vanilla* vine indeed resemble black flowers, and that is probably where the name comes from.

But why does the fruit of this orchid have such a powerfully enchanting smell? It has been widely assumed that *Vanilla* seeds are not dispersed by the wind, as is the case for the vast majority of orchids. That is a sound hypothesis given that, unlike the tiny, dust-like, transparent seeds of most other orchids, the seeds of *Vanilla* are relatively heavy, rounded and covered by a hard, black seed coat. There is another key difference, of course. Seeds of certain *Vanilla* species are packed into a notoriously fragrant fruit. Biologically speaking, the natural fermentation process of the fruit that leads to the fragrance production must have a function. But what? Even though you may think this has been well established, there is actually little definite evidence and no consensus among scientists. Different reports in the literature point the finger at animals as diverse as ants and bats, bees and reptiles, birds and marsupials, and occasionally monkeys.

Our ongoing studies on *Vanilla* in Costa Rica have revealed a most extraordinary means of natural seed dispersal. Vanillin has long been known to attract male euglossine bees. It is one of the aromatic compounds used in baits to attract the bees. In the neotropical region, Euglossini are often observed approaching harvested vanilla beans as they mature. The bees brush the surface of the fruits with their forelegs, a behaviour we know corresponds to fragrance collection. However, and despite their importance in pollinating flowers, bees are very rarely known to disperse the seeds of plants. They are unable to open or carry the large *Vanilla* fruit, and therefore have been mostly disregarded as seed dispersers. But the fruits of certain *Vanilla* species — such as *V. planifolia* (see Figure 4.6.4) and *V. odorata* — naturally open up on the vine when they ripen. Growers know they have to collect and process the beans when they are still green to avoid this dehiscence. A dehiscent *Vanilla* fruit exposes its black seeds and — when left on the vine — the

**Fig. 4.6.4.** *Vanilla planifolia* flowering and fruiting *in situ* at Cahuita National Park in Costa Rica.
© APK.

**Fig. 4.6.5.** Natural seed dispersers of *Vanilla planifolia*. **A.** The fragrant 'black flower' attracts male Euglossini bees, that remove seeds while collecting vanillin. **B.** Indehiscent fruits drop to the ground where they are eaten by mammals, such as the spiny rat *Proechimys semispinosus*.
© APK.

highly aromatic 'black flowers' are visited by the Euglossini (see Figure 4.6.5). It is while the males collect the fragrances from the open fruit that they remove seeds. We recorded several *Eulaema* and *Euglossa* species getting the tiny black seeds stuck to their bodies as they brush up the vanillin (see Figure 4.6.5A); even more seeds drop to the ground as the bees work. The attractiveness of the fruits and importance of vanillin were demonstrated experimentally in the field. Dozens of euglossine bees were caught in bottle traps containing mature, dehiscent *V. odorata* or *V. planifolia* fruits. Interestingly, commercially sold vanillin was also effective in attracting the bees, but at most sites, the extract attracted fewer bees than the natural fruits.

What does this mean for those indehiscent (non-splitting) *Vanilla* species? How are their seeds dispersed? This is where the story becomes really remarkable. We noticed that the fruits of *V. pompona* and certain *V. planifolia* relatives — including the commercially grown *V. × tahitensis* hybrid — become mature on the vine and don't open at all. Instead, they drop to the ground, becoming brownish and aromatic as they ferment. Those fruits just seem to lay there and unless they are damaged or opened, bees are not interested. So, we decided to place motion-activated camera traps in front of the fruits. The results have been illuminating. Several different animals can be seen passing in front of the camera. Most of them ignore the *Vanilla* fruits entirely. Several species of birds — which are known for having a poor sense of smell — seem not to notice the brown fruits among the litter at all. Similar reactions are observed from armadillos, felines and lizards. Animals with a good sense of smell, such as the gray four-eyed opossum, white-nosed coati, and even dogs, can be seen sniffing out the aromatic fruit, but are clearly repulsed. However, one animal showed up multiple times in our cameras at all sites, nibbling on the fruits and returning until they finished eating them whole. The spiny rat, *Proechimys semispinosus*, a common rodent in Central and South America, was caught repeatedly enjoying the fleshy aromatic fruits of both *V. planifolia* and *V. pompona* (see Figure 4.6.5B). It and the common opossum *Didelphis marsupialis* were the only animals that showed up on multiple occasions on our cameras (a few other rodent species were also recorded consuming the fruits).

This was an extraordinary discovery, but does animal ingestion necessarily lead to *Vanilla* dispersal? By experimentally feeding the spiny rat and common opossum with *Vanilla* fruits, we were able to prove that the tiny black seeds passed through their digestive systems and showed up in the faeces. Through an ongoing collaboration with Jyotsna Sharma from Texas Tech University, we have now been successful in germinating *Vanilla* seeds from the digested fruits *in situ* in Costa Rica. This suggests that the rodents and marsupials are able to disperse the orchid's seeds by eating the aromatic fruits, retaining the pulp and passing the seeds without harming them. Despite not having been previously shown to disperse *Vanilla* seeds, anyone who studies rodents can testify that they are commonly caught using vanillin as bait! Which means that vanillin is not only the queen of flavours to us humans, it also plays a key role in attracting at least two very different groups of seed dispersers in *Vanilla* orchids: bees and mammals.

The experiments showed that euglossine bees were quite keen to visit dehiscent *Vanilla* fruits in bottle traps, but indehiscent fruits did not interest them much. Curiously, after leaving the fruits out in the field for a few weeks, they attracted another insect. Crickets! We caught several individuals of varying sizes belonging to the genus *Idiarthron*, a member of the Tettigoniidae family, commonly known as bush crickets. From the bite marks, it is clear that the crickets fed on the aromatic fruits of *V. planifolia*. The tiny black seeds were also found in their faeces. Perhaps the crickets are merely accessory seed dispersers, but what is a fact is that they can feed on *Vanilla* fruits. In the very same way, they have already been shown to feed and disperse the seeds of other orchids with fleshy fruits and rounded seeds with a hard seed coat, such as *Apostasia* and *Yoania*, mentioned previously. We have also found bite marks on the fruits of the fragrant *V. trigonocarpa* that suggest they were eaten by a bat. This brings us back to the initial observations by Ruschi, who saw bats interacting with *V. chamissonis* — another aromatic species from Brazil. Perhaps the bats he saw were eating the fruits rather than pollinating the flowers, which is more consistent with their nature. In any event, *Vanilla* ecology seems to be extremely complex and much remains to be discovered and tested. How about the *Vanilla* species that lack vanillin? What is dispersing their seeds? No one knows.

The word 'vanilla' is commonly used to describe things that are boring and plain. Vanilla sex generally refers to conventional sexual practices that conform to basic expectations. The word has even gained a culturally negative connotation. "I am not vanilla! I've done lots of crazy things!" one of the characters of the American television sitcom *Friends* objected angrily. The origin of this idiomatic usage of the word probably comes from ice cream vendors in the United States, who offered their primary flavouring 'plain vanilla' as the cheaper, basic option. But as we have seen, there is nothing plain about *Vanilla*, especially when it comes to the complex and still only partially understood ecological relationships needed for it to reproduce. I say, all hail the lady in black!

**LADY IN BLACK**
Pollination in *Vanilla*

SCAN ME
https://youtu.be/XOim1HWaZsQ

*Vanilla pompona* is pollinated by *Eulaema* bees through a double mechanism of fragrance collection and food deception. The male *Eulaema* collect fragrances on the sepals and petals, but will only visit the lip when hunger strikes. They enter the tube with their tongues extended looking for nectar, and leave quickly when they realise the flower is nectarless. However brief, the visit is enough for the bee to get the orchid's pollen smeared on its back. *Vanilla hartii*, on the other hand, produces copious amounts of nectar that attracts several floral visitors. Large bees are unable to enter the floral tube to pollinate the flower and are nectar robbers. Hummingbirds can also be observed piercing the nectary externally and robbing the flowers of their nectar. It is only smaller species of *Euglossa* that fit the floral tube and remove the orchid's pollen as they exit. *Vanilla planifolia* is known as the black flower for its highly aromatic fruits that split when mature. The open fruit is visited by male Euglossini bees that collect the fragrant compounds and disperse the orchid's tiny seeds in the process. Indehiscent fruits fall to the ground where they are consumed by mammals, mostly rodents.

The videos and photographs shown here are owned by Adam P. Karremans, Charlotte Watteyn and Daniela Scaccabarozzi and have been reproduced and edited by the author with permission. Videos published as Supplementary data in Watteyn *et al.* 2021 (https://doi.org/10.1111/btp.13034) have been reproduced with the kind permission of *Biotropica* and Wiley.

## 4.7 WHAT IF

> *"This co-existence with the dinosaurs makes me wonder if small dinosaurs might have pollinated these early orchids, given that modern orchids are pollinated by such a wide range of animals."*
> MARK CHASE in the foreword to *Extraordinary Orchids* by SANDRA KNAPP (2021)

Even though orchids have developed a seemingly infinite number of contrivances to succeed in becoming pollinated, the act of pollen transfer itself is almost exclusively carried out by either insects or birds — at least in modern times. Consequently, the diversity of organisms that pollinate orchid flowers may not appear as overwhelmingly diverse as the intricate mechanisms themselves. But that doesn't mean there are no reports of unusual pollinators in such a ubiquitous and versatile family of plants. In closing this *Misfits* chapter, we will discuss the two major 'what ifs?' among orchid pollinators: mice and lizards.

Flying mammals — such as the bats discussed in the previous story — are well known for their pollinator services to many flowering plants. Conversely, non-flying mammals have only rarely been documented as pollinators. Marsupials, rodents, shrews, lorises, monkeys and lemurs have been shown to pollinate some 60 or so flowering plant species across some 20 different families. Among the rodents, only mice have been suggested as pollinators in the Orchidaceae family. Even though they have been proven to pollinate the flowers of certain plants effectively, they are odd among orchid pollinators and quite unexpected, for good reasons. Plants that are regularly pollinated by rodents — such as the South African sugarbushes in the genus *Protea* (family Proteaceae) and the hedgehog lilies in genus *Massonia* (family Asparagaceae) — are known to produce large amounts of nectar and have nutty or yeasty smells. Their flowers or inflorescences also tend to be large, the anthers and stigmas are easily accessible, and their pollen is abundant. These features are not regularly found in orchid flowers, which are much more commonly adapted to pollination by small insects. As far as I can tell, there is a single 'preprint' study claiming that mice were found to be the exclusive and consistent pollinators of the orchid *Cymbidium serratum*. The study was never formally published, for reasons unknown to me. Perhaps after careful scrutiny the authors withdrew their hypothesis, or maybe additional proof was requested by their peers. Be that as it may, there seems to be good reason to be sceptical. However, the extraordinary case deserves mentioning here given that the study is cited in orchid pollination literature.

Without exception, all the well-documented pollinators of *Cymbidium* species are bees in the Apidae family (see Figure 4.7.1). The most commonly cited pollinator seems to be the Asian honeybee — *Apis cerana* — which is attracted to the food-deceptive *Cymbidium* flowers through fragrances. The bees transfer the orchid's pollinaria while exploring the base of the lip and contacting the column apex with their backs while exiting. The flowers of *C. serratum* are structurally not particularly different from those of many other species in the genus. The floral morphology would suggest a pollinator that is similar to those of its sisters, but instead, pollinaria of *C. serratum* are allegedly removed while the rodent munches on the lip. Although it is not impossible for this species to be adapted to such a different pollinator — unique in size, shape and behaviour — it is rather difficult to imagine how the *Cymbidium* pollinarium could be placed with the precision required for it to be later transferred by such an animal. Mammal pollination — by both flying and non-flying animals — can of course not be ruled out completely in the Orchidaceae family, especially considering that we lack pollinator information for about 90% of all orchids. However,

Fig. 4.7.1. *Cymbidium serratum* (**A**) a species allegedly pollinated by mice and *C. cyperifolium* (**B**), which like many other congeners, is pollinated by the Asian honeybee *Apis cerana*.
© Ron Parsons (A) and Hong Liu (B).

**Fig. 4.7.2.** *Myrmecophila thomsoniana*, a rare orchid species endemic to Grand Cayman
© Christine Rose-Smyth.

flowers pollinated by such animals would need to be large and offer copious nectar. Explicit floral modifications would certainly be expected to allow specific placement of the pollinarium on a furry mammal. Until proven otherwise, the idea of bat and mice pollination in orchids is perhaps best taken with a pinch of salt.

Another very odd group of potential pollinators is reptiles. We may never know if small dinosaurs ever really pollinated any of the early orchids — as Mark Chase wondered in his foreword to Sandra Knapp's *Extraordinary Orchids*. But we do know that birds pollinate several different orchid groups, and there is a possibility that some of their non-flying modern relatives do too. It is well established that reptiles — specifically lizards — visit the flowers of several different plant species, and they are occasionally important pollinators. This is especially true on islands, where lizards can have high population densities and other

**Fig. 4.7.3.** The isolated condition of *Myrmecophila thomsoniana* may have led to the development of a unique pollination mechanism, being occasionally pollinated by cetoniid beetles. Rare footage also shows an *Anolis* lizard carrying the orchid's pollinarium after feeding on extrafloral nectar.
© Christine Rose-Smyth.

pollinators may be limited or absent. Cases of lizards either aiding the pollination of or actually pollinating flowers have been found on small, offshore islands in Brazil and New Zealand, on the Balearics, Corsica and other Mediterranean islands, on Madeira and Mauritius, and on Tasmania. On La Réunion, the endemic day gecko *Phelsuma borbonica* was observed feeding on the nectar of *Angraecum bracteosum* and *A. cadetii*. The day gecko did not effectively pollinate any of the flowers, but it did remove pollinaria on one occasion — but ended up eating it. Even though both orchid species are known to be pollinated by other agents, floral visits by these lizards present the possibility that reptiles may eventually contribute to orchid pollination. In fact, one ongoing study on *Myrmecophila thomsoniana* (see Figure 4.7.2) — a rare orchid species endemic to Grand Cayman, an oceanic island in the Caribbean — may offer a paradigm shift.

Unpublished studies by Christine Rose-Smyth suggest that as a result of the isolated conditions under which *Myrmecophila thomsoniana* grows on the island, it has adapted to attract a range of rather unusual floral visitors and pollinators. The ecological isolation of oceanic islands can cause a reduction in the diversity of available pollinators, which may promote the evolution of generalist systems and adaptation to unusual pollinator groups. Initial observations suggest just that. Rose-Smyth's ongoing studies reveal that even though bees are still able to pollinate the flowers of this species, the most important group of pollinators is actually cetoniid beetles. Several beetle species have been recorded transferring pollinaria. But they are not the only unusual pollinators. She has also documented occasional pollinaria removal by a bird species and the lizard *Anolis conspersus* (see Figure 4.7.3). In this extraordinary case, the blue-throated anole has been observed to feed on both the sweet, extrafloral nectar offered by *M. thomsoniana*, and on the ants that are attracted by the sugary liquid. The anole was twice recorded removing pollinaria during this feeding process. Given that the pollinaria are placed on top of the animal's head (see Figure 4.7.4), there is a real possibility of transferring them to a stigma. These lizards may also jump onto and climb the inflorescence to reach the nectaries, which occasionally disrupts the work of pollinating beetles. It can therefore indirectly affect pollination not only by impeding the beetles from accessing the flowers, but also by stimulating those that have already removed pollinaria to seek other *M. thomsoniana* flowers.

These findings suggest that the rare *Myrmecophila thomsoniana* may have evolved towards generalist pollination strategies to cope with the particular ecological conditions on the island of Grand Cayman. Despite being only preliminary, these results are in line with observations that lizards play a role in dispersing seeds and pollinating the flowers of other island plants while feeding on them. The possibility of reptile pollination is an exciting new avenue of research in orchid pollination — these plants truly are full of surprises!

**Fig. 4.7.4.** The lizard *Anolis conspersus* carrying the pollinarium of *Myrmecophila thomsoniana*.
© Christine Rose-Smyth.

*Stelis villosa*, bears unusual dangling appendages of unknown function on the sepals.
© Henry Oakeley.

CHAPTER 5

# REDESIGN

In the preceding chapters, we have reviewed the diversity of animals that are known to pollinate orchid flowers and how they are persuaded to do so, from the very basic rewarding systems to some of the most intricate deceptive strategies among flowering plants. We have also scrutinised the diversity of pollinator groups that orchids employ to do their bidding, from the more traditional groups such as bees and flies, to less frequently recruited animals — including ants, thrips and lizards.

From a strictly anthropomorphic point of view, orchid flowers appear to have taken as much advantage of animals as they possibly can. They have selected as their pollinators those organisms that seem to be best suited for the job and disregarded those that aren't. They provide rewards only when strictly necessary, and rely on deception when needed. Orchids have figured out how to reproduce in the most cost-efficient way possible, luring the most effective pollen carriers at the lowest price possible.

But in the end, most orchid species still depend on their pollinators to succeed. Animals have their own agenda and can be unpredictable or unreliable at times. Orchids — on the other hand — are not in the business of leaving something as important as the survival of their species completely up to the whims of pollinators. By making certain morphological and physiological changes, they have found ways to increase their odds of reproductive success. These particular changes are an improvement on the basic orchid flower or a particular pollination strategy, a 'redesign', as it were, that has been made to ensure fertilisation at all costs. Particular 'improvements' on the mechanism of pollination are the subject of this chapter.

The extensively invasive *Oeceoclades maculata*, a selfer aided by rain. © APK.

## 5.1 COMFORTABLY NUMB

The Greek historian and warrior Xenophon of Athens — a student of Socrates — famously told the story of how, on its way back to Greece after defeating the Persians, the army he led feasted on local honey along the shores of the Black Sea. The troops became disoriented and sick, suffering from nausea and diarrhoea, and were no longer able to stand up straight. The army was rendered completely defenceless. However, the next day none of his men had died, and they slowly came back to their senses. After two or three days, they were able to get up and continue their travels. But not everyone has had such luck after consuming this honey. A few centuries later, the Roman army, led by Pompey the Great, was stationed in the same area in Northeastern Turkey. After delighting on the pots of honey left by King Mithridates of Pontus and his men, the Romans fell ill, exhibiting the same symptoms described by Xenophon. Well aware of the effects the honey would have on the invading army, this time the Persians took advantage. Unable to properly defend themselves, more than a thousand Roman troops were killed. The Persians had not altered the honey themselves, so why was it making the soldiers so ill?

The sweet honey these men feasted upon is naturally hallucinogenic. If ingested in large amounts it is indeed noxious and can be quite dangerous. Today it is known as mad honey and it continues to be produced and consumed in Turkey, Nepal and a few other parts of the world. Secondary products derived from plants — such as honey — can contain chemical compounds that, depending on their concentration and application, can be considered medicinal or poisonous. What gives mad honey its particular properties are a group of neurotoxins known as grayanotoxins\*, which naturally occur in the beautiful *Rhododendron* flowers as well as a few other genera in the Ericaceae family. Honey that is made by bees collecting the nectar and pollen of *Rhododendron* flowers contains the grayanotoxins and is used both in traditional medicine and as a recreational drug.

You may be wondering what neurotoxins in honey have to do with pollination in orchids. Well, the presence of toxins in the nectar

---

\* The name grayanotoxins honours Asa Gray, who is said to have been the most important 19th century American botanist, and with whom Darwin corresponded frequently.

of flowers is not as uncommon as you might think and, as a matter of fact, it is known to occur in orchids. Floral nectar is mostly made up of glucose, fructose and saccharose, but it may also contain amino acids, lipids, organic acids, vitamins, enzymes, antioxidants, mineral ions and secondary metabolites. In flowers, nectar has a single role: luring and feeding pollinators. So why does toxic nectar, which can be potentially harmful to the pollinator, occur at all? That is still a matter of debate. The function and possible consequences of toxicity in nectar are not well understood, nor are its extent and diversity in the world of flowers. There are several hypotheses about the function of toxic nectar, including pollinator fidelity, protection against nectar robbers, antimicrobial effects, and pleiotropy (one gene that influences two or more unrelated traits) — which suggests that poisonous nectar is simply an indirect consequence of the production of toxic secondary compounds by the plant for defence. Here, we will focus on the drunken pollinator hypothesis. This particular hypothesis proposes that intoxication causes pollinating insects to become sluggish, and therefore more easily drawn to pick up pollinaria and less likely to remove them from their bodies.

It is not uncommon to hear rumours about pollinators becoming sluggish after visiting orchid flowers. In Stig Dalström's magnum opus *The Odontoglossum Story*, he mentions that large bumblebees visiting an *Odontoglossum* species in Colombia were seen to "fall out of the flower and remain motionless on the ground for an hour or more." While in Costa Rica, my colleagues and I observed that flies feeding on the nectar offered by the flowers of *Specklinia* — in time — "become slower and less aware of their surroundings", and may remain completely immobile on a single spot for up to an hour. Whether any of these orchids somehow intoxicate their pollinators remains to be proven, but there is one group of orchids in which such intoxication has become a well established fact. It all began with observations made on a terrestrial orchid from Denmark. In 1974, Bernt Løjtnant noted that the flowers of *Epipactis helleborine* and *E. purpurata* (see Figure 5.1.1) were pollinated by wasps that became inebriated and sluggish after repeatedly consuming their nectar. In a short paper written in Danish, the author put forward an engaging hypothesis: the poisonous nectar is the tool by which the orchid manipulates the insect. The intoxication makes the wasps slow, sloppy and less aggressive. Addicted to the nectar, they crawl from flower to

**Fig. 5.1.1.** Helleborines are common terrestrial orchids in Europe. **A–B.** *Epipactis hell*eborine, the broad-leaved helleborine is a common and widespread species. **C.** *E. helleborine*, photographed in New York state where it has become naturalised. **D–E.** The less frequent *E. purpurata*.
© Jean Claessens (A–B, D–E) and John Gange (C).

flower accumulating pollinaria and are less likely to groom themselves to remove them.

*Epipactis helleborine* is a common and widespread orchid species in Europe. It can be found growing in a variety of habitats, including woods, bushes, peat bogs, meadows and dunes, as well as many anthropogenic ones such as parks, roadsides and cemeteries. The broad-leaved helleborine — as it is popularly known — is so versatile it has now colonised North America, becoming a weed in Wisconsin and other regions. Flowers of this species are somewhat inconspicuous, and tend to be variable in the shape and especially the colour of their sepals and petals. They vary from light green, yellowish-white, pink to almost purple flowers, generally in dull shades. *Epipactis purpurata*, the less common of the two sister species, is morphologically quite similar but prefers heavily shaded woods with little competition. Irrespective of the colour, insects — especially wasps — have often been seen to visit the *Epipactis* flowers, feeding on the nectar and becoming intoxicated. Strong reactions, such as coming to a complete standstill, have been observed in beetles, whereas larger insects such as bees and wasps become sloppy and fly in an erratic fashion.

**Fig. 5.1.2.** A common wasp, *Vespula vulgaris*, carrying several pollinaria of *Epipactis purpurata*. Becoming drowsy as it visits multiple flowers (**A**), picking itself up after falling to the ground (**B**).
© Jean Claessens.

So researchers set out to understand what the nectar of *Epipactis* contains exactly, and their discovery was even more bizarre than expected. What they found was that the nectar does indeed contain ethanol (alcohol). But even more incredible, the ethanol seems to be produced by microorganisms that thrived in the sugar-rich medium after being introduced into it by the pollinators themselves. Floral nectar is commonly infested with microorganisms, but very little is known about those living in orchid nectar. The nectar of *Epipactis* was found to be inhabited by many different kinds of yeast and bacteria, especially members of the Enterobacteriaceae family. Several of the bacteria discovered in the nectar of *Epipactis* where completely new to science. The presence of microorganisms in the nectar increases its amino acid content, as well as enriching it with byproducts of fermentation such as alcohol. This has been long suggested to be the reason for the sluggish *Epipactis* pollinators. Wasps have been shown not to discriminate between flowers that are contaminated or uncontaminated with yeast, which results in a quicker spread of the microorganisms from flower to flower. They are also not deterred by the nectar's high alcohol content and become drunk as a result. However, in an incredible twist of events, one study found that fermentation in *Epipactis* is not that common under natural conditions and perhaps the presence of alcohol cannot fully explain the sluggish behaviour of pollinators. Instead it seems that the nectar also contains compounds that have strong narcotic properties, including morphine.

*Epipactis* orchids have one more trick up their sleeve. Even though they are occasionally visited by other insects, the flowers of *E. helleborine* are usually pollinated by wasps, mainly social wasps such as *Vespula vulgaris* (see Figure 5.1.2) and *V. germanica*. The mechanism of attraction has only recently been discovered. Researchers found that the social wasps that pollinated the flowers typically prey on caterpillars to feed their developing larvae. The wasps use a combination of visual and olfactory cues to locate their prey, specifically the volatile chemicals that plants emit when they are being eaten. Using gas chromatography, researchers showed that *E. helleborine* releases the green-leaf volatiles that plants such as cabbages emit when they are infested by herbivores. By mimicking a plant's defence mechanism, this orchid succeeds in attracting pest control in the form of parasitic wasps, and gets them to pollinate its flowers as a bargain.

So we now know that *Epipactis* flowers deceive their pollinating wasps by smelling like a caterpillar-infested cabbage. Instead of food for their growing larvae, the wasps are given a cocktail of sweetened narcotics, keeping the insects comfortably numb as they do the orchid's bidding. I bet the common, dull-flowered *Epipactis* growing in your garden is suddenly not that boring any more!

**COMFORTABLY NUMB**
Poisonous nectar and inebriated pollinators

SCAN ME

https://youtu.be/wzyFIE0JnFQ

Among the highly specialised modifications we may find in orchid pollination mechanisms, toxic nectar is one of the most curious. Upon observing the behaviour of floral visitors, naturalists noticed that insects would become drowsy or sluggish after consuming the nectar of certain orchid flowers. After carefully analysing the nectar content of *Epipactis* species, researchers discovered that in fact the nectar may contain alcohol as a side product of its natural fermentation by microorganisms brought by the pollinators themselves. However, they also found that the flower itself produces substances such as morphine, which intoxicate the pollinators. Apparently, the sluggish wasps that pollinate *Epipactis* are more likely to visit multiple flowers and less likely to remove the pollinaria from their bodies.

The videos and photographs here are owned by Jean Claessens (https://europeanorchids.com/) and have been reproduced and edited by the author with permission.

## 5.2 THUNDERSTRUCK

*"Bien curious aussi est une petite Orchidée à fleurs blanches qui se trouve par milliers sur les arbres. Pour cette espèce également, la floraison a lieu en mème temps pour tous les spécimens, mais au lieu ode se produire une fois par an, elle se produit une fois par mois, done douze fois par an."*
MELCHIOR TREUB (1887)

**Fig. 5.2.1.** The pigeon orchid, *Dendrobium crumenatum*, in Peninsular Malaysia, showing its usual gregarious flowering.
© Ong Poh Teck.

European travellers to South East Asia were fascinated by the region's exuberant plants. The Dutch botanists visiting Buitenzorg in Java — today, Bogor in Indonesia — noted a curious feature of an orchid that commonly grew there. One of the botanists was Melchior Treub, a graduate of the University of Leiden. In 1887, Treub wrote about an unusual small orchid with white flowers that is found on trees by the thousands, all of which flower simultaneously once a month (his exact words are reproduced above). He was the first naturalist to notice synchronic mass flowering events in orchids, and the orchid he was referring to was *Dendrobium crumenatum* — also known as the pigeon orchid (see Figure 5.2.1).

In 1895, the Belgian botanist Jean Massart elaborated on Treub's observations: "Here is *Dendrobium crumenatum*, one of the most common Orchidaceae around Buitenzorg. There is no tree that bears only a few tufts. Some days all are decorated with large sprays of white *Dendrobium* flowers; the same evening, the petals lose their turgor. Then, for weeks, we would search in vain for a flower, until all of a sudden, one fine morning, we were again amazed to see that the trees retaliated for their ephemeral adornment. This curious periodicity to which M. Treub first drew attention is not at all explicable. What makes synchronism even more mysterious is that the *Dendrobium*, torn from their support by thunderstorms, and lying around in the grass — those which are potted and cultivated in conditions as disparate as possible — even those which are imported to Buitenzorg from other islands of the Indian Archipelago — all bloom on the same day as those which remained quiet on the tree where they were born. When you examine the buds five to six days before they bloom, you notice that they are far from being equally developed. But the differences disappear on the following days: the most advanced buds increase slowly — those which were late are more hasty; and all bloom the same morning, as if in response to a wave of a magic wand."

The extraordinary phenomenon that Massart so passionately describes is known as synchronic mass flowering and it is key to the successful reproduction of *Dendrobium crumenatum* — as well as several other orchids. But let's first talk about the longevity of the orchid flower.

Orchids are widely grown for their flowers. The lifespan of orchid flowers varies immensely. From a few hours, as in some *Dendrobium*, *Palmorchis*, *Sobralia* and *Vanilla*, to several months, as in certain *Grammatophyllum*, *Phalaenopsis* and *Vanda*. Long-lasting flowers are certainly a desirable trait among orchid enthusiasts. Orchid collectors are well aware that even though it can take several years for an orchid to reach maturity, and that it may only bloom once a year, the vast majority of orchid flowers start to wither only a few days after opening. The brief exuberance of their flowers is in part what makes them unique. In fact, some orchid flowers don't even make it through a single day. This may be offputting to growers, but it also represents a challenge for reproduction in nature. Given how specialised pollination systems are, it seems quite obvious that the shorter the flowering period, the less likely it is that a pollinator will visit the flower, and the less likely it is that the pollinator will find other flowers open at the same time. Consequently, ephemeral flowers may not only be a disappointment to growers, they also represent a true challenge for the successful transfer of pollen.

But orchids always have a trick up their sleeves. They have significantly improved the odds of successful reproduction of ephemeral plants by flowering gregariously. This means that diverse individuals in the same area will flower simultaneously. An extreme type of gregarious flowering is synchronic mass flowering, which involves multiple individual plants of the same species blooming on the exact same day. This unusual strategy has been reported in several orchid groups on diverse continents, but remains poorly understood. Synchronic mass flowering has been reported in Asia in the genera *Bromheadia*, *Dendrobium*\* and *Thrixspermum*; in Central and South America in the genera *Palmorchis*, *Psilochilus* and *Sobralia*; and in the widespread genus *Triphora* (see Figure 5.2.2). In Costa Rica, I have personally observed massive synchronic flowering of the ephemeral *Myoxanthus parahybunensis*, *Octomeria valerioi* and *Vanilla trigonocarpa*, three genera for which this particular flowering syndrome has not been previously reported in the literature.

Gregarious flowering is perhaps most famously known to occur in bamboo. The massive clumps of bamboo are rarely seen in bloom, up to the point that many of us forget that they actually do flower. The reason that we may have missed it — besides the fact that they have relatively small, grass-like flowers — is that a bamboo plant blooms only once in its lifetime. After that the plant withers. An entire clump of bamboo will flower once and then die off. Given that bamboos may live up to 32 years before they reproduce and perish, strong selective pressures ensure that enough of them flower at that very same time to ensure cross-pollination. Their periodicity is an inherited character, so all plants of the same age will bloom gregariously no matter where they are and under what conditions they are grown. Hundreds of thousands of tiny flowers therefore appear simultaneously, and as suddenly as they appeared, they disappear again. Mass flowering in bamboo is 'rhythmic'. In other words, it is genetically rather than ecologically determined, happening once at a pre-established time. This is not the case in orchids. Floral development in orchids is induced by an external trigger, and unsurprisingly the first studies of flower induction and development were carried out on the gregariously flowering *Dendrobium crumenatum*.

To flower, an orchid plant must reach a certain stage of maturity and this differs among orchids. In general terms, it has been estimated

---

\*   Including species of the former *Diplocaulobium* and *Flickingeria*.

**Fig. 5.2.2.** Synchronic mass flowering is reported in diverse orchid genera worldwide. **A.** The gorgeous *Sobralia labiata* is a one-day flower from Costa Rica. **B.** *Myoxanthus parahybunensis* flowers gregariously throughout the Neotropics. **C.** The North American *Triphora trianthophoros* is pollinated by a halictid bee. APK (A), Ron Parsons (B) and John Gange (C).

that orchids can take four to seven years to reach the adult stage, but they may take longer. This duration is genetically determined, but some orchids only flower under specific ecological conditions. High-elevation orchids may grow well but never bloom in the lowlands, whereas warm growers may not flower when transported to cooler habitats. Studies in West Africa show that some orchids may be categorised as long-day, short-day or neutral-day plants, whereas others are influenced by low or fluctuating temperatures. A long-day plant is one that requires nights that are shorter than a specific minimum to bloom, while short-day plants are those that require nights to be longer than a certain threshold. Neutral-day plants are those that are unaffected by these changes in day and night lengths. Plants that are adapted to tropical conditions — where variations in day length are subtle — seem to be more sensitive to smaller differences in the hours of darkness compared with their temperate counterparts. Temperature fluctuations — which may sometimes be difficult to tease apart from other factors such as day length — are also known to induce flowering in some orchid species.

Fire has been shown to influence pollination substantially in the fire-prone regions of North America, Australia, South Africa and the Mediterranean. In Australia, fire stimulates flowering in terrestrial orchids such as *Caladenia*, *Diuris*, *Microtis*, *Prasophyllum* and *Thelymitra*, all of which have been cited as fire-dependent for flowering. Similarly, *Disa* and *Pterygodium* species in South Africa flower more abundantly on sites that have been previously burned. Flowering and pollination in such orchids increases immediately after an area has been burned, and declines as post-fire plant succession proceeds. Incidentally, this supports the use of controlled burns as a tool in conservation. But fire can't explain synchronic flowering in tropical epiphytes. What is stimulating them to bloom simultaneously?

It was once believed that the flowering of certain orchids responded to the phases of the moon, though this has never been proven. Authors have also noted that the pigeon orchid blooms after thunderstorms and heavy rainfall, and initially suspected that those events could be the trigger. Flower development and anthesis (opening and closing) is now known to respond to rain-induced cooling in *Dendrobium crumenatum* and several of its close relatives. In this species, the floral buds develop up to a certain point and then stop. At that stage, as development ceases and the bud becomes dormant, most floral organs having reached maturity. It is not until a sudden drop in temperature of around 5°C, or a prolonged, gradual cooling, that all the buds resume growth and the flowers open in concert. Eight to ten days after the induction, the whole population of *D. crumenatum* blooms in a sudden and brief spectacle of beautiful white aromatic flowers. During that period, the fresh weight of the flowers increases by a whopping 50-fold. The opening and senescence of the *D. crumenatum* flower is regulated by changes in its carbohydrate and water status, and in the metabolism of its cell wall, that begin in young flower buds. The cell wall starts to disassemble during the floral bud development stage — evidence for the early onset of a senescence program. In a sense, the flowers are dying well before they open. Despite its short-lived flowers, *D. crumenatum* is able to reproduce effectively. It is pollinated by the Asian honeybee *Apis cerana*. The bees visit the flowers in the early morning, making multiple quick visits to a large portion of the available flowers (see Figure 5.2.3).

But how do other orchids manage to synchronise? At Lankester Botanical Garden in Costa Rica, plants of *Myoxanthus parahybunensis* produce several thousand flowers over a single period of just a few days every year. In less than 10 days, they develop hundreds of floral buds per plant, and then bloom. We have observed them five years in a row now and, independently of their provenance and growing conditions, all individuals flower gregariously sometime in November or December. Even plants in the field — in different regions of the country — flower simultaneously with those in the various greenhouses at the garden. We suspect that shorter day lengths or lower temperatures may be the trigger, but preliminary analyses of complex meteorological data from the garden has yet to reveal any pattern. Studies by my colleagues on the flowering of *Sobralia* — another orchid with obvious synchronistic flowering — have also failed to find any discernible trigger pattern. A sudden change in humidity or drought regimes and drastic drops in temperature or prolonged periods of cooler temperatures have been proposed as possible triggers for gregarious flowering.

*Coffea arabica* — which we all know and love — is another plant that flowers synchronously. Whole coffee plantations are suddenly adorned with tiny white flowers once or twice per year, in a spectacle that lasts only a couple of days. In the Costa Rican countryside, thousands of

*C. arabica* plants bloom at exactly the same time, signaling the beginning of the rainy season and filling the air with their sweet smell. Rainfall has been considered the prime stimulus that induces flowering in these plants. Rapid rehydration of the plants after a dry period triggers both blooming and the growth of new shoots. It may well be that the particular stimulus differs from one species to another. It is also possible that a combination of stimuli trigger different steps and that flowering comes at the end of a sequence of changes.

The mechanisms behind the notoriously precise synchronicity in the Orchidaceae largely remain a mystery. However, that they have improved their chances of becoming pollinated by being gregarious goes without question.

**Fig. 5.2.3.** The pigeon orchid, *Dendrobium crumenatum*, with its pollinator *Apis cerana*.
© Ong Poh Teck.

## 5.3 TRIGGER

> *"Perhaps the most interesting study connected with the structural peculiarities of Orchids is that of the varying means by which ... fertilisation by insect agency is secured. The wonderful co-adaptation of all the parts of the flower to effect this end ... can never fail to excite our admiration and surprise."*
> THOMAS F. CHEESEMAN (1872)

We tend to think of plants as being immobile. They seem to be stuck to one spot, incapable of the freedom of movement of animals. But that doesn't mean they are completely motionless. Plants are of course well known for growing and climbing, altering the position of leaves and even snap-trapping insects to feed upon. Charles Darwin was fascinated by movement in plants, writing profusely on the subject in three of his books. *On the Movements and Habits of Climbing Plants* (1875), *Insectivorous Plants* (1875), and *The Power of Movement in Plants* (1880) still inspire research on these topics and are a testament to Darwin's deep-rooted interest in the evolution of plant mechanics.

Darwin is also well known for having discovered several of the transformations that the floral organs of orchids undergo. But it was the English botanist John Lindley who first described the presence of irritable lips in the Orchidaceae family. In 1832, while examining a species of *Pterostylis* brought to his attention by Robert Brown, he wrote: "I observed a kind of convulsive action of the labellum." In 1853, he expanded on the subject, cautioning that "among many other remarkable peculiarities the irritability of the labellum must not be passed over in silence." Lindley noted that like *Pterostylis*, many *Bulbophyllum* species have a mobile labellum, and that the delicately hinged lips of *Caleana*, *Drakaea* and *Spiculaea* were all easily moved by the action of a gentle breeze or visiting insects. However, the irritable lip of *Pterostylis* is unlike those of the other genera in one key aspect. In the flowers of *Bulbophyllum*, *Caleana*, *Drakaea* and *Spiculaea* — as well as numerous other orchids — the lip moves through the action of water, the wind or an insect. These movements are all passive. What makes *Pterostylis* different is that its lip is actively motile. Its trigger lip, or snap-trap lip, is not only movable, but also sensitive to touch. A light touch causes the narrow hinge to contract, swiftly raising the lip towards the column and catching visiting insects.

In the first edition of his orchid book, Darwin lamented not having been able to examine any orchid species with a sensitive lip. This prompted New Zealand botanist Thomas Cheeseman to draw up his own accounts on the pollination of local *Pterostylis*. He noted that the fleshy lip of the plant was delicately hinged at the base by a thin membrane. The slightest touch was sufficient to flip the lip upwards towards the column, a movement that Cheeseman proved was crucial for the plant's reproduction. The careful observer noted that the movement of the lip would enclose a small insect in the hood of the flower. Pollen transfer occurred as the prisoner escaped through the only available exit, by crawling through the passage made by the appendages of the column. To confirm his suspicions, Cheeseman trapped diverse small insects in the *Pterostylis* flower — just as Darwin had done when studying slipper-orchids — and observed some of the insects successfully exiting with the orchid's pollinarium attached to their bodies. He then examined flowers with a closed lip, looking for naturally caught prisoners. He soon found hostage flies caught within the flowers, most having perished after being unable to escape their imprisonment. Cheeseman sent his observations on *Pterostylis* to Darwin, who appreciated the interesting findings and fully agreed with his conclusions. The English naturalist quoted Cheeseman's notes extensively in the second edition of his orchid book. Despite the early interest of these elite naturalists, the means by which *Pterostylis* species attract their pollinating insects would remain a mystery for more than a century after the discovery of the trigger-lip mechanism.

There are a few hundred named *Pterostylis* species, and many more remain unrecognised. They are tiny terrestrial orchids found only in Australasia, especially Australia where more than 200 species are known to occur. Floral morphology is variable among the different species, but most feature greenish flowers and a well-developed hood — giving rise to their popular name 'greenhoods'. Even though botanists have been aware for over a century that greenhoods are pollinated by flies (see Figure 5.3.1), how they attracted the pollinators was unknown until quite recently. In 2014, a group of Australian researchers led by Ryan Phillips discovered that *Pterostylis sanguinea* — like several other Australian orchids — is sexually deceptive. What makes *Pterostylis* unique — besides the snap-trap lip mechanism — is that pseudocopulation is performed by a tiny fungus gnat from the

Fig. 5.3.1. Greenhoods usually feature highly sensitive snap-trap lips. *Pterostylis sanguinea* was recently found to be sexually deceptive, here shown visited by a male fungus gnat by (**A**). The striking *P. barbata*, endemic to the south-west Western Australia (**B**). Untriggered and triggered *P. picta* (**C–D**).
© Mark Brundrett (A) and Jeremy Storey (B–D).

Mycetophilidae family, rather than a bee or wasp (see Chapter 2, section 2.2 for more on sexually deceptive Australian orchids). Ryan and his team observed that when visiting *P. sanguinea*, the male fungus gnat courted the flower by curling its abdomen and then engaged in vigorous copulatory probing over the upper surface of the lip. This energetic attempt to copulate triggers the lip to snap shut, trapping the fly inside the hood chamber. In a matter of seconds, the male fungus gnat goes from having a great time to being a prisoner of the flower. The only way to escape is through a tunnel made by the lobes of the column, where the pollinaria are successfully transferred.

Follow-up studies have now determined that pollination by sexual deception occurs in at least four more *Pterostylis* species, and although perhaps not the only strategy in the genus, it is likely to be widespread and commonplace among the greenhoods. Using live bait flowers — where flowers are cut from elsewhere and then artificially presented to pollinators in a suitable habitat — researchers succeeded in attracting male fungus gnats of the Mycetophilidae and Keroplatidae families. The insects were observed showing behaviours that are consistent with courtship and copulation, including fanning their wings and extending and curling or rocking their sexual organs. The insects aligned with the lip and probed the base with their sexual organs in an attempt to copulate. Chemicals are crucial in attracting the flies and eliciting this behaviour. The power of sexual attraction over the male flies is such that — on occasion — male flies have been observed pulling down a previously triggered *Pterostylis* lip and desperately attempting to copulate with it.

Snap-trap lips give pollination by sexual deception a whole new dimension. But trigger lips also occur in at least four independent lineages within the fly-pollinated neotropical subtribe Pleurothallidinae — popularly known as pleurothallids. But their ecology is poorly understood and the exact mechanism by which these orchids are pollinated remains a mystery. Among them are the members of the former genus *Acostaea* (today included within *Specklinia*). The seven closely related species live as tiny epiphytes in the humid forests of Central America and the Andes. They grow amid mosses on small twigs in cool, humid forests, where they are easily missed on account of their size. The minute flowers bear a characteristically broad hood and have an intricate lip that are remarkably similar to those of *Pterostylis*. The complex labellum is lobed, with a distinctive, elongate, more or less warty or hairy, central callus (a raised lip lobe). A gentle touch of the callus will trigger the lip (see Figure 5.3.2).

The first to note the snap-trap mechanism of *Specklinia* was Auguste Endrés, who in 1867 scribbled "labellum very irritable ... the least touch, or a slight gust of wind causes it to clap close to the column, never relaxing into its former position" below his illustration of a *S. colombiana* specimen from Costa Rica. Endrés was a meticulous researcher and his drawings and observations of *S. colombiana* were made with precision. Unfortunately they remained unknown to the general public until they were picked up and published by pleurothallid guru Carlyle Luer in 1987. The lip of *S. colombiana* is extremely sensitive and indeed very easily snapped by disturbances, especially when the base of the callus is touched. It is especially sensitive during the warmer hours of the day. In the early mornings and late afternoons, the lip is open and can be triggered, but is much less sensitive. Interestingly, no matter at what time the lip is triggered, it only resets overnight, opening up again the following morning in the cool and moist conditions before the sun comes up. After a few resets, the flower seems to lose sensitivity. I have not been able to trigger a single flower on more than three consecutive days, and some triggered flowers appear to become permanently closed, with the lip not resetting at all.

There are no published pollination studies on *Specklinia colombiana* or any of its relatives. A photograph taken by my colleague Sebastián Vieira Uribe shows a fly caught inside the flower of a triggered *S. trilobata*. It is hard to say whether the fly was indeed in the process of transferring pollinaria. But, given its size and position, it does seem quite likely that such a fly could act as a pollinator. Unfortunately this tells us little about the mechanism of attraction to the flowers. Sexual deception, where male flies attempt to copulate with the lip, is a definite possibility given the superficial similarity to the sexually deceptive *Pterostylis* flowers. But there are other possibilities. *Specklinia caulophryne* — the anglerfish *Specklinia* — is a species endemic to Colombia that bears a unique, rod-like callus. This odd species looks as if it has placed a hook with bait in front of its gaping flower. What poor insect may it be waiting to snap up?

Among the pleurothallids, there is another group of species with trigger lips that warrants a mention — the extraordinary genus

Fig. 5.3.2. **A–B.** Untriggered and triggered *Specklinia colombiana*. **C–D.** Untriggered and triggered *S. trilobata* with a fly. **E–F.** Untriggered and triggered *S. caulophryne*.
© APK (A–B) and Sebastián Vieira Uribe (C–F).

*Porroglossum*. In 1887, Mr. W. Bean, foreman of the orchid house at the Royal Botanic Gardens Kew, discovered the trigger-lip mechanism while inspecting the flowers of a specimen of *Masdevallia muscosa* (today *Porroglossum muscosum*). Endrés's earlier observations on *Specklinia* were unknown at the time, warranting *Porroglossum* the distinction of being the second orchid known to bear a trigger-lip mechanism — after *Pterostylis* — and the first such orchid native to the American Tropics. The plant that flowered at Kew was further investigated by Francis Wall Oliver, who prepared a detailed account of the snap-trap phenomenon. Oliver, of the Jodrell Laboratory at Kew, noted that the movement is a sudden and rapid folding-up of the labellum that causes the apex to move towards the column. He noted that the upper surface of the lip's triangular blade was particularly sensitive: "The movement is called forth by the gentlest touch of a hair or insect's foot on the median crest of the blade." Within a couple seconds after stimulation, the blade snaps upwards. The movement itself results from a sudden loss of turgidity in the cells of the contractile region.

*Porroglossum* species only grow in the Andes in Venezuela, Colombia, Ecuador, Peru and Bolivia. Today we recognise about 50 or so species, and all of them have sensitive, snap-trap lips (see Figure 5.3.3). Unpublished studies of the trigger-lip mechanism shows that there is significant variation between species in the speed and acceleration of the movement. Using high-speed videography, researchers have found that the movement of the lip varies from 0.04 to 5.50 millimeters per second, concluding that differences may be caused both by genetic variation and environmental factors. Either way, it is clear that this trigger-lip mechanism is much slower than that in either *Pterostylis* or *Specklinia*. In *Porroglossum* species, the lip closes gradually in a couple of seconds rather than abruptly in fractions of a second. After some 30 minutes, turgor in the contractile region is restored and the lip returns to the receptive position. Another interesting observation made by Oliver is that the lip of *Porroglossum* closes up and assumes a 'sleeping' position at night, reopening at dawn. These features differ from those observed in *Specklinia* — which only closes when triggered, snaps shut in fractions of a second, and opens only once per day — but their biological significance remains unknown. Unfortunately, as with many other pleurothallids, no pollination studies of any *Porroglossum* species have been published to date. Flies of the Drosophilidae family have been photographed visiting

**Fig. 5.3.3.** The extraordinary neotropical genus *Porroglossum* is well-known for its trigger lips. **A.** The highly sensitive *P. muscosum*. **B–C.** Untriggered and triggered *P. mordax*.
© Ron Parsons (A) and Sebastián Vieira Uribe (B–C).

the flowers of both *P. muscosum* and *P. hystrix* and may act as pollinators. But why the flies are visiting the flowers, what they do during these visits and exactly how the trap mechanism works, remain a mystery.

Darwin never witnessed the trigger mechanism of *Porroglossum*, *Pterostylis* or *Specklinia* himself. Given his passion for plant movement he would surely have experimented with them, taking careful notes of what triggers the lip, how fast it snaps shut and how long it takes to reset. But most of all, he would have been mesmerised by the thought of an orchid flower that uses sex to kidnap flies, cruelly catching and imprisoning them as they attempt to copulate with the lip, only releasing the hostages once a bargain, in the form of pollen transfer, has been struck.

**TRIGGER**
Triggered movement and pollination in orchids

SCAN ME

https://youtu.be/gIh0kkVsMCg

Some orchids have bettered their chances of getting pollinators to do their bidding by actively trapping them in their flowers. One such group of orchids are the members of the Australasian genus *Pterostylis*. Many species of *Pterostylis*, popularly known by the name of greenhoods, have highly sensitive lips that trigger with touch. When the male fungus gnats that pollinate these flowers through sexual deception touch the crest of the lip, it immediately closes trapping the insect. It is while escaping the hood that the fly transfers the orchid's pollinarium. In the neotropical regions, a few fly-pollinated members of the Pleurothallidinae subtribe also have trigger lips. The most famous are the species of genus *Porroglossum* and those of the former genus *Acostaea* (now *Specklinia*). No pollination studies in these orchids have been published, but the lip mechanism suggests certain similarity to *Pterostylis*.

The video of *Pterostylis* is owned by Daniela Scaccabarozzi and has been reproduced and edited by the author with permission. The videos of *Specklinia* and *Porroglossum* are owned by Adam P. Karremans. Videos published as Supplementary data in Phillips *et al.* (2013) (https://doi.org/10.1093/aob/mct295) have been reproduced with the kind permission of *Annals of Botany* and Oxford Academic Press.

## 5.4 PATIENCE

*"Malaxis in its fertilisation is an uninteresting form; Listera and Neottia, on the other hand, are amongst the most remarkable of all Orchids from the manner in which their pollinia are removed by insects, through the sudden explosion of viscid matter contained within their rostellums."*
CHARLES DARWIN in *Fertilisation of Orchids* (1862)

We have learned that orchids go to extreme lengths to ensure pollination and guarantee reproduction. In the following stories, we will read that in some cases they are willing to resort to self-pollination as a short-term solution to adverse conditions — such as a lack of pollinators. And yet, there are many mechanisms that prevent selfing in orchids as well. Outcrossing is thought to increase genetic diversity and consequently fitness, whereas self-pollination reduces genetic diversity and can result in inbreeding depression (decreased survival and fertility of offspring). Many plants have therefore evolved complex mechanisms to prevent self-pollination. One way to secure outcrossing is to separate the sexes. In 'dioecious' plants, one individual produces male flowers and another female flowers — guaranteeing that seeds will have genes from two sources. Orchids are 'hermaphroditic', meaning that the flowers of the vast majority of species have both a male and female function (see Chapter 3, section 3.10 for exceptions). Many plants have figured out other ways to have both a male and female function, be self-compatible and still reduce the chances of selfing. One of the most common strategies is known as 'dichogamy'. In dichogamous flowers, the two sexes are separated in time — at one stage the flower is functionally male and at another stage it is functionally female. Several orchids bear flowers that have a male function at anthesis (opening) and only become functionally female later on (when the male function comes first, this is known as protandry). It has now been shown experimentally that inflorescences with protandrous flowers do indeed have reduced self-pollination. This suggests that the evolution of protandry in orchids is driven by the important consequences it has on mating success.

Protandry brings us to the well-known common twayblade — *Neottia ovata* — and its sister species the lesser twayblade — *Neottia cordata*. Darwin was fascinated by these orchids, discussing them in great detail in his orchid book — under their former generic name *Listera* — and specifically addressing the 'remarkable' findings published by Joseph Hooker a few years before. The diligent naturalists noted that when the flowers of *Neottia* open, the pollinia rest loosely on top of a very broad rostellum. In this orchid, the rostellum, a lobe of the stigma that separates the anther from the stigma, is so large that it partially involves the pollinia and almost completely blocks access to the stigma. Only after the pollinia are released does the rostellum transform into an extended flap, fully exposing the stigma. So even though this orchid flower is hermaphroditic, functionally it is first a male and then a female. But this is not the reason why Darwin and Hooker were captivated by *Neottia* flowers. In addition to protandry, these orchids exhibit yet another fantastic trick. Most orchids have dry pollinaria that only become stuck to the animal after the glue-like, viscous liquid produced by the rostellum gets smeared on them. In other orchids, the viscous material formed by the rostellum has become part of the pollinarium itself, being physically attached and removed as a unit by the pollinator. But Hooker noted that the rostellum of *N. ovata* is unique in that it contains and violently expels the viscous matter. "So exquisitely sensitive is the rostellum, that a touch from the thinnest human hair suffices to cause the explosion" wrote Darwin. The tip of the rostellum has pressure-sensitive trigger hairs that upon the gentlest touch, instantly shoot out a viscous droplet. In a recent follow-up study, James Ackerman and Michael Mesler observed that this "quick-drying cement" is shot onto the pollinator, and immediately after, the margins of the rostellar flap flex and release the pollinia.

There are numerous diverse means of protandry among orchid flowers. In some species of *Pleurothallis*, the temporal unavailability of the stigma is achieved in a slightly different manner, but with similar implications. When the flowers of *P. eumecocaulon* open, the apex of the column is completely covered by the anther cap. The rostellum, which impedes self-fertilisation by separating the male and female organs, sticks outwards, covering the stigma completely. Just as in *Neottia*, once the pollinarium has been removed, the rostellum flap twists upward more than 90 degrees, fully exposing the stigma (see Figure 5.4.1). However, in this case the stigma also expands, noticeably extending outwards. The transformation occurs overnight and at the end, the stigma has become a large spherical outgrowth that completely dominates the column apex. So a flower that was functionally male one day will become functionally female the next. Several other species of *Pleurothallis* share this extraordinary feature. It is easy to spot because flowers with the anther cap present will have the triangular rostellum flap sticking out below, whereas those with an erect rostellum have an inflated stigma underneath.

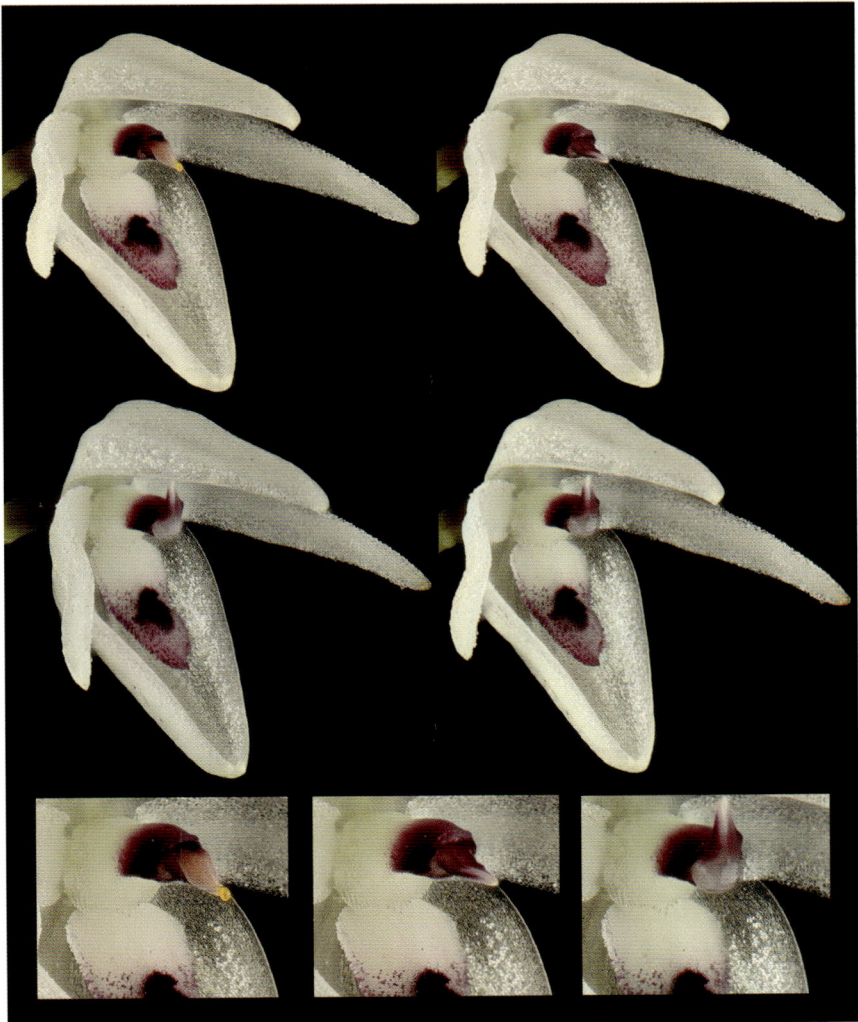

**Fig. 5.4.1.** Protandry in *Pleurothallis eumecocaulon*. Once the anther cap has been removed, the rostellum flips and the stigma becomes exposed and enlarged. © APK.

There are several other instances of protandry reported among the Orchidaceae. It is a well-known feature of certain terrestrial orchids, such as the members of tribe Cranichideae — specifically in subtribes Goodyerinae, Prescottinae and Spiranthinae. In the genus *Spiranthes*, when the flower opens the column is bent downwards and the stigma is inaccessible. After 2–30 days the column changes position and exposes the stigma, which is now receptive. The multi-flowered inflorescences of species belonging to this genus are spirally displayed and open in a strict base-to-top sequence, which means that functional female flowers are always below the functional males. The bumblebees that pollinate the flowers always inspect them in an ascending motion (see Figure 5.4.2).

Fig. 5.4.2. *Spiranthes lacera* being pollinated by a *Bombus* species in Vermont. The spirally arranged flowers open in a strict base-to-top sequence with functional females always below functional males, thus avoiding self-pollination.
© John Gange.

This means that they first deposit pollen on the older, functionally female flowers at the bottom, and as they move upwards, they pick up pollen from the functionally male flowers above. This sequential flowering strategy prevents selfing, despite both male and female flowers being present at the same time, and promotes outcrossing. In some of the smaller-flowered members of the subtribe Oncidiinae — specifically *Macradenia*, *Macroclinium* and *Notylia* — the elongate column has a slit-like stigma that is initially extremely narrow and afterwards inflates, broadening longitudinally. In Puerto Rico, *Lepanthes* species have been found to have an age-related change in receptivity that could be interpreted as a form of protandry. In this particular case there is no visible change, but fruit set increases significantly in older flowers.

Another group of protandrous orchids are the members of the genus *Mormodes*. As previously discussed, in *Mormodes* flowers the column is always twisted to the side at an angle of about 90 degrees. The twist in the column stalk completely impedes access to the stigma. At this stage the flowers have non-operational female organs and are functionally male. It is only after the removal of the pollinarium that the column begins to stretch and straighten out. The process of fully stretching the column is slow. In *Mormodes colossus* the column is found unwound, exposing the receptive stigma, the day after the pollinarium has been removed. Protandry in *Mormodes* differs from that in *Neottia* and *Pleurothallis*, not only in that it is the column itself that transforms, but also in that there is a second mechanism impeding self-fertilisation — namely, temporal transformation of the pollinarium. Upon removal, the elongate pollinarium of *Mormodes* coils up completely, resembling a sushi roll. The coiling is so forceful that the pollinia themselves are pressed to the side, hanging outside the roll. Immediately after, the pollinarium begins to slowly uncoil, and after a couple of minutes there is enough space for the pollinia to return to their central position in the pollinarium. From this moment on, the pollinarium slowly grows and stretches, becoming completely straightened after 30–45 minutes (see Figure 5.4.3). The transformation of the pollinarium does not end there. For the next few hours, it dehydrates and the accessory structures shrink to about half the size, becoming hard and dark. The pollinia themselves remain virtually unchanged during this process. Their size and colour are the same, but by the end of the process they are fully exposed, sticking out on the very top of the pollinarium. A similar uncoiling process has also been found in the genera *Cycnoches* and *Dressleria*, both close relatives of *Mormodes*. By contrast, in the unrelated genus *Vanda* the pollinarium flattens over time. Becoming disk-like a few minutes after it has been removed from the flower (see Figure 5.4.4). This temporal transformation of the pollinarium has an important function: it prevents autogamy and promotes outcrossing. The pollinia will not reach the stigma unless they are fully exposed on a straight pollinarium, and this requires time. By then, the bee is likely to have left the inflorescence where it originally picked up the pollinarium.

Like many other quirks of the orchid flower, the transformation of pollinaria was noted and extensively discussed by Darwin in his orchid book. In the literature, pollinaria transformation is also referred to as bending. Darwin used the term "movement of pollinia" instead, as indeed movements are not limited to merely bending of the pollinarium. But, considering that the different components may coil, uncoil, twist, shrink and change colours, my personal take is that words such as 'transformation' or 'reconfiguration' more correctly describe the complexity of changes that a pollinarium may undergo after removal from the flower, especially as these changes seem to be permanent. Darwin discussed pollinaria transformation in the first chapter of his orchid book, and his drawing of an *Orchis mascula* pollinarium bending downward on a pencil has been extensively used to illustrate this phenomenon. He carefully noted that this "beautiful contrivance" allowed the pollinarium to reorient in such a manner that it could later be placed in the stigma.

Pollinaria transformation can take from a few seconds to several hours. Darwin predicted that it evolved as a mechanism to avoid self-pollination. It has now been established that the time spent by the pollinator on an inflorescence and the time it takes for the pollinia to reorient are indeed correlated, and invariably the latter exceeds the former. Observations on the pollination of the food-deceptive genus *Orchis* in Europe, for example, indicate that the bee spends less time on the inflorescence than it takes the pollinarium to transform. This effectively prevents 'geitonogamy' — self-pollination among different flowers within the same individual plant. Very long reconfiguration times could make sure that self-pollination is always avoided, but could also be detrimental in terms of lost cross-pollination opportunities. Therefore, mating chances are in fact maximised if the transformation occurs shortly after the pollinator has left the original source, and not much longer after. Interestingly, reconfiguration times can vary between

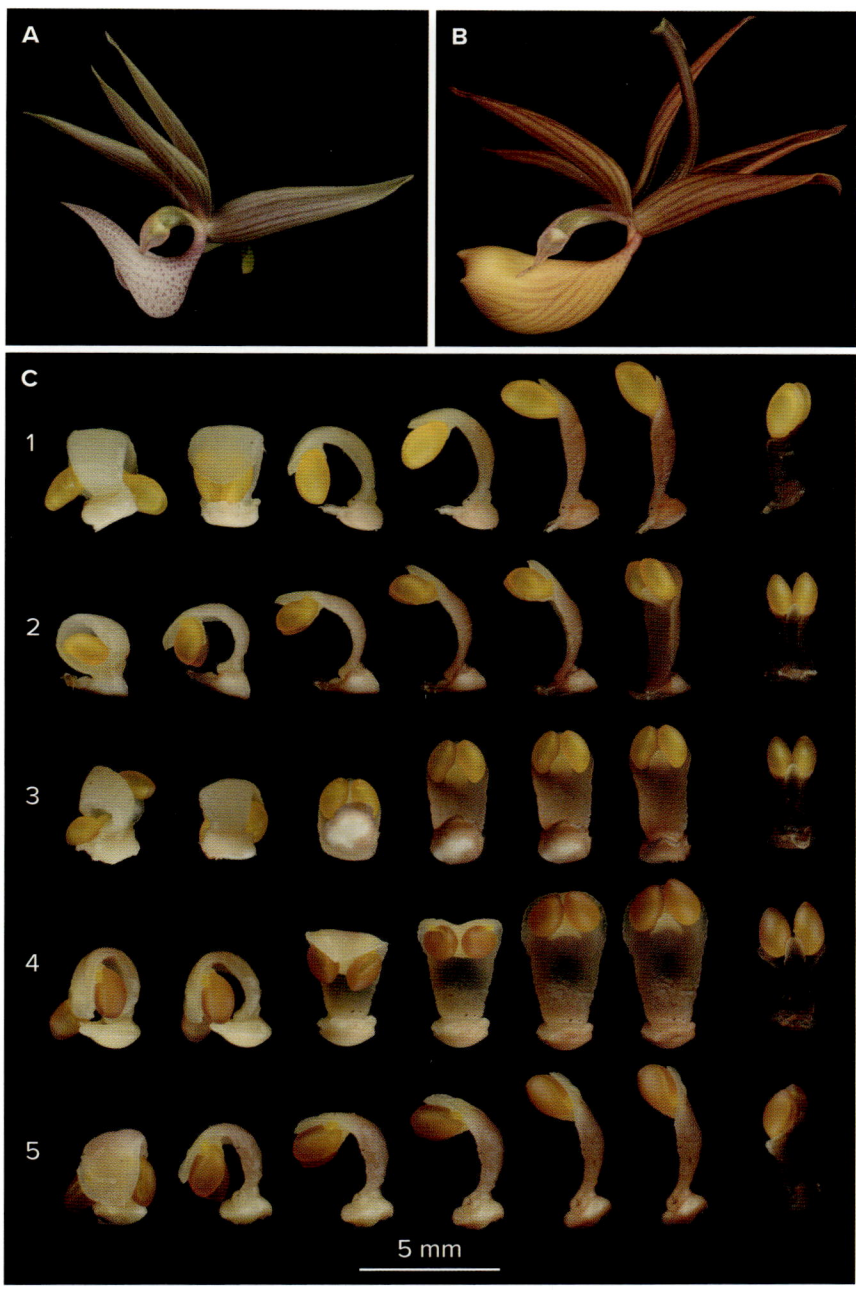

**Fig. 5.4.3.** Protandry in *Mormodes*. **A–B.** Twisted column of *M. colossus* and *M. fractiflexa* respectively. **C.** Temporal transformation of the pollinarium of *M. fractiflexa* (1), *M. horichii* (2–3) and *M. colossus* (4–5) after removal from the anther.
© APK.

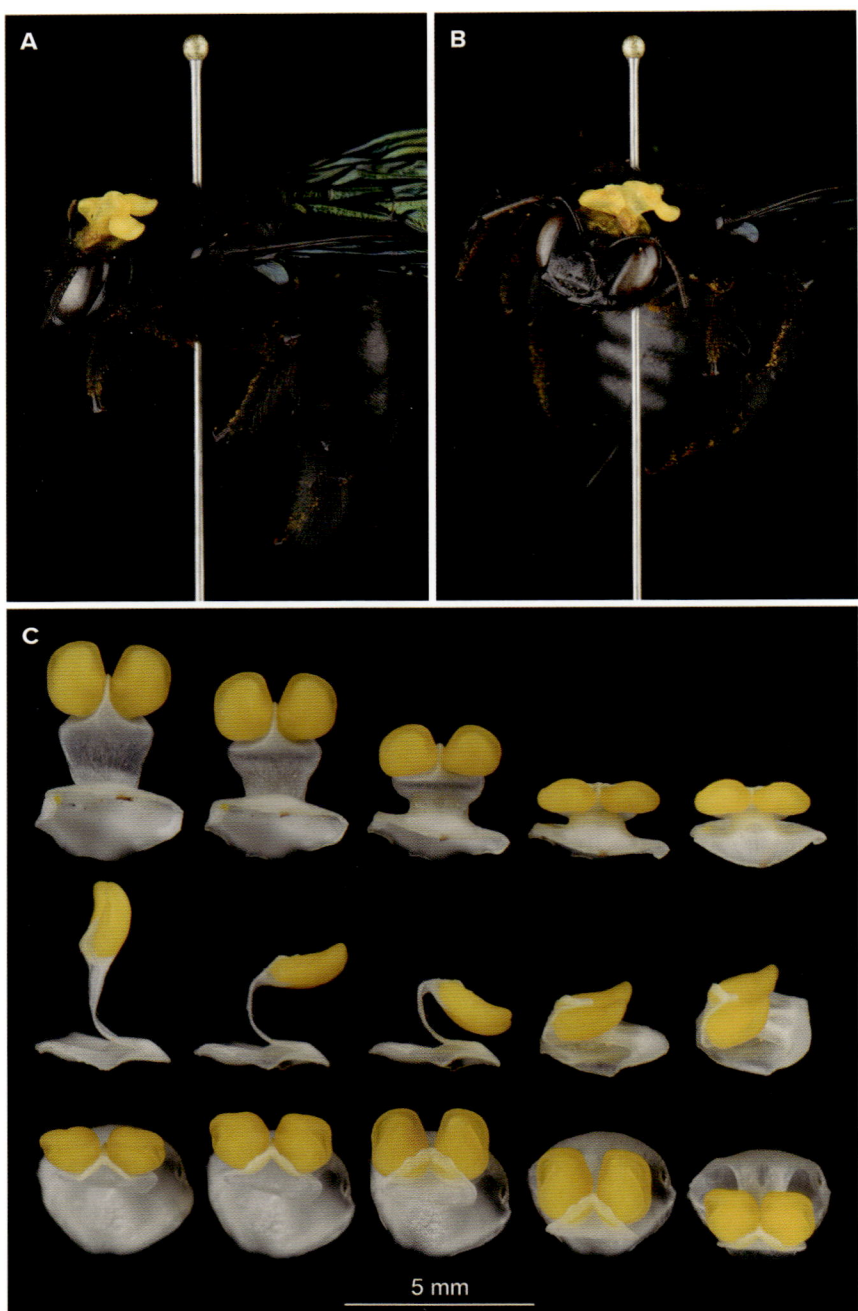

Fig. 5.4.4. Pollinarium transformation in *Vanda tricolor*. **A–B.** A carpenter bee carrying two fully transformed pollinaria. **C.** The gradual transformation of the pollinarium following removal from the anther.
© APK.

close relatives, for example in two subspecies of *Eulophia parviflora*. One subspecies is pollinated by a slow-moving beetle and its pollinarium takes an average 100 seconds to transform, whereas another that is pollinated by fast-moving bees takes just 28 seconds to reconfigure.

Most of the pollinaria transformations result from reconfiguration of the accessory structures, especially the stipe (a slender stalk often connecting the pollinia and caudicles with the viscidium). In the Catasetinae — at least in *Catasetum*, *Cycnoches* and *Mormodes* — the movement is also accompanied by a contraction and noticeable darkening of the stipe. Shrinking has also been suggested to occur to the pollinia themselves. Do orchid pollinia really shrink? To quote Jerry Seinfeld: "Like a frightened turtle!". One well-documented example is that of several Brazilian species of *Bulbophyllum*. Accessory structures are very much reduced in the pollinaria of *Bulbophyllum*. Pollinaria transformation in these orchids entails a reduction of about 50% of the pollinia size through dehydration. After only a couple of minutes, the pollinia have shrunk enough to be inserted into the narrow stigmatic cavity. The requirement of a period of shrinkage of the orchid pollinia in order to fit the stigma had already been noted by Dressler in the 1960s, but it remains largely anecdotal and understudied. Given that it has been reported to occur in several unrelated genera, such as *Trigonidium* — in the Maxillariinae — and *Gongora* and *Stanhopea* — in the Stanhopeinae — it seems likely to be much more widespread in the Orchidaceae family than is currently known.

Trigger stigmas, protandry and pollinaria transformation require further investigation in the Orchidaceae. We rarely think of flowers and their organs reconfiguring over time, but careful observation of living flowers and their interaction with pollinating agents has led to the discovery of intricate mechanisms that influence the pollination process. Curiously, pollinarium transformation also occurs in the Apocynaceae — now including the milkweed family Asclepiadaceae — some of which, like orchids, package their pollen into single functional units.

> *"In investigating the obscure subject of generation, additional light is perhaps more likely to be derived from a further minute and patient examination of the structure and action of the sexual organs in Asclepiadeæ and Orchideæ, than from that of any other department either of the vegetable or animal kingdom."*
> ROBERT BROWN (1833)

**PATIENCE**
Protandry and pollinaria transformation in orchids

**SCAN ME**
https://youtu.be/poakyuhRJhQ

Protandrous orchids are those that have a temporal separation between their male and female functions. Early naturalists noticed that besides protandry, the flowers of *Neottia* had a curious mechanism of placing their pollinaria on the visiting insect. Touching the sensitive rostellum of this orchid causes the stigmatic fluid to be expelled explosively, cementing the pollinia onto its pollinator. In the neotropical genus *Mormodes*, the column is initially twisted to the side, blocking the stigma while the male anther and pollinia are still in place. Once removed, the column straightens out progressively until the flower is functionally a female. In this orchid, the pollinaria also require transformation in order to be placed into the stigma. The uncoiling of the pollinarium takes around 30 minutes, thus preventing self-pollination. Pollinaria transformation is a little studied phenomenon that occurs in many orchids around the world.

The videos and photographs reproduced here are owned by Jean Claessens (https://europeanorchids.com/) and Adam P. Karremans, and have been reproduced and edited by the author with permission.

### 5.5 NOTHING ELSE MATTERS

*"Nature tells us, in the most emphatic manner, that she abhors perpetual self-fertilisation."*
CHARLES DARWIN in *Fertilisation of Orchids* (1877)

We have learned about the great lengths that plants go to in order to avoid selfing and promote cross-fertilisation — as Darwin so adamantly stated. And yet, at the summit of the El Ángel paramo (a variety of alpine tundra found at high elevations in the Andes Mountain Range in South America) in Ecuador one finds that the five pleurothallid species that grow there are all conspicuously and profusely selfing. The site — which is more than 3,500 metres above sea level — features extreme conditions. It is dominated by dwarf vegetation, strong winds and cold temperatures. The orchids at El Ángel belong to several different genera — *Brachionidium*, *Draconanthes*, *Lepanthes* and *Stelis* — in subtribe Pleurothallidinae, and all produce flowers that are auto-pollinated and mostly fully cleistogamous — that is, selfing while remaining in bud. Self-pollinating flowers are not unusual in *Brachionidium* and *Draconanthes*, which are typically found growing at high elevations, but *Lepanthes* and *Stelis* are more common at mid elevations and their species are predominantly out-crossing. The likelihood that all five pleurothallid species growing at a single site are 'autogamous' is very low, unless a strong selecting force favours this particular condition.

Autogamy or auto-pollination is a type of self-pollination in which flowers are pollinated with their own pollen autonomously — without the aid of an external vector. Even though fruit production is assured with autogamy, reduced genetic variability and inbreeding depression may be important trade-offs. Inbreeding depression refers to the expression of genes that cause a reduction in the fitness of the offspring. Avoidance of these negative side effects is a major selective force acting on the evolution of plant breeding systems. In hermaphroditic plants — those with male and female organs in a single flower — such as most orchids, there are many adaptations to reduce self-pollination, ranging from morphological adaptations to functional strategies. Given the complex floral traits and intricate contrivances developed to enhance the chances of cross-pollination, one might expect autogamy to be rather unusual in the orchid family. However, most orchids are in fact self-compatible — meaning flowers are able to fertilise their ovules with their own pollen — which is probably an important prerequisite for the evolution of autogamy. Autonomous selfing is in fact widespread in the Orchidaceae, and the phenomenon apparently evolved independently many times in the family. The most recent and extensive review by Ackerman and collaborators found that autogamy has been reported in 573 species of orchids worldwide, that is in 19% of the orchids for which pollination data are available. Therefore, selfing could be occurring in about one out of every five orchid species. and is not at all rare in the family. So when do orchids resort to autogamy?

In previous chapters we have learned about the immensely diverse schemes by which orchids are able to ensure cross-pollination. From the promise of sex or food, to the mimicry of pheromones or fetid odours, from offering shelter to punishing the visiting insect, and from nectar toxicity and floral synchronicity, to organ sensitivity and temporal transformation — all these remarkable floral adaptations have the sole purpose of securing pollen transfer from one individual to another. Despite that, many orchids have modified their highly specialised floral structures in such a way that they have dispensed with the need for an external pollinating agent. It seems paradoxical for a flower that has gone through such remarkable adaptations to ensure cross-pollination, only to afterwards develop autogamy. "The whole case is perplexing in an unparalleled degree, for we have in the same flower elaborate contrivances for directly opposed objects," Darwin wrote in *Fertilisation of Orchids*, baffled after studying the selfing local orchid species. How can we explain autogamy in orchids? German botanist and entomologist Hermann Müller offered a simple and straightforward answer: "Cross-fertilisation is better than self-fertilisation, yet self-fertilisation is infinitely better than absence of fertilisation and consequent sterility." In essence, a plant will seek to guarantee reproduction at all costs. Nothing else matters!

Cross-fertilisation does not always lead to higher fitness when compared to self-pollination, and many of the most common and widespread orchid species are selfers. In fact, in some cases selfers have been known to outlast their pollinator-dependent relatives. Autogamy is obviously closely related to high levels of fruit set, and average fruit set in diverse autogamous orchids was estimated at 77% in one study — significantly higher than in allogamous (cross-fertilised) species. Many of the most successful orchid colonisers are autogamous. Roadsides and other intervened areas soon become colonised by self-pollinated orchids such as *Habenaria monorrhiza* in Central and South America (see Figure 5.5.1), *Platanthera aquilonis* and *P. huronensis* in North America (see Figures 5.5.2 and 5.6.2), and *Epipactis helleborine* in Europe and Asia (see Figure 5.1.1). This suggests that selfers can cope better with new or unstable conditions and, at least in the short term, are not at any disadvantage due to their lower level of genetic variability. A deficiency of pollinators is often documented in orchid populations, and autonomous self-pollination ensures reproduction. Self-pollination seems to be more likely to occur in species that are on the periphery of their geographical or elevational ranges. Self-pollinators are also

**Fig. 5.5.1.** *Habenaria monorrhiza* is a common roadside weed in Central and South America, it is an autonomously self-pollinated orchid that quickly colonises intervened areas. Note the transition from flower to fruit.
© APK.

**Fig. 5.5.2.** *Platanthera huronensis* colonising a roadside in Colorado.
© John Gange.

more common on islands than in continental regions, and at higher, rather than lower, latitudes. This is congruent with the hypothesis that autogamy develops to ensure reproduction when pollinator services are unpredictable, infrequent or lacking altogether. Selfing was estimated at 10% in Barro Colorado Island in Panama, 15% in North America, 25% for Puerto Rico, and 27–50% in Europe.

We often think of orchids as being either strictly self-pollinated or strictly allogamous, when actually many species have both selfing and non-selfing populations or individuals. Certain orchids probably have obligatory autogamy — meaning they depend exclusively on selfing to reproduce — but there are many studies that report cases of 'facultative' (occasional) autogamy at the end of a flower's life span. This facultative autogamy may be quite important because it allows the flowers to first attempt cross-pollination, but reproduction is assured through autogamous means if all else fails. Occasional crossing in such cases may be sufficient to infuse enough genetic variability to diminish any effects of inbreeding depression from recurrent self-pollination. Perhaps the same concept may apply to the occasional open flowers observed in populations of 'cleistogamous' orchids (which self-pollinate while still in bud). Cleistogamous pleurothallids — such as those at the El Ángel paramo — frequently produce a few open flowers, which may allow occasional crossing.

Autogamy is often accompanied by an underdevelopment or overdevelopment of one or more floral organs. Such monstrous, 'peloric' flowers may aid in the detection of auto-pollination. Pelorism typically refers to an abnormality whereby the flowers appear to revert to an earlier, more primitive form. Darwin was interested in this phenomenon and is commonly credited with having been the first to apply the term. As early as 1864, pelorism had already been described in the flowers of multiple orchid species. The most commonly described condition is the development of three columns or three anthers in a flower, rather than a single one as is normally found in advanced orchids. In pleurothallids, one may find pelorism in the form of abnormal flowers that bear three anthers, or an undifferentiated lip that has become petal-like, or undifferentiated petals that have become sepal-like — all of which are normally accompanied by autogamy. I wonder if it is autogamy that causes redundant floral organs to become undifferentiated as they are no longer under selective pressure? Or if it is the other way around? The monster develops first and it later becomes autogamous to cope with the limited functionality of its imperfect organs. Be that as it may, pelorism is a curious phenomenon that requires further study.

Pelorism may have taxonomic significance, but more often the morphological changes in peloric individuals, populations or species are so striking that they can cause significant confusion about the relationships among plants. In Costa Rica, certain populations of the common *Masdevallia cupularis* have cleistogamous individuals that produce flower buds that develop into fruits without ever opening. The undeveloped buds are morphologically aberrant, featuring long cylindrical tails on their petals and lip. Occasionally, an autogamous flower opens fully, exposing these odd floral features. The morphology of the petals and lip are so unlike those of typical flowers of *M. cupularis* that botanists described one autogamous individual as a different species — *M. smallmaniana*. Based heavily on floral morphology, the two *Masdevallia* species have even been assigned to different sections within the genus. Curiously, the inconsistency in the development of these autogamous *Masdevallia* prompted the recognition of two more species, one with shorter and broader petals — *M. driesseniana* — and another with cleistogamous flowers — *M. rostriflora*. All three were described from a single individual in cultivation. My hypothesis is that they are all peloric forms of *M. cupularis*, in which the petals and lip have reverted to a sepal-like shape (see Figure 5.5.3). Within the same population, one may find individuals with typical *M. cupularis* flowers, individuals with malformed cleistogamous flowers — like *M. rostriflora* — and individuals with malformed open flowers — as in *M. driesseniana* and *M. smallmaniana*.

The uncertainty that peloric individuals can cause in orchid taxonomy is nowhere better exemplified than in slipper orchids. Slipper orchids characteristically feature a pouch-like lip with two fertile anthers. Therefore, when the pouch-less *Phragmipedium lindenii* was first discovered in 1846, it was naturally classified as a distinct genus under the name of *Uropedium*. The odd combination of having a long, flat lip and three fertile anthers immediately set *Uropedium lindenii* apart from any other known slipper orchids at the time. But all its other plant and flower features were so similar to those of *Phragmipedium* species that it was eventually concluded that *U. lindenii* was a peloric member of that genus. This was later confirmed using DNA studies, and it is today widely accepted that *P. lindenii* is an autogamous species in

Fig. 5.5.3. *Masdevallia cupularis*. **A.** Typical form with normal petals and lip. **B.** The peloric form which has sepal-like petals and lip and was described as a different species, *M. smallmaniana*.
© Wiel Driessen.

which the flowers have reverted to a petal-like lip shape — a primitive condition that is also reflected in their bearing three anthers. Strangely, history would repeat itself more than a century and a half later when a peloric *Selenipedium* species from French Guiana was recognised as a distinct genus under the name *Apedium*. This slipper orchid also features flowers with a petal-like lip and three fertile anthers, but there seems to be little discussion that its overall features are characteristic of species belonging to genus *Selenipedium*.

Certain floral changes are directly involved in the various mechanisms of self-pollination in orchids. These include structural modifications such as the lack of a rostellum, falling of friable pollinia onto the stigmatic surface, over-secretion of the stigma, and different movements of the perianth, stigma, anther or pollinarium. In most autogamous orchids, the rostellum — which typically separates the male and female organs in orchids — doesn't develop, develops incompletely or disintegrates during flowering. The removal of this physical barrier is an important step in promoting close contact between the pollinia and the stigma. To cite a few examples, in Costa Rica it is common to find that flowers of the widespread selfer *Stelis deregularis* lack a rostellum altogether. When — and if — the flowers open, the pollinia are already in contact with the stigma. In the partially and fully mycoheterotrophic species (plants that obtain their food from parasitism upon fungi rather than from photosynthesis) of *Cymbidium* in Japan, the degradation of the rostellum has been suggested to be an adaptation to low insect visitation in their hostile habitats. An understudied phenomenon is self-pollination through pollinarium transformation. The mechanisms by which pollinaria transform, as discussed in the previous section, are usually considered to be an effective way to prevent selfing, and they regularly function as an adaptation for outcrossing. Paradoxically, pollinaria transformation has also been exploited to facilitate the evolution of selfing through spontaneous autogamy.

*Dactylorhiza fuchsii* is a deceptive orchid species with a generalist pollination syndrome that suffers from an unpredictable pollinator environment. It is a curious orchid that on the one hand promotes cross-pollination by being food-deceptive, and on the other hand enables the possibility of autonomous self-pollination when pollinators are scarce or absent. Researchers have found that in *D. fuchsii*, the pollinaria that are not removed by insects gradually lean out of the anther by

Fig. 5.5.4. *Holcoglossum amesianum* a facultative self-pollinating orchid from China.
© Ron Parsons.

means of the caudicle (the granular process to which the pollinia are attached). This is caused by the pressure of the pollinia on the long and flexible caudicle, which bends and become convex. Through gravity and dehydration, the pollinia become twisted to the side and slowly droop until they reach the stigma. This effectively self-pollinates the flower through the transformation of the pollinarium. Researchers in Yunnan, China, found that under harsh conditions, the anther cap of *Holcoglossum amesianum* naturally turns upwards and then falls, allowing the pollinarium to bend into the stigma (see Figure 5.5.4). This mechanism of self-fertilisation apparently occurs in the dry season when pollinators are scarce. Pollinarium transformation is also responsible for what researchers have called 'accidental autogamy' in the European orchid *Gymnadenia conopsea*. This orchid, which is mostly pollinated by lepidopterans (butterflies and moths), may undergo spontaneous self-pollination in a very small percentage of the flowers produced in any given population. Autogamy in this case is believed to be a side product of the mechanisms that enable the pollinarium to reconfigure in order for insect-mediated pollination to occur.

Despite previous notions, autogamy is indeed widespread in the orchid family and may be advantageous, at least in the short term. Several diverse mechanisms leading to self-pollination have been identified, many of them accompanied by functional or non-functional floral modifications. Such changes in the structure of the orchid flower need to be carefully scrutinised before their biological and taxonomical significance can be inferred. The orchid itself is only interested is ensuring its reproduction, by any means possible. The selfing of some orchids may even be caused or facilitated by abiotic factors, such as water and wind. Their influence on the pollination of orchid flowers is discussed in the final two sections of this chapter.

## 5.6 RIDERS ON THE STORM

Britain is stereotypically known for having terrible weather, plagued by a cold, cloudy and rainy climate. But despite its reputation, rainfall in the United Kingdom is usually intermittent and not very heavy. In fact, with just 1,154 mm of rain on average per year, Britain is not that rainy at all, relatively speaking. There are of course local variations, and the West Highlands of Scotland get doused annually with closer to 3,000 mm of rain. And yet, there are areas in Costa Rica where twice as much rain can fall in a year, and in certain periods rainfall can be detected every single day in a year. Not surprisingly, a map of World Bank data shows that the rainiest countries in the world are centered along the equator in the tropical regions of America, Africa and Asia. By now, you are probably wondering what the rainiest place on Earth actually is? Believe it or not, this is a highly disputed matter with several different locations competing for this particular distinction. Mawsynram and Cherrapunji — two small towns in northeast India — have held the record, and actively quarrel for the title of 'wettest place on Earth'. Cherrapunji was declared the world's wettest place when it recorded 12,262 mm of rainfall in 2002 — more than 10 times the British average. But in recent years, Cherrapunji lost its position to Mawsynram. With an average annual rainfall of 11,872 mm and a maximum of 26,000 mm registered in 1985, Mawsynram holds the title, according to the *Guinness Book of World Records*. The claim is disputed by the Colombian municipalities of Lloró and López de Micay. Lloró reported an average rainfall of 12,717 mm yearly between

1952 and 1989, while López de Micay reported 12,892 mm per year between 1960 and 2012. The latter is therefore potentially the wettest inhabited place in the world. To make this odd competition even more peculiar, accusations fly over how data are recorded at each of these sites. The Cherrapunjis question the reliability of the data obtained from Mawsynram, while the Colombian sources present contradictory measurements for the same time and place.

How does this relate to orchids? Well, orchid species thrive in the tropical regions where rainfall is extremely high. Bad weather is widely known to limit plant reproduction and is typically considered a threat to the delicate flowers of orchids. Torrential rain, coupled with strong gusts of wind, may damage the flowers or inflorescences, dilute rewards such as nectar, and dislodge or reduce the viability of pollen. Bad weather may also pose a problem for pollinators, limiting their activity and capacity to find and reach flowers. It is thus to be expected that certain adaptations protect the flowers against rain and wind damage. Studies show that plant species living in very wet areas are more resistant to water damage than those adapted to dry areas. As precipitation increases, so too does the number of plant species with downward-facing flowers. Orchid species proliferate under humid conditions and have not only adapted themselves to high rainfall, but have in fact figured out a way to profit from seemingly adverse climates.

Previous stories have dealt with the multiple strategies by which orchids use animals as a vector for pollen transport — pollination systems that are mediated by living organisms. Here, we will discuss pollination through abiotic vectors. In a nutshell, abiotic pollination refers to the fertilisation of flowers without the involvement of animals. The main forms of abiotic pollination in plants are pollination by wind, known as anemophily, and pollination by water, known as hydrophily. As anyone that suffers from hay fever — or pollen allergy — will have noticed, pollination by wind is not uncommon in plants. Many plants produce abundant amounts of pollen, with the minute size of pollen grains allowing them to be easily swept away in a breeze. Even though much of it is lost, enough pollen finds its way to other flowers. This would be near to impossible in orchids because their pollen is packed into a few heavy sacs, called pollinia, and they do not produce large amounts of dust-like pollen. Water can carry orchid pollinia, but this reduces its viability and the pollen can't easily get to orchid flowers.

Even if orchid pollinia could be efficiently transported by wind and water, it would be virtually impossible for them to reach the small and hidden stigmas of orchid flowers without a more precise vector. The only way for orchids to be pollinated by abiotic vectors — as far as I know — is through assisted self-pollination.

Self-pollination — as we learned from the previous section — refers to the fertilisation of flowers with their own pollen. In most selfers this occurs mechanically, without the necessity for an external agent. But abiotic vectors, such as water, wind or gravity, can also assist self-pollination. That is exactly what has been found to happen in the flowers of certain orchid species. The flowers of *Acampe rigida* are self-compatible — that is, they may be fertilised by their own pollen (see Figure 5.6.1). However, this Chinese epiphytic orchid will not self-pollinate on its own. The cup-shaped flower is positioned upwards, and it is raindrops that dislodge the anther cap. This causes the pollinia to bounce upwards, after which they are pulled by a string-like structure, causing them to slingshot back into the stigma. We can theorise why a

**Fig. 5.6.1.** *Acampe rigida* is a self-compatible orchid. The cup-shaped flowers positioned upwards allow raindrops to dislodge the anther cap, causing the pollinaria to bounce and slingshot back into the stigma, effectively self-pollinating it.
© Ron Parsons.

self-pollinating mechanism that involves an external factor, instead of a well-developed autonomous mechanism, would evolve in these orchids. The key may be in the possibility of allowing cross-pollination when weather conditions are favourable. These orchids bloom in the wet season, and heavy rains deter pollinators. There are a few rainless days during the wet season, and no rain results in no selfing. It is possible that not having a strict selfing strategy would eventually allow a biotic agent to pollinate these flowers when the climate allows it.

Rain-assisted self-pollination, or ombrophily — a special kind of hydrophily has been reported in a few other orchid species, including the terrestrial orchid *Liparis loeselii*. Contrary to what has been found in the *Acampe*, this species does not exclusively rely on the raindrops to fertilise itself. Here the selfing occurs autonomously when the anther cap degenerates, thus allowing the pollinia to descend into the stigma by themselves. Nevertheless, it has been shown that drops of water will accelerate the process by disturbing the anther cap and knocking the pollen masses out of the anther and onto the stigmatic surface. In *Cyrtopodium polyphyllum*, it has been suggested that rain assists self-pollination by accumulating in the stigmatic cavity, dissolving the stigmatic substance and then evaporating. This species differs from *A. rigida* and *L. loeselii* in that, under normal conditions, it is pollinated by bees. The rain-assisted selfing only occurs under high levels of precipitation and air humidity, when pollinator visits become scarce. Something similar has been found in the invasive terrestrial orchid *Oeceoclades maculata*. The species, originally restricted to Africa, has spread extensively throughout the Neotropics (see Figure on page 254). At the periphery of its distribution it self-pollinates and, as in the species of *Liparis*, rain assistance increases fruit set but is not strictly necessary. In Brazil, where the species initially appeared two centuries ago, *O. maculata* also outcrosses through butterfly pollination. However, this occurs only occasionally, on rainless, sunny days, and the species mostly self-pollinates, with or without the assistance of rain. A curious variant has been found in *Platanthera aquilonis*, a widespread North American terrestrial orchid (see Figure 5.6.2). Water droplets from dew that collect in the center of the flowers can transport the pollen masses. Once the droplets evaporate they deposit the pollinia onto the stigma. Rain, mist or fog may also play a similar role for this species in areas where these conditions are prevalent.

Fig. 5.6.2. *Platanthera aquilonis*, a rain-assisted autogamous orchid, at flowering (**A**) and fruting after being fully self-pollinated (**B**).
© John Gange.

In Japan, one researcher noticed that the naturally fructiferous *Liparis kumokiri* would rarely set fruit under greenhouse conditions. Its flowering season coincides with the early summer rainfall, which produces heavy showers that can suppress reproduction in other plants. This led to the hypothesis that the species does not self-pollinate autonomously, and that selfing only occurs when the pollen is splashed by rainwater. To test this hypothesis, the author covered the plants with screens to exclude pollinators and subjected some of the plants to artificial raindrops by watering them with a shower-head. Under natural conditions, the flowers of this orchid are visited by several dipterans which land on the lip and feed on the secreted nectar, but none were observed removing pollinia. Around three days after the flowers have opened, the anther cap degenerates and raises, liberating the pollinia. Water droplets, be it from rain or otherwise, knock the anther caps from the flower and cause the pollinia to drop onto the stigma, self-fertilising the flowers. The fact that flowers placed under the screens produced no fruits proves that external factors are required for fertilisation. Perhaps what is most interesting is that artificial rain significantly increased the fruit set of the flowers, but autonomous autogamy — actual rain-assisted selfing in the field — was still higher. This suggests that the disruptive effect of rainfall could be compounded by the vibrating effects of the wind.

## 5.7 BLOWIN' IN THE WIND

> *"In some species ... the labellum is furnished with a beard of fine hairs, and these are said to cause the labellum to be in almost constant motion from the slightest breath of air. What the use can be of this extreme flexibility and liability to movement in the labellum, I cannot conjecture, unless it be to attract the notice of insects…"*
>
> CHARLES DARWIN in *Fertilisation of Orchids* (1877)

Darwin noticed that the lips of certain orchids — specifically those of species belonging to the genus *Bulbophyllum* — continuously tremble in the wind. He could not imagine what the function of this quality could be, but, as the first true evolutionary biologist, he was a faithful believer in the functionality of all floral adaptations and therefore suspected it could somehow be involved in pollinator attraction. What exactly may the evolutionary purpose of the highly sensitive vibrating lip be? The answer, my friend, is blowin' in the wind.

In spite of what Darwin instinctively believed, it has never been shown that the trembling lip of *Bulbophyllum* species lures its pollinators (see Figure 5.7.1). What has been found is that they have an important function in wind-assisted pollination. Unlike other plants, wind pollination — strictly speaking — is not known to occur in orchids. Orchid pollen is found in masses that are too heavy to be blown by the wind from flower to flower. However, wind, like rain, has been shown to assist in pollination. The lip of *B. penicillium*, an orchid from Myanmar and China, emits a rotting fruit odour and incessantly undulates in the wind. A group of local researchers studying the pollination of this species stated that the movements are "similar to that of a caterpillar's undulating crawl" and are meant to "attract and use potential pollinating insects." But what they in fact found is that the small *Drosophila* flies that pollinate the orchid are unable to reach the column apex when stationed on the lip, due to their size. The continuous movement of the lip lifts the pollinator to the column apex, eventually touching the anther and stigma and effectively removing the pollinaria. It is unlikely that the lip in any way resembles a caterpillar. Rather, the long, hollow hairs in which it is completely covered make it so much more sensitive to minor air movements.

The lips of a few Brazilian species belonging to genus *Bulbophyllum* are delicately hinged to the column, and tremble in the wind. The flowers are pollinated by minute flies that remove the pollinarium when they are pressed against the column by the trembling lip. This was first described in 1978 in a study on the pollination of *Bulbophyllum* in the Serra do Cipó, in the state of Minas Gerais, Brazil. The orchid was observed to be visited and pollinated by female flies of the genus *Pholeomyia* — in the Milichiidae family — which would approach the flower with erratic movements and head straight to the base of the lip upon landing. There, they have been seen feeding on nectar. However, the weight of the fly is not enough to cause the lip to tilt, and so the assistance of an air current is needed to press the lip and insect onto the column. Further studies on other *Bulbophyllum* from the campo rupestre grasslands in Brazil suggest that this mechanism of wind-assisted pollination may be common in species found in open areas where they are exposed to continuous wind currents. They too are pollinated by flies of the Milichiidae family, the adults of which are known to be kleptoparasitic (a form of parasitism in which an insect feeds on the hemolymph of another animal's dead prey), while the larvae feed on decaying plant matter and animal faeces or other detritus. Odours were shown to play an important role in the pollination mechanism of these *Bulbophyllum* species, but it is not clear whether they are sapromyiophilous — pollinated by mimicking decaying flesh — as is the case for others in this genus. Even though the lip movement was suggested to draw the insects' attention, it was only definitively shown to participate in mechanically aiding the insects in successfully removing the pollinaria. Strong winds were in fact suspected of hindering the insect's landing and diminishing the olfactory attraction of the orchid flowers.

Delicately hinged, trembling lips are also found in several other fly-pollinated orchid genera, including *Anathallis*, *Lankesteriana* and *Trichosalpinx* — which, as discussed in Chapter 3, section 3.6, have hit on a similar pollination strategy through evolutionary convergence.

**Fig. 5.7.1.** Many *Bulbophyllum* have hairy flowers, some species have a highly sensitive plumose lip that trembles with the slightest breeze. *B. tremulum* (**A**), *B. jolandae* (**B**) and *B. dayanum* (**C**).
© Ron Parsons.

They are all members of the exclusively neotropical Pleurothallidinae subtribe, which is the most species rich and rapidly evolving group in the orchid family with more than 5,500 species recognised today. Like *Bulbophyllum*, pleurothallids are mostly pollinated by flies. Together they represent about a third of the whole orchid family, and — in what must be the most large-scale example of convergent evolution in orchids — many of their species have small, dark, hairy flowers with mobile lips. Flowers of certain members of these genetically unrelated groups are virtually identical in every aspect. Were it not for their geographical separation and the presence of pseudobulbs in *Bulbophyllum*, which are strictly absent in all Pleurothallidinae, they would constantly be mistaken for one another. Species of the genera *Anathallis*, *Lankesteriana* and *Trichosalpinx* have also converged in having lips that constantly vibrate in the wind, and it would not be a stretch to believe that they too benefit from its assistance for pollination. However, an alternative explanation for *Trichosalpinx* is that the vibration of the ciliated lip might produce a visual effect similar to that of a trapped prey. *Trichosalpinx* could attract female Ceratopogonidae flies by mimicking the tactile properties of an invertebrate host prey with its hairy surfaces, visually in its dark purple colour, and mechanically in the movement of the lip.

Some members of the genus *Bulbophyllum* and members of several genera in subtribe Pleurothallidinae bear movable appendages on floral organs other than the lip (see Figure on page 252). These are especially mind-blowing in species belonging to *Bulbophyllum* section *Epicranthes*, where the petals of many species are bizarrely ornate (see Figure 5.7.2). The appendages are often suspended by thread-like stalks that move with the slightest air current. Others are rigidly attached, long and conspicuously ornate, making the flowers look like they are wearing halloween costumes. One of the most striking of these is found in the eccentric *Bulbophyllum macrorhopalon* which bears three tassel-like appendages hanging from each petal. The six appendages dangle in the wind from little threads like a mobile hung over a baby's crib. The many long, hairy appendages of *B. macneiceae* and *B. tarantula* — as their names suggest — make them look like the fruiting bodies of slime moulds or the legs of spiders, respectively, while those of *B. davidii* and *B. johannulii* give the appearance of a corpse that has been taken over by a parasitic fungus. These are not the kinds of accessories that would help you in attracting and captivating a partner.

Fig. 5.7.2. The flowers of certain members of *Bulbophyllum* section *Epicranthes* are intricately adorned wind movable appendages of unknown function. *B. cimicinum* (**A**), *B. flavofimbriatum* (**B**), *B. macrorhopalon* (**C**), *B. davidii* (**D**), and *B. nocturnum* (**E–F**).
© Andre Schuiteman (A, C, E), Ron Parsons (B, D) and Rogier van Vugt (F).

*Bulbophyllum nocturnum*, another member of this curious assembly, is perhaps the only orchid known to have exclusively nocturnal anthesis. Several other orchids are known to have nocturnal pollination, but the flowers of *B. nocturnum* only open at night. They open at 10 pm and last only until the early morning hours — something that I was able to witness at the Hortus Botanicus in Leiden, thanks to Rogier van Vugt, the greenhouse head who kindly took me to see it after closing hours on a winter's night. This rare, short-lived species features long, grey appendages that hang in the centre of the flower. To my eyes, the flowers look like they have been overgrown by a slimy grey mold. The appendages are quite conspicuous and yet their function is unknown. Like many of the other species in the flamboyant *Epicranthes* group, *B. nocturnum* is endemic to New Guinea, where plants are found very sporadically and in remote areas. One may suspect that these appendages play an important role in pollination, and although they may attract pollinators, such a function has never been demonstrated.

As has been amply pointed out throughout this book, the only way to definitively solve the mystery of the function of these ostentatious accessories is to make field observations. Despite the commendable efforts to culture these exuberant tropical orchids in Europe, such ecological studies can only be carried out successfully where these plants grow naturally in the wild.

The role that wind plays in the pollination syndromes of orchids certainly requires further investigation. Besides those discussed here, many other orchid flowers or flower parts are under constant movement due to the wind, and their function — if any — is not yet known. In Costa Rica, *Muscarella strumosa* thrives in windy areas. The flowers are borne on a long, wiry inflorescence, and they continuously move. The generic name *Muscarella* — meaning small fly — is a reference to the impression that the flowers give of flies hovering above the plant. It is curious that most species in the genus share this feature, but whether it has any impact on pollination remains to be shown. In the previous examples, air currents are assisting the insect in successfully fertilising the orchid flowers, but one may also think of the wind as assisting self-pollination without animal mediation — as rain or dew do. The importance of wind-assisted self-pollination in orchids is not well understood.

The only published study I could retrieve regarding wind-assisted autogamy in orchids is a 2002 paper by Jean Claessens and Jacques Kleynen on *Ophrys apifera*, which they later illustrated magnificently in their book *The Flower of the European Orchid: Form and Function*. This European orchid has long been known to be autogamous, but the details of how this takes place were not clear. What the researchers found is that after the pollinia drop from the anther, they are held by the caudicles, and dangled in front of the stigma. They are able to swing back and forth, and are eventually blown into the stigma by a gust of wind. The process is aided by hollow spaces in the caudicles, which weaken their structure and allow them to move more freely. Wind-assisted self-pollination seems odd, but in this case proved to play a crucial role in assisting the autogamy. I have seen this in a Costa Rican *Ornithocephalus* growing under greenhouse conditions. Species belonging to the genus feature a long, straight column and rostellum,

**Fig. 5.7.3.** The wind-assisted self-pollinating orchid *Ornithocephalus lankesteri*. Flowers showing the pollinaria, which are still attached to the apex of the rostellum, inserted in the stigma; note the ovaries swelling.
© APK.

bearing a correspondingly elongated pollinarium. In *O. lankesteri*, the anther cap shrivels a few days after anthesis. As the pollinia drop out of the anther, the pendulum-like pollinarium does not fall to the ground but rather remains hanging from the tip of the long rostellum, held by a sticky pad called the viscidium (see Figure 5.7.3). The air currents make the pollinarium oscillate, and it is eventually swung into the stigma.

How many orchid species may actually be profiting from wind and rain assistance? The answer, my friend, is that anemophily and ombrophily are likely to act as auxiliary pollination mechanisms much more often than is currently believed. Wind and rain may be especially important in assisting self-pollination when conditions are adverse, but — like most other phenomena that require extensive fieldwork — they need to be studied more carefully.

**BLOWIN' IN THE WIND**
Wind-mediated pollination in orchids

**SCAN ME**
https://youtu.be/L1cGLX64NfM

The delicately hinged lips of many orchids tremble continuously in the wind, this feature is especially frequent in species of *Bulbophyllum* and certain Pleurothallidinae. The trembling of the lip has been suggested to be attractive to the pollinating insect, but it may simply be mechanically aiding the pollination process instead. The lip is not the only floral organ that may be prone to movement by the action of the wind. Flowers of the genus *Muscarella* are borne on a long, wiry inflorescence which causes them to constantly move in windy areas. Certain other orchids have odd appendages of unknown purpose that are easily moved by action of the wind. Finally, some orchids may profit from wind-assisted self-pollination in which pollinaria are blown into the stigma.

The videos reproduced here are owned by Adam P. Karremans, Kurt Metzger and Jean Claessens (https://europeanorchids.com/), and have been reproduced and edited by the author with permission.

The supergeneralist *Neottia ovata* being pollinated by *Apis mellifera*. © Jean Claessens.

CHAPTER 6

# **FALLACIES**

*"As with most of the sciences whose origins we can trace to Ancient Greece, skepticism and the demand for first-hand evidence would eventually come to dominate botany, but when it came to orchids — as we shall see — myths, folklore, and superstition would persist a lot longer than with other plants. Indeed, orchids were to continue accumulating myths for two thousand years after Theophrastus first wrote about them.'"*
JIM ENDERSBY in *Orchid: A Cultural History* (2016)

Perhaps we have an innate fascination with the mystique that surrounds the orchid flower and its intricate strategies to ensure perpetuation. A passionate affair with the mysterious, as it were. Or maybe it is simply the challenge of grasping their overwhelming complexity. For whatever reason, the perpetuation of mistaken beliefs is extremely common when it comes to the orchid flower, especially with regards to sex and reproduction.

From the very genesis of the name *Orchis* — commonly attributed to Ancient Roman or Greek mythology — orchids have invariably been associated with lust and beauty. But even the apparent origins of this myth, copied, rewritten and repeated thousands of times for over three centuries, have been called into question. In his *Orchid: A Cultural History*, Jim Endersby points out that just as the 15th century advent of printing allowed for a rapid dissemination of literature — with both facts and errors — the internet has encouraged the same process at an even faster and greater scale, to the point that it is more difficult to disprove a previously widespread belief than to propose a brand new hypothesis altogether. Errors tend to become 'facts' by mere repetition, regardless of contradiction and refutation — a phenomenon known as proof by assertion.

Certain aspects of the orchid flower's reproductive ecology have been repeated so frequently that they have become part of our common folklore. Some ideas have been stated so often that we seem to forget they require fact-checking at all. When we dig deeper into the origin of such factoids, we may discover that there is very little evidence to support them. From the widely accepted fantastical pollination strategies such as pseudo-antagonism and pseudo-parasitism, to the conventional wisdom regarding the overwhelming dominance of bees and the high degree of pollinator specificity, from mantises and crab spiders to fossils, orchid pollination — both in scientific literature and popular culture — is far from immune to delusion. Most of the recurrent fallacies involving the ecology of orchid flowers and their pollinators have been discussed in previous chapters. Those that did not fit anywhere else are discussed here.

The orchid mantis, *Hymenopus coronatus*, ambushes and feeds on pollinating insects © James O'Hanlon.

## 6.1 STRAY CAT STRUT

*"A flower, a red orchid, that catches and feeds upon live flies. It seized upon a butterfly while I was present, and enclosed it in its pretty but deadly leaves ... That flower and this plant must have a nervous system closely allied to that of animals."*
JAMES HINGSTON in *The Australian Abroad. Branches from the main routes round the world* (1879).

James Hingston's report of an orchid flower that could actively catch and feed upon living insects surely stirred the imagination of his readers. But what this neurotic English travel writer had witnessed in Java (Indonesia) was not an orchid. In fact, it was not a flower of any sort. That which could seize a butterfly using its 'pretty but deadly leaves' was in fact a praying mantis. The charming orchid mantis (see Figure on p. 310).

Mantises are popularly known as praying mantises because of the prayer-like posture of their folded, raptorial forelimbs. They are agile generalist predators that belong to the insect order Mantodea, and typically use camouflage to catch a diversity of prey. Praying mantises commonly combine greenish or brownish colour patterns with cryptic body structures to blend into the surrounding vegetation. Among the praying mantises, orchid mantises — *Hymenopus coronatus* — are unique in their colourful bodies and large, leaf-like leg expansions. Their four hind legs bear large white, pink or yellow expansions that resemble petals, whereas the colour patterns of the broad abdomen are reminiscent of an orchid's prominent lip. Rather than ambush their prey by blending with their surroundings, the orchid mantis has transformed its whole body into a beautiful, yet deadly, blossom. But is the orchid mantis really mimicking an orchid flower? And if so, which one?

Since their discovery, these mantises have been associated with orchid flowers. A riveting account was famously given by Alfred Russel Wallace — co-discoverer of evolution by natural selection — in 1889: "The most curious and beautiful case of alluring protection is that of a wingless Mantis in India, which is so formed and coloured as to resemble a pink orchis or some other fantastic flower. The whole insect is of a bright pink colour, the large and oval abdomen looking like the labellum of an orchid. On each side, the two posterior legs have immensely dilated and flattened thighs which represent the petals of a flower, while the neck and forelegs imitate the upper sepals and column of an orchid. The insect rests motionless, in this symmetrical attitude, among bright green foliage, being of course very conspicuous, but so exactly resembling a flower that butterflies and other insects settle upon it and are instantly captured. It is a living trap, baited in the most alluring manner to catch the unwary flower-haunting insects." Around the same time, several travellers noted the resemblance between the mantis and the most beautiful orchid flowers. But none of them pointed at any specific orchid species.

In modern times, the orchid mantis has been typically associated with the large, colourful flowers of commercial *Phalaenopsis* orchids. But most of those are the result of artificial hybrids and their floral features are the product of a selection process by humans. They are not found in nature. Wild *Phalaenopsis* flowers may be quite spectacular, and some perhaps similar enough to the insect's attire to seem to have been mimicked. Yet, orchid mantises have never been encountered in the vicinity of *Phalaenopsis* flowers in nature. In fact, no large-flowered orchids have been found in association with the orchid mantis in their natural habitat. If the orchid mantis is mimicking a large orchid flower, shouldn't the flowering orchid be found close by?

As early as 1900, authors began to become suspicious of the alleged orchid mimicry. The Scottish entomologist Nelson Annandale pointed out: "I do not know of any specific orchid which it may have simulated; orchids of sufficient size and brilliancy of colour are rare, if not unknown, in lower Siam." He had a point. If the orchid mantis was mimicking a specific orchid flower, you would expect the real orchid itself to be found in the same area. Orchid species with such flowers are unusual in the natural habitat of the mantises, such as Thailand (then known as Siam), and given the rarity of the mantises, their ecology is poorly documented and remains virtually unknown. Consequently, the association between the two remained largely speculative (see Figure 6.1.1).

In 2014, a group of Australian scientists undertook a series of experiments to test whether the orchid mantis was in fact mimicking a flower and — if so — which one. To answer those questions the authors carried out a field experiment where they placed living mantises together with common local flowers, and observed pollinator behaviour. They also compared the spectral reflectance — the amount of energy a

Fig. 6.1.1. Although not found together, the beautiful *Phalaenopsis deliciosa* is one of the few species in the genus that grows in several of the same countries as the orchid mantis.
© Ron Parsons.

surface reflects at a specific wavelength — of different body parts of *Hymenopus coronatus* with that of the flowers of diverse plant species known to co-occur in the same habitat. In other words, they measured the colour of the orchid mantis and that of local flowers. The results were stunning. Based on the reflectance study, the authors were able to show that pollinating insects cannot see the difference between the orchid mantis and local flowers. In fact, the field experiments suggested that the pollinators were more attracted to *H. coronatus* than to the actual flowers. The researchers also assessed whether the mantis was mimicking an orchid flower or any other flower model specifically. By looking at morphological similarities between the mantis and flowers of different plants from its natural habitat, they discovered something quite extraordinary. By comparing the shape (a technique known as geometric morphometrics) and colour (vision models) of the mantis with that of local flowers, they showed that in fact the orchid mantises do not resemble one particular flower more than another, but rather their colour and shape varied within the range exhibited by many flowers. In other words, from the perspective of bees, flies and birds, the orchid mantis looks like a flower. Any flower!

The reason the mantises are more attractive than the actual flower — outperforming them in the choice test — is that the mimicry may not be limited to only visual cues. The floral resemblance is most apparent in juveniles. Researchers from Japan discovered that juvenile orchid mantises release pheromones that attract the oriental honeybee. This proves that the orchid mantis's ornaments are not meant to camouflage it, but to act like a flower. Like a stray cat strutting down a dirty alleyway, the mantis proudly displays itself like a beautiful flower regardless of its surroundings. In the wild, *Hymenopus coronatus* is often found on a range of plant substrates, including flowers, leaves, stems and bark; but it can also be found simply on the ground. Behavioural studies confirm that these mantises are generalists in their immediate substrate choice and do not prefer perching on flowers as compared to green foliage — so carefree and wild. Nevertheless, like non-rewarding orchid flowers, the mantis benefits from the magnet effect of nearby rewarding flowers and is more likely to be visited by insects when sited among patches of high floral density. Although the orchid mantis is attractive enough on its own, it benefits from the presence of other flowers because they increase the abundance of insects and enhance its deceptive signals.

The orchid mantis represents a unique case in nature as it is the only animal known to mimic flowers. Like other praying mantises, the orchid mantis is a generalist predator. It is a highly successful predator that specialises in mimicking flowers, catching floral pollinators using its raptorial forelimbs. It uses a form of generalised food deception, as no specific flower model is mimicked. Incredibly, it has one more trick up its sleeve. Praying mantises are known to have sexual size dimorphism — the female of the species is consistently larger than the male. In the orchid mantis this size dimorphism is even more pronounced than in other mantis species, with the female being several times larger than the male. Scientists have now confirmed that female gigantism in the orchid mantis is actually a result of the flower mimicry. The reason the female is so much larger than the male is that this allows her to display the floral mimicry without competing with the male for food resources. She will feed on a variety of larger, pollinating insects, while he is a generalist that preys on smaller insects that are not necessarily associated with flowers.

## 6.2 EVERYWHERE

Ecological and evolutionary studies that look into the complex interactions between flowers and their visiting insects typically focus on pollination. But pollinators are not the only animals that interact with flowers and their visitors. In fact, a whole community of different organisms take part and — directly or indirectly — affect plant reproduction. Among the most common organisms that can have an effect on plant reproduction we can list the predators of pollinators, floral herbivores and seed predators. Herbivores, in addition to parasites of floral parts, fruits and seeds, play a significant role in the successful reproduction of orchids, but these interactions are outside the scope of this book. Here we will discuss those animals that target the pollination process directly.

Highly adaptive as nature always is, predators have figured out that flowers can be a good source of prey. Animals that feed on pollinators are commonly called ambush predators or anti-pollinators. As far as I can tell, the term anti-pollinator was first proposed by Mariano Ospina in 1969 when describing the presence of ambush predators camouflaged on or around the flowers of certain orchids in Colombia. In his 1969 and 1972 accounts, Ospina describes the presence of small white spiders hidden on the lips of *Epidendrum ciliare* and *Cycnoches chlorochilon*. According to the author, the spiders patiently waited for visiting insects that they could ambush. But the most shocking of his observations must surely be that of an animal lurking among the inflorescences of *Elleanthus xanthocomus* while he was traveling on the Atrato River in Chocó, northwestern Colombia. As the author approached the bright yellow flowers of the hummingbird pollinated orchid he was cautioned. "*¡Cuidado… hay una culebra!*" (Careful, there is a snake!) And indeed, hidden among the rich vegetation waited a *Bothrops atrox*, ready to strike. The fer-de-lance, as it is commonly called, is a very aggressive and venomous creature that feeds mainly on small mammals and birds. It is generally terrestrial, but will climb trees when necessary to catch prey. The term 'anti-pollinator' is a direct reference to the antagonistic behaviour these animals can have towards pollinators and other floral visitors. The term has a negative connotation and most of us may subconsciously root for the pollinator and dismiss predatory animals. Ambush predators are certainly not enjoyable for pollinating animals, but are they really negatively affecting the plant and its reproductive success?

**Fig. 6.2.1.** Crab spiders, such as those on *Cephalanthera rubra* (**A**) and *Anacamptis morio* (**B**) in Europe, are found lurking in orchid inflorescences all around the world, ready to ambush insects visiting the flowers.
© Jean Claessens.

In Brazil, a snake was observed catching and feeding upon a bat that was pollinating the flowers of the leguminous tree *Parkia nitida*. Perhaps this explains why, despite the copious amount of nectar offered, bat visits to the flowering tree are shorter than expected. In his studies on certain neotropical trees, Alwyn Gentry referred to insectivorous birds as anti-pollinators. He believed that the high nectar production and synchronic mass-flowering of certain trees were part of a complex strategy to attract an excess of insects, that in turn would be irresistible to predators. But why would these trees plot against their own pollinators by encouraging the presence of anti-pollinators? Al's hypothesis was that, by attracting their natural predators, the pollinators would be dispersed, which would consequently promote cross-fertilisation by forcing the insects to distance themselves from the original tree. In fact, studies on the western monkshood showed that the presence of predators of their pollinating bees significantly reduced fruit-set.

Surprisingly, there are virtually no studies on the diversity and possible effects of anti-pollinators in orchids. In North America,

Carl Luer reported the presence of well-known predators, including ambush bugs (Phymatinae), assassin bugs (Reduviidae), and especially crab-spiders (Thomisidae). These small spiders mimic the colour of the floral parts with such precision that they are commonly overlooked when the flower is photographed and only spotted later after careful inspection (see Figure 6.2.1). In Indonesia, spiders were seen weaving their webs in front of the flowers of *Dendrobium crumenatum* and *Ceratochilus biglandulosus* in an attempt to catch the pollinating bees that hover over the flowers. Crab spiders were also reported to catch non-pollinating flies visiting the flowers of *Brassia antherotes* in Colombia, and in Patagonia they were found not to affect the reproductive success of *Chloraea alpina*, given that their prey were mostly non-pollinating floral visitors.

In southern Europe, Royal Botanic Gardens, Kew botanists Dave Roberts and Richard Bateman observed praying mantises hunting hoverflies on the inflorescences of *Orchis italica* and crab spiders likewise lurking on the flowers of *Serapias neglecta*. Realising that such ambush predators were also commonly found on the inflorescences of orchids in the United Kingdom, they later carried out one of the few experimental studies on anti-pollinators in orchids I could find. They compared the preference of ambush prey for conspecific orchids, specifically the nectar-rewarding *Gymnadenia conopsea*, versus the deceptive, nectar-less *Dactylorhiza fuchsii* and *Anacamptis pyramidalis*. Crab spiders occupied 2.5–4.9% of the orchid inflorescences and — somewhat unexpectedly — the presence of a floral reward (for the pollinators) was found not to result in a higher frequency of these predators. Rather, the density of crab spiders was shown to be site- and year-dependent, and perhaps their frequency may be influenced by other factors such as ease of concealment, nature of the dominant pollinators and ease of movement between inflorescences. Interestingly, Roberts and Bateman arrived at a similar conclusion to that of Gentry, speculating that the presence of ambush predators may in fact make rewarding orchid species effectively deceptive: shortening insect visitation times and promoting cross-pollination.

Admittedly, I have personally not paid much attention to anti-pollinators lurking among orchid flowers in the past. But careful inspection produced immediate results. Anti-pollinators are indeed everywhere! In Costa Rica it is relatively common to observe spiders building webs or hiding in or around orchid inflorescences, evidently hoping to ambush visiting insects. At Lankester Botanical Garden, the inflorescences of abundant and prolific species, such as *Epidendrum radicans* and *Stelis gelida*, are among the preferred hiding places for different spider species. The elongate, erect inflorescences of the former are ideal for spinning a web (see Figure 6.2.2), while the tiny greenish-white flowers of *S. gelida* are a good place for the crab spider to go undetected. So too are the flowers of other non-native orchids grown outside the greenhouses, including white-flowered *Coelogyne* and *Dendrobium* species (see Figure 6.2.4). Meanwhile, an insectivorous *Anolis* lizard looked irritated as I photographed it while it was — surprisingly well — camouflaged on the long inflorescence of an *Oncidium stenotis* (see Figure 6.2.3). Upon further inspection of video footage of diverse pollination events of native orchids, I have been able to witness how predators attempt to catch visitors. Such close scrutiny allows for interesting additional observations regarding ecological interactions in and around orchid flowers.

**Fig. 6.2.2.** The multiple erect inflorescences of the butterfly pollinated *Epidendrum radicans* offer a perfect spot for spiders to place their web.
© APK.

Fig. 6.2.3. An *Anolis* lizard looks back angrily at the photographer who has spotted it camouflaged on the inflorescence of *Oncidium stenotis* in Costa Rica. © APK.

In section 4.1 (Chapter 4) I described the proclivity of certain orchids to attract ants through the use of extrafloral nectaries. On one hand, the orchids provide food and shelter, while on the other hand, the ants patrol the different plant parts, ferociously defending their food source against herbivores. These insects are very efficient and indiscriminate. They chase away intruders, including pollinators. Inspection of our video footage of *Vanilla* pollination in Costa Rica showed that as the Euglossini bees land on the flowers, they clearly startle and excite the ants, which quickly rush towards the labellar tube. However, the bee visits are brief. By the time the ants arrive on the scene, it is too late. The bees have already left. Even though most bees clearly have enough time to remove pollinia successfully, could the ants also act as anti-pollinators by chasing away potential pollinators? In this sense, plants may pay an ecological cost for the defence provided by ants, in terms of reduced pollinator visitation times and frequency. However, a study on the cactus genus *Opuntia* suggests that the growth benefits provided by the ants exceed the costs. In fact, the plant does not seem to suffer any reduction in fitness. Are there any unexpected consequences? Going back to Gentry's hypothesis of anti-pollinators in mass-flowering synchronic trees, could the violent behaviour of the ants in *Vanilla* and other orchids be favouring outcrossing? It is an intriguing hypothesis that needs to be put to the test.

Anti-pollinators are an integral part of plant-pollination systems and more studies are needed to assess their prevalence, diversity and impact on orchids. Are anti-pollinators universally found on orchid flowers? Or, do they show preferences for certain orchids, and if so, for which ones and why? What other animals are using the orchid flowers as hunting grounds? Are they detrimental to the orchid's reproduction success, as we might expect? Or are they in fact promoting outcrossing and therefore increasing the orchid's fitness? Many such questions still remain unanswered. It seems logical to expect anti-pollinators to be more common in multi-flowered inflorescences with long-lasting flowers where concealment is also facilitated. But perhaps the number of floral visitors and how long they stay also play a role. Further field studies are needed.

Fig. 6.2.4. Crab spiders ambush insects visiting flowers of exotic orchids in Costa Rica. **A.** A bee caught on *Dendrobium cymbidioides*. **B.** A fly caught on *Coelogyne intermedia*.
© APK.

**EVERYWHERE**
Anti-pollinators

**SCAN ME**
https://youtu.be/TN4nHNY8Oz4

Orchid flowers attract many visitors and not all of them are harmless. Anti-pollinators, such as ambush predators, are often found lurking on the inflorescences of orchids all around the world. They are ready to strike, and when other floral visitors, including pollinators, approach the flowers they catch and feed on them. Despite being relatively common, anti-pollinators have not been found to reduce orchids' reproductive success. On the contrary, they may promote outcrossing.

The videos and photographs here are owned by Adam P. Karremans and have been reproduced and edited by the author with permission.

## 6.3 WORTH FIGHTING FOR

Despite being one of the most celebrated and commonly referenced pollination strategies attributed to orchids, pseudo-antagonism, otherwise known as territorial defence or pseudotrespassing, remains largely mysterious. Pseudoantagonism refers to a unique pollination strategy based on the exploitation of the antagonistic behaviour of territorial male bees towards enemies. The flowers supposedly mimic the bees' rivals. Pseudoantagonism is often cited as one of the most extraordinary means by which orchid flowers become pollinated. However, aggressive behaviour towards orchid flowers is an extremely rare event and — given the recent passing of Bob Dressler and Cal Dodson — seems to have no living witnesses remaining. Pollination through territorial defence has been exclusively reported to occur in members of the neotropical orchid subtribe Oncidiinae (genus *Oncidium* and its close relatives). These orchids typically bear long, swaying, multi-flowered inflorescences, and the flowers are thought to mimic swarms of intruding insects dancing in the wind that are reportedly attacked by male bees of the genus *Centris*. But what do we know about this alleged aggressive behaviour?

*Centris* (Centridini) are solitary, ground-nesting bees. The females excavate tunnels under flowering host trees and in grassy areas, where males have established territories. Nests tend to be widely dispersed over a given area. Males and females forage on a wide variety of host plants, mostly trees, during the dry season, which typically coincides with their flowering periods. The aggressive territorial behaviour of *Centris* bees is well known. Males select certain areas for their territorial displays, possibly close to nesting areas or flyways used by foraging females. These territories serve as interception areas. Males mark their territory with odours, forming a circular or parabolic odour screen. Scent-marked territories provide a defined and defendable space for the males, and these territories may be important mating arenas. Territories are typically set up in grasslands, but the males of some *Centris* species establish them on flowering trees and lianas. They select and mark the flowering crowns of certain tree species that are adapted to pollination by large bees. Within the crowns, the male bees may scent-mark the flowers, twigs and leaves to establish their territory. Usually, any given tree is patrolled by a single male. They continuously patrol the tree by hovering and occasionally perching on it. Sometimes the male bee leaves the crown for a few seconds or minutes to survey nearby flowering trees. In the late morning hours, they disperse in search of nectar, and abandon their previously defended territories. Finally, in the mid-afternoon, the male bee seeks an overnight sleeping site, only to start the whole process of marking and patrolling a territory all over again early in the morning the following day.

*Centris* bees are quite ritualistic in their behaviour and they are known to interact with territorial intruders in a very characteristic way. They patrol an air space with a radius of up to 1.5 metres and their territorial activity occurs exclusively during the morning hours. The quarrelsome males behave the same towards any intruder: following a quick initial inspection, they aggressively chase them away for a distance of up to 2–3 metres from their territory. After the pursuit, the bees return to their marked territories and either perch immediately or first hover and then perch. With other males of the same species, the bee exhibits a dominance behaviour. A dominant male may replace another in a previously marked territory. During this process the males are very aggressive, chasing and colliding with invaders. Males also trail other airborne objects that invade their

Fig. 6.3.1. *Oncidium hyphaematicum*, a species allegedly pollinated through pseudo-antagonism.
© Ron Parsons.

territories — including blowing leaves, sticks and stones. For the male *Centris* bee, it is true that some things are really worth fighting for. But could orchids have figured out a way to profit from the bee's innate macho behaviour?

The first ever report of a bee antagonising orchid flowers is probably that of Dodson and Frymire. On 15 January 1960, they described witnessing a *Centris* bee striking the flowers of *Oncidium hyphaematicum* (see Figure 6.3.1) at Montecristi, in the Manta province of Ecuador. The authors noted: "This bee ... would sit on a stick near an erect inflorescence of the *Oncidium* and would drive away any other bee that passed nearby. Occasionally this bee would fly out and hover in front of the inflorescence of the *Oncidium* and when the wind moved the inflorescence the bee would dart in and strike a flower." A few months later on the road to Quevedo, north of Guayaquil, the authors observed this astonishing interaction once again. This time it involved a small *Centris* and the flowers of *O. pardothyrsus* (now a synonym of *O. planilabre*, see Figure 6.3.2). "The bee would hover in front of the flowers and the pendant racemes and when the wind moved them, would fly in and strike one of the flowers." On describing these events, Dodson and Frymire brought up the possibility of the interaction being a consequence of territorial protection by male bees. Pollinating the 'intruding' *Oncidium* flowers by striking them and attempting to drive them away was a completely novel phenomenon in floral ecology.

In 1966, van der Pijl and Dodson proposed the term pseudo-antagonism for this unusual pollination strategy. In their book *Orchid Flowers* the authors meticulously described the previous observations regarding *Centris* bees attacking the flowers of two *Oncidium* species in the coastal zone of Ecuador. "When the flowers of an *Oncidium* located in the territory of the bee move, the bee attacks them. If the breeze continues the bee strikes flower after flower ... The action is extremely precise. The bee does not land on the flower but merely strikes it." Their vivid description of this provocative behaviour, accompanied by revealing photographs of a *Centris* bee carrying a pollinarium while approaching a flower of *Oncidium* head first, was extremely convincing. Given the minute size of the viscidium — the sticky end of the pollinarium — and the reduced stigmatic area, the authors emphasised the millimetric

**Fig. 6.3.2.** Dodson and Frymire observed *Centris* males striking the flowers of *Oncidium planilabre* (formerly *Oncidium pardothyrsus*) in Ecuador.
© Ron Parsons.

precision that the bee would need to strike the flower to effectively pollinate it. "The viscidium attaches to the frons of the bee, between the compound eyes, and the stipe rapidly depresses, holding the pollinia extended directly in front of the bee," they wrote.

The exploitation of the defensive behaviour of territorial *Centris* bees by *Oncidium* species has since been upheld by some and disputed by others. On the one hand, some authors have elaborated upon the initial observation and hypotheses. Jim Ackerman suggested that the interaction could be mutualistic. Rather than being exploited, the bees could benefit from the hitting practice and become better territory defenders. John Alcock — who described the event as "head-to-flower combat" — colourfully added "markings on the orchid flower encourage the male to collide with the column." On the other hand, others have offered alternative hypotheses. Mark Chase — renowned Oncidiinae specialist — stated that the observations of male bees attacking flowers were unquestionable. But he argued that, given how unlikely it is for the bee to contact the viscidium or stigma — let alone both — it was highly doubtful that the behaviour could result in pollination. Rather, Chase expected them to be pollinated by bees that collect oils, in a strategy that involves mimicry of Malpighiaceae flowers, as has been shown in several species belonging to the genus. Castro and Singer later suggested that perhaps the males patrol the inflorescences waiting for females, and that while the males defend their territory the females pollinate the flowers. Yet, no studies have ever been able to prove and confirm pollination of *Oncidium* species by *Centris* bees through pseudoantagonism.

The flowers of many Oncidiinae look extremely similar to each other. But thanks to DNA studies it has now been amply shown that species with almost identical flowers are not necessarily close relatives. We know today that this floral similarity is the result of convergent evolution as a result of floral mimicry. Many Oncidiinae flowers are mimicking the flowers of another plant — members of the Malpighiaceae family. There are more than one thousand species of neotropical Malpighiaceae and they share a floral morphology theme that attracts oil-collecting bees. The mimicry of malpighiaceous oil-flowers appears to be a recurrent pollination strategy among many orchids in the subtribe Oncidiinae. Some of them are known to produce oil-rewards that are biochemically similar to those of the Malpighiaceae, while others appear to be

**Fig. 6.3.3.** Donald Dod observed male *Centris* bees approaching and occasionally striking the flowers of the sexually deceptive *Tolumnia henekenii* in the Dominican Republic.
© Robert-Jan Quené courtesy of Orchids Limited.

deceiving their pollinators by resembling an oil-producing flower and offer no reward at all. Elaiophores, or oil-secreting glands, are present in the flowers of many Oncidiinae species. Solitary bees — especially females — gather the oils and use them to nurture their larvae. The oils adhere to their legs through capillary action while the bee scratches the surface of floral segments, especially the highly developed lip callus and lateral lobes. It is during this gathering process that these bees remove or deposit the orchid pollinaria. It seems inexplicable that the highly characteristic *Oncidium*-like flowers, which are the outcome of an evolutionary convergence due to the mimicry of Malpighiaceae flowers, would at the same time serve pollination strategies as contrasting as pseduo-antagonism and oil-collection.

I remained — in that stubborn naivety that comes from inexperience — highly sceptical of the idea that *Oncidium* orchids could trigger territorial *Centris* bees to attack them aggressively in the belief that they are competing males or other intruders. But I was even more suspicious of the thesis that this chaotic interaction could result in the effective fertilisation of the *Oncidium* flowers — especially given the small size and complex shape of the column, and the precision that would be required to successfully remove and place the pollinarium within a stigma. Such a behaviour had never been fully documented and explained. Yet, orchid pollination guru Cal Dodson observed *Centris* bees antagonising *Oncidium* flowers in Ecuador, and his observations were reinforced by Bob Dressler, who saw the same behaviour occurring in Panama. Then there were the curious observations by Donald Dod in the Dominican Republic. When studying the pollination of the bee orchid *O. henekenii* (now a synonym of *Tolumnia henekenii*, see Figure 6.3.3), Dod noted two types of behaviour. One was a "hit and fly maneuver" and the other "was a definitive mounting of the flower, caressing the lip with the hind feet and tapping movements of the abdomen down toward the lip of the orchid." The bee orchid has since then been regarded as one of those orchids that are pollinated through sexual deceit, or pseudocopulation. One may wonder if what Dod saw were simply failed attempts at copulation or perhaps also territorial behaviour by the male *Centris*. Be it as it may, the swift attacks described were certainly reminiscent of the observations made by both Dodson and Dressler. This meant three researchers had independently observed antagonistic behaviour, so there had to be more to it than only imagination. But even if the bees did attack the flowers, could they remove pollinaria in the process? And if they did, why would an orchid develop such a seemingly unreliable pollination strategy? Questions flooded my mind.

One of the most common *Oncidium* species in Costa Rica is the robust *O. stenotis*. The trees at Lankester Botanical Garden are covered in them and fruits can be seen hanging here and there. In November, the long inflorescences bear hundreds of little yellow flowers, showering their host plants in gold — which gives them their local name of '*lluvia de oro*' or golden rain. So there I sat, looking up at the trees. Staring like a mad man at one inflorescence, then another, and yet another, for hours at a time. Once in a while I would see a bee or a wasp approach, calmly hovering over the flowers. I never witnessed any aggressive behaviour towards the flowers, but then again, neither did I observe pollinaria removal. It was not until the beginning of 2020 that my luck unexpectedly changed. Karen Gil and I had just been filming the pollination of a bucket orchid at the Monte Alto nature reserve in Guanacaste province, and we were heading home in the university's pickup truck. It was a warm morning in February. By then — the middle of the dry season — many trees in the dry forest have completely shed their leaves and the grasslands are quite desiccated. Suddenly, a large patch of yellow flowers swaying in front of a shriveled tree caught my eye. It was a large Oncidiinae, a healthy specimen of *Cohniella ascendens* (previously known as *Oncidium ascendens*, see Figure 6.3.4) in full bloom. Something appeared to be attacking the flowers!

We got as close as possible, climbing into the pickup's cargo bed. I couldn't believe my eyes. There it was, the mythical antagonistic event was occurring right in front of us. The *Centris* bee seemed to aggressively chase the bright yellow *Cohniella* flowers as if to intimidate them. The bee hovered and charged at full speed, zig-zagging from one side of the inflorescence to the other. Its movements were extremely fast, each lasting fractions of a second, and it was very hard to understand — with the naked eye — what was happening exactly. The bee sprinted back and forth so quickly that I couldn't be sure if it had even touched the flowers, let alone if it was carrying any pollen. Luckily, as we had been filming earlier that day, the recording equipment was at hand. We immediately set a phone and small camera on record, and just as we were taking out the professional equipment, the bee left. The whole event took about a couple of minutes, and the bee's movements had been so

**Fig. 6.3.4. A.** The golden yellow flowers of *Cohniella ascendens* mimic oil-rewarding Malpighiaceae. **B.** A stingless bee gets the tiny pollinarium stuck to the head as a result of the oil-collecting behaviour.
© APK.

erratic and speedy that we were not really sure what we had or hadn't recorded. Back home we were able to inspect everything carefully. By playing the video recordings back in slow motion, we could clearly see the bee's actions. It repeatedly charged the inflorescence, then — just before coming in contact with the flower — it stopped midair and turned away. Not once did the bee actually hit any of the flowers, although it came very close several times. An even closer look shows smaller flying animals approaching the flowers — perhaps it's them that the bee chases away. There was absolutely no doubt about it: *Centris* may aggressively patrol Oncidiinae flowers! But there is no way this behaviour can lead to pollinaria removal or deposition, and it remains to be shown if it's actually the flowers or other visiting insects (or both) that elicit the territorial behaviour of the male *Centris*.

Earlier that same day — at Monte Alto — I had filmed a small *Trigona* bee visiting the flowers of another plant of *Cohniella ascendens*. The bee diligently visited each flower, systematically scraping the base of the callus every time. It was obviously showing the oil-collecting behaviour that has been described as a pollination strategy in many Oncidiinae species. I did not think too much of the tiny bee at the time. But after watching the *Centris* attacking the flowers of that same orchid species, I decided to reexamine those videos carefully too. And there it was, staring right back at me! In one of the scenes, the little *Trigona* bee places itself facing the column, and as it proceeds to scrape the oily surface of the callus, it touches the viscidium with its head and removes the extremely discrete and inconspicuous pollinarium. The flower is a very precise piece of machinery, with the pollinarium accurately becoming placed on the front of the bee's head, delicately sticking out — as can be seen in the photograph originally published by Dodson of a *Centris* with the pollinia of *Oncidium hyphaematicum* to accompany his discussion of pseudoantagonism. A large *Centris* bee recorded while visiting the flowers of *O. ornithorhynchum* in Colombia also shows a very calm, oil-collecting behaviour. The orchid's pollinaria can be observed accumulating on the bee's head as it peacefully inspects each flower (see Figure 6.3.5).

Oncidiinae inflorescences are certainly patrolled by some male *Centris* bees, but pseudoantagonism as a pollination strategy is unlikely — if not impossible. Rather, their flowers are pollinated through other means, such as oil-collection, fragrance collection or sexual deceit. Certain species do induce territorial behaviour, and can therefore occasionally be observed seemingly being antagonised. Perhaps, given the oil-seeking behaviour of female *Centris* bees, male bees actively establish their territory in or around the inflorescences of the oil-producing Oncidiinae as part of their own strategy to find a mate. As Jim Ackerman elegantly phrased it in a personal communication: "The male bees are fooled by the *Oncidium* as much as the female bees. Only that the boys are hot for the girls, and hang out where the girls go." An exciting hypothesis that remains to be tested.

**Fig. 6.3.5. A–B.** A female *Centris* accumulates several pollinaria on the head as she collects oils on the flowers of *Oncidium ornithorhynchum* in Colombia. © Sebastián Moreno.

## WORTH FIGHTING FOR
Pseudoantagonism in the Oncidiinae

**SCAN ME**

https://youtu.be/7liKit5lmbo

Despite being undocumented, pseudoantagonism is one of the most frequently mentioned pollination strategies among orchids. The hypothesis is that certain Oncidiinae species elicit the antagonistic behaviour of male *Centris* bees, and pollinate the flowers while attacking them. Antagonistic behaviour in the vicinity of Oncidiinae flowers has been observed by at least three independent researchers in the Neotropics and a recording of such an event is shown here for the first time. However, it is unlikely to result in the transfer of pollinaria. Rather, these Oncidiinae mimic the oil-producing flowers of the Malpighiaceae family and are pollinated by bees collecting oils.

The videos and photographs here are owned by Adam P. Karremans, Karen Gil and Sebastián Moreno and have been reproduced and edited by the author with permission.

## 6.4 PARANOID

In *Orchid Flowers*, van der Pijl and Dodson introduce the term 'pseudoparasitism' to describe another obscure pollination strategy based on deceiving insects. Pseudoparasitism refers to a pollination strategy that relies on insects that are attempting to sting and parasitise their victim. Much less known about this strategy than about its cousins, pseudocopulation — where a flower is pollinated by inciting an insect to copulation — and pseudoantagonism — where the flower elicits territorial behaviour by male insects —; pollination through parasitism is highly speculative and remains cryptic. Besides the lack of strong supporting evidence, pseudoparasitism also suffered from the fact that the authors themselves were not convinced about it. In their own words: "This is an unproven category. We only include it because it seems as logical as pseudocopulation and pseudoantagonism."

**Fig. 6.4.1.** *Calochilus campestris*, the odd bearded orchid *in situ* in New South Wales, Australia.
© Ron Parsons.

However, it is now clear that this hypothesis was based on misconstruing an observation of a wasp interacting with the flowers of the Australian orchid *Calochilus*.

In 1946 — upon studying the pollination of *Calochilus campestris* — school headmaster Fred Fordham suggested that the dark, yellow-banded scoliid wasp *Campsomeris tasmaniensis* was attempting to sting the flower (see Figure 6.4.1). Female scoliids parasitise beetle grubs in the soil. They paralyse each victim by stinging before laying an egg on it. Species of *Calochilus* — known popularly as beard orchids — are among the more unusual and attractive of Australia's orchids. As the name suggests, their lips are prominently covered in long, dense, reddish or bluish hairs. They are pollinated by male wasps of the scoliid genus *Campsomeris* through sexual deception (see Figure 6.4.2). It has been suggested that the flowers release sexual pheromones that elicit the male wasp's copulation attempts. It is during the process of stroking the lip with its genitalia that the insect removes the pollinarium with its head. Australian ecologist Colin C. Bower and Peter Branwhite correctly pointed out that Fordham's wasps had in fact later been determined to be males. Males cannot oviposit. They lack a stinger, and therefore also the capacity to sting. What van der Pijl and Dodson

**Fig. 6.4.2.** *Calochilus stramenicola* being pollinated through pseudocopulation by a male Scoliid wasp.
© Jeremy Storey.

missed is that Fordham's account includes a statement by entomologist Keith Collingwood McKeown of the Australian Museum, Sydney. In the foreword, McKeown, who identified the wasp specimens as *C. tasmaniensis*, is cited as having said: "This species is widely distributed along the south-eastern coast of Australia. The males of the flower wasp (family Scoliidae) are much smaller than the females." This precludes the possibility of pseudoparasitism. Bower and Branwhite point out that what Fordham believed to be "*stabbing movements, as if... stinging the hairless ribbon*" were in fact the copulation attempts by the males. This definitively puts the matter of pollination through pseudoparasitism in orchids to rest — at least for now!

Pseudocopulation, pseudoantagonism and pseudoparasitism are similar in that they are pollination strategies based on the deception of an insect into believing the flower is an organism that it isn't. In this way, the orchid flower is able to elicit a particular behaviour that ultimately results in pollination. Captivating as these strategies may be, only pseudocopulation has actually been proven to occur in orchids.

## 6.5 NICE GUYS FINISH LAST

*'I hate myself, I hate clover, & I hate bees'*
CHARLES DARWIN in a letter to JOHN LUBBOCK (1862)

Darwin — of course — didn't really hate bees. The English naturalist was well aware of the crucial part these insects play in the ecosystem and was simply frustrated with trying to understand them. Bees are highly charismatic and have an excellent reputation. Their popularity has surged in recent years, and we have all probably heard of their essential role as flower pollinators. In fact, in orchids — as in many other plants — bees are generally considered the most important pollinator group and adaptation to bee pollination has been identified as a key evolutionary driver. It goes without saying that bees pollinate the flowers of an extremely high diversity of Orchidaceae, and among them are some of the most appealing species that we all know and love. Pollination systems involving bees are also among the most well studied and best understood. Orchid scientists and enthusiasts alike have a clear notion of the importance of bees and a higher sensitivity towards them, perhaps much more so than for other pollinator groups. This relevance of bees as pollinators filters into casual conversation and scientific literature. But what is the actual importance of bees in relation to other pollinator groups such as wasps, flies, moths, butterflies and birds?

The perception of the great importance of bee pollination in orchids was initially conceived at a time when very little was known about pollinating agents across the family. In *Orchid Flowers* (published in 1966), van der Pijl and Dodson estimated that 60% of all orchids were pollinated by Hymenoptera — 55% bees and 5% wasps. Only 40% of the Orchidaceae were believed to be pollinated by other agents — 15% flies, 8% moths, 3% butterflies, 3% birds, 3% autogamy, and 8% mixed agents. These numbers suggested that bees alone pollinated more orchids than all the other pollinator groups combined, and that they pollinated almost four times as many species as the second most important group — flies. This idea of bee dominance has persisted over time and is repeated virtually anytime orchid pollination is discussed. However, it was the result of a general notion — albeit an educated one — rather than of an accurate estimation of any sort. On the one hand, bees, like some of the large-flowered orchids they pollinate, are admired and revered. On the other hand, flies, like some of the orchids they pollinate, are often unappealing and disliked. To prove my point, let your online search engine finish the phrases 'Bees are' and 'Flies are' automatically for you, and compare the results. We may culturally adore bees and dislike flies, but the latter are much more important as pollinators than is typically believed.

The most recent account of pollination records in the Orchidaceae — an extensive database by Ackerman and colleagues — still finds that Hymenoptera are the most widespread and common orchid pollinators. Current data suggest that 57.5% of all Orchidaceae are pollinated by bees, and bees are the only pollinator group found to service all five orchid subfamilies. The number of orchids now confirmed to be pollinated by bees has grown from 222 reported in *Orchid Flowers* back in 1966, to 1,006 today — which is more than four times as many species. However, the number of fly-pollinated orchids has grown tenfold since van der Pijl and Dodson's account, currently taking up 21.9% of the Orchidaceae. This steep increase in the importance of flies is not surprising. Bees pollinate a larger number of different orchid groups, but when it comes to species numbers most bee-pollinated genera are relatively small. The most species-rich genera in the orchid family are *Bulbophyllum*, *Dendrobium*, *Epidendrum*, *Lepanthes* and *Stelis* — each with over a thousand species. Out of the big five, only *Dendrobium* may include significant bee pollination; *Epidendrum* seems to be mainly pollinated by moths and butterflies (but see 'Never enough', Chapter 2, section 2.7); while *Bulbophyllum*, *Lepanthes* and *Stelis* are fly-pollinated. In fact, fly pollination predominates in most of the species-rich and rapidly diverging orchid groups, namely subtribes Pleurothallidinae (5,500+ species), Dendrobiinae (3,400+ species) and Malaxidinae (1,255+ species). Butterflies and moths are no minor contributors either. They are reported to pollinate at least a few thousand orchid species, most notably the genera *Epidendrum* (1,800+ species) and *Habenaria* (850+ species), as well as many other terrestrial orchids, and many species of subtribe Angraecinae (600+ species). But records are available for an extremely small proportion of these large orchid groups. Pollinator data are available for just 1% of the species of *Lepanthes* and *Stelis*, 3.5% of the *Habenaria* and around 5% of the *Bulbophyllum* and *Epidendrum* (see Figure 6.5.1 and

Fig. 6.5.1. *Bulbophyllum sicyobulbon* being pollinated by a drosophilid fly. © APK.

**Fig. 6.5.2.** A fungus gnat pollinates the flowers of a *Stelis* species in Colombia. © Luis Eduardo Mejía.

Figure 6.5.2). These numbers suggest that orchids pollinated by Diptera and Lepidoptera are significantly understudied. But why?

The Ackerman database includes pollination records for about 10% of all orchid species. Even though this may sound like a small sample, with around three thousand records, it is the largest database on orchid pollinators ever compiled. The records are, however, unevenly distributed. Tropical species are much less well represented than their temperate counterparts, and the large-flowered orchids are better studied than the tiny ones. The latter are probably the most important reasons why fly-pollinated orchids have fewer pollination records: flies are typically tiny, like the flowers that they pollinate and the pollinaria they remove. Observations on the pollination of smaller orchids by tiny flies are more uncommon than those of bigger flowers visited by large bees simply because they are more difficult to see and understand. Another reason for the discrepancy among these groups is probably the fact that Hymenoptera pollination regularly occurs during the daytime and is therefore easier to study than nocturnal pollination by Diptera and — especially — Lepidoptera (see Figure 6.5.3).

This bias towards having better data for more charismatic, diurnal pollinators that pollinate larger, more accessible orchids is obviously not easy to overcome. But given that pollinator groups are quite consistent within genera with similar floral morphology, we can safely expect that the vast majority of the other 95% or more of *Bulbophyllum*, *Lepanthes* and *Stelis* species are going to be pollinated by Diptera, while the *Epidendrum* and *Habenaria* species are mainly going to be shown to be pollinated by Lepidoptera. Luckily, as data are available for at least a few species in most orchid genera, we are able to estimate with relative certainty what will be found in the remaining 90% of the Orchidaceae. Extrapolations of pollinator information have indeed been carried out and the result is that the patterns of the relative importance of pollinators change significantly. Perhaps one of the most interesting findings is that Hymenoptera and Diptera become more or less equally important as orchid pollinators, each being responsible for pollinating more than 8,000 orchid species.

A few years ago we estimated that the hyper-diverse fly-pollinated subtribe Pleurothallidinae was increasing at an average rate of 85 to 90 newly discovered species per year — that is about half of the new orchid species discovered annually. This tendency persists today, and in

**Fig. 6.5.3.** The hawkmoth *Oligographa juniperi* pollinating *Habenaria epipactidea* in South Africa.
© Craig Peter.

fact fly-pollinated orchids consistently dominate species discovery. Just as an example, of the 212 new orchid species added to the Orchidaceae family in 2020, around 100 belong to Diptera-pollinated groups, around 50 to Hymenoptera-pollinated groups, and around 35 to Lepidoptera-pollinated groups. This means that, by mere species discovery alone, the number of fly-pollinated orchids grows twice as fast — on average — as the number of bee-pollinated orchids. Given the tendency towards fly pollination in the most species-rich orchid groups, and the accelerated growth in both species numbers and pollination data, it seems likely that flies will eventually be found to be as important as bees when it comes to pollinating orchids, perhaps even displacing them at the top. Sometimes, nice guys finish last!

### 6.6 SOMEBODY TOLD ME

*"Conventional wisdom also holds that the orchids are highly specific in their pollination relationships, implying that most orchid species are limited to single pollinator species. Such highly specific relationships do occur, but they are probably the exception rather than the rule"*
ROBERT DRESSLER (1993)

For as long as I can remember, I have 'known' that orchids are exceptional because each species has a highly specialised relationship with a unique pollinator. Through the years I kept hearing about how their species-specific interactions made them special. I never really stopped to think about what the evidence for such an extraordinary specialisation in the orchid family was. With time I came to realise that we ignore the pollinators of the majority of known orchid species, and that species-specific interactions are rather rare in many well-documented pollination systems. How specialised are orchid-pollinator relationships really?

Even though orchids have very particular pollination mechanisms, most studies find that their flowers are often visited by a range of animals. Unsurprisingly, not all floral visitors are able to remove and deposit pollinia, but in most cases pollination is not restricted to a single species. Also, it is well known that a single pollinator species

**Fig. 6.6.1.** *Pleurothallis helleri* attracts many floral visitors belonging to diverse orders and families of arthropods, including butterflies, beetles, wasps and flies, but only the tiny Ceratopogonidae (Diptera) are able to remove the pollinarium. © APK.

may be shared among different orchid species, because animals carrying the pollinarium of one orchid species are occasionally observed visiting the flowers of another orchid species (see Fig. 1.3.1, Chapter 1). From an evolutionary perspective, it is clearly an advantage not to rely on a single species of animal for reproduction, as its absence could entail the plant's demise. Nevertheless, the orchid also benefits from and requires a certain degree of precision, to avoid wasting pollen — as well as other valuable resources — on animals that are poor or inconsistent vectors. This results in a trade-off between restricting access to some animals while allowing the 'right ones' to access the floral resources. So are floral visitors always pollinators?

In Costa Rica, the bright yellow flowers of *Pleurothallis helleri* attract diverse animals of multiple families within diverse orders. The flowers are visited by a plethora of flies, bees, wasps, beetles, butterflies and arachnids. Curious as to which actually could pollinate them, I collected every arthropod that visited a single plant to look for the tiny pollinaria on their bodies. During a period of about one week, more than 80 specimens were collected on the flowers of this particular orchid. Most specimens belonged to diverse families of flies, but beetles, butterflies, wasps and even a few spiders were also captured. In total, 16 out of the 80 individuals collected bore a pollinarium. Rather unexpectedly, without exception, all the specimens carrying pollinaria were tiny flies belonging to a single fly family, the Ceratopogonidae (see Figure 6.6.1). So despite their general appeal to arthropods, the flowers of *P. helleri* clearly have a well-designed mechanism of pollinarium placement that sorts the floral visitors from the effective pollinators. Similar observations have been made in species of the genera *Earina* and *Dendrobium* in New Zealand. They are visited by insects of several different families of flies, beetles, wasps and bees, but only a small subset of the fly families actually removed pollinaria, and only these were regarded as putative pollinators by the authors. But how common is it for orchids to specialise in a single family of pollinators — such as the Ceratopogonidae in *P. helleri* — or to several families in one order — such as the flies pollinating *Earina* and *Dendrobium* species?

The largest dataset on pollinators and floral visitors in the Orchidaceae has been produced by Ackerman and colleagues. This database includes pollination records for 2,770 orchids — covering roughly 10% of the known species — and is by far the largest and most comprehensive compilation made for the family. Perhaps one of the most interesting results from this study is that only 6.6% of the orchid species — 113 out of 1,700 species for which this information was available — are known to be pollinated by more than a single order of pollinator — for example either Diptera, or Hymenoptera, or Lepidoptera. Just 15% of all orchid species are known to be pollinated by more than a single family of pollinator — for example Apidae (Hymenoptera), or Ceratopogonidae (Diptera), or Sphingidae (Lepidoptera). Scarcely 28% are pollinated by more than one genus. Perhaps most strikingly, the list reveals that 56% of all orchid species represented have a single species of pollinator, and fewer than 10% of all orchids for which pollination data is available have more than five known pollinator species in two or more different families. But the list also reveals that some orchid species are 'supergeneralists' with dozens of known pollinator species. The top three supergeneralists are: *Neottia ovata* (see Figure 6.6.2), with 162 recorded pollinator species, belonging to 102 genera in 34 families distributed among

Fig. 6.6.2. The king supergeneralist *Neottia ovata* may be pollinated by bees, beetles or wasps, among many other arthropods. **A.** An ichneumonid wasp carrying pollinaria. **B.** The beetle *Cantharis rustica* pollinating the flowers.
© Jean Claessens.

Fig. 6.6.3. The supergeneralist *Anacamptis pyramidalis* pollinated by the butterflies *Zygaena purpuralis* (**A**) and *Melitaea parthenoides* (**B**).
© Jean Claessens.

five arthropod orders; *Epipactis palustris* (Figure 4.1.5), with 136 pollinators in 96 genera from 41 families belonging to four orders; and *Anacamptis pyramidalis* (see Figure 6.6.3), with 75 pollinators in 48 genera from 14 families in four orders. Perhaps rather surprisingly, only half of the ten most generalist orchids provide food rewards in the form of nectar; the other half are food deceptive.

*Neottia ovata*, otherwise known as the common twayblade, is king among the orchid supergeneralists with 162 recorded pollinators. It is an inconspicuous forest plant that is common throughout western Europe, and reaches as far as Siberia in Asia. The green flowers offer a sweet scent and nectar is produced on the wide base of the lip, and in lesser quantity in the central, longitudinal groove. The availability of nectar entices a range of pollinators. Insects pollinating its flowers follow the sweet trail of nectar from the central groove towards the base of the lip. As the animal reaches for nectar it touches and triggers the rostellum, removing the pollinarium (the very particular mechanism was discussed in Chapter 5, section 5.7). *Neottia ovata* is most regularly visited by ichneumonid wasps, sawflies and beetles, but pollinators may also include crickets and spiders. Interestingly, the species may pay a price for its highly successful reproductive strategy. Despite outcrossing very effectively, it has been determined that populations of this species have very low genetic variability. The reason appears to be a phenomenon known as biparental inbreeding, where high pollination efficiency and close proximity results in mating between the same individuals every season, and therefore in the continuous production of closely related descendants. This seems to be causing exceptionally low or absent germination of plants from seeds!

Another well-known supergeneralist is *Cypripedium calceolus*. The lady's slipper — as it is commonly known — is famous as the most beautiful European orchid and, consequently, is highly sought after by collectors. Unlike the common twayblade, *C. calceolus* is a deceptive orchid. Like other slipper orchids (see Chapter 2, section 2.10), the pouch-like lip traps small flying insects and forces them to crawl out through a pair of basal orifices — effectively transferring pollinia in the process. But unlike the fly-specialised slipper orchid species discussed before, *C. calceolus* is predominantly pollinated by Hymenoptera, mostly bees of the genera *Andrena*, *Halictus*, and *Lasioglossum*, and in lesser proportions also by Diptera. Studies on the pollination of this orchid across Europe demonstrate that it has low pollination specificity. And yet, despite being a supergeneralist, several of its populations — in Britain, Denmark and Poland, for example — show pollinator limitation. In fact, it has been proposed that the reasons for the high diversity of pollinators in *C. calceolus* may actually be the unavailability of pollinators in time and/or space in some areas. In other words, *C. calceolus* may be adapted to more diverse landscapes and a broader temporal activity than any specific insect that pollinates it.

Specialisation in orchids can be defined in different ways. Just above half of all orchid species seem to be serviced by a single pollinator species. We may also think of specialisation in the sense of phylogenetic groupings. For example, 85% of all orchid species are being serviced by a single pollinator family, while 93.4% are pollinated by a single order of animals. Therefore, no matter the definition of specialisation, orchids indeed tend to be highly specialised in terms of their pollinators, with most relying on a single functional group and having several pollinator species that are similar in size, shape and behaviour. However, there are supergeneralists among the orchids, too. They teach us that specialisation may not be a sign of weakness or disadvantage — as one might expect. Notwithstanding their generalist strategy, populations of *Neottia ovata* show very low genetic diversity, while those of *Cypripedium calceolus* suffer from low numbers of individuals and pollinator deficiency resulting in low fruit set and slow — if any — growth of the population. Perhaps one can imagine why Britain has placed its last known wild individual of *C. calceolus* under protective guard. For a few weeks each spring, a lone officer zealously safeguards the orchid round-the-clock, fending off thieves in the hopes that Europe's most beautiful orchid may one day thrive and recolonise the country.

## 6.7 DIAMONDS AND RUST

It is commonly believed that, given their incredible species diversity and absurd complexity of pollination strategies, orchids are among the most evolved living beings on Earth. But their actual age remained a mystery until quite recently. The reason for this uncertainty is that — despite their universal presence today — orchids are virtually absent from the fossil record. This is of course not evidence for divine creation, nor spontaneous generation. Most authors agree that the scarcity of orchid fossils is probably the result of certain intrinsic characteristics of members belonging to the family. Essentially: 1) that they predominantly occur, both in the present and presumably in the distant past, in the wet tropics, where decay is quick; 2) the dominance of soft herbaceous plant tissues make orchids prone to decomposition; 3) the preference for epiphytic habits, which keeps them farther from the aquatic conditions that typically lead to fossilisation; 4) the reduction of loose pollen grains to pollinia, and; 5) the small seeds that are difficult to detect and easily degradable.

Nevertheless, a few orchid fossils have been claimed through the years. Unfortunately, none of them was considered very reliable. The first orchid fossils ever described were *Palaeorchis* and *Protorchis*, proposed by Abramo Bartolomeo Massalongo on the basis of specimens discovered in the calcareous Eocene deposits at Monte Bolca in northeastern Italy in 1857. Even though Massalongo placed the fossils initially in the Orchidaceae, he had difficulty assigning them with confidence. These uncertainties led him to place them in a new taxon, the Protorchidaceae. According to Massalongo, the Protorchidaceae were close relatives of orchids and aroids. Later authors have strongly debated whether these fossils are in fact part of the orchid family, with a more or less accepted conclusion that nothing about them definitively proves they are orchids. Their features may represent any monocot plant with a tuber and juvenile leaves.

The Massalongo relics were followed by *Antholithes*, an 'orchid fossil' claim based on a specimen from the Lower Oligocene found in Florissant, Colorado, and published by Cockerell in 1915. However, it suffered a similar fate to its predecessors. Initially attributed to the Orchidaceae on account of the flower's superficial resemblance to the lip of *Cypripedium*, it was later regarded as *incertae sedis* — meaning of uncertain placement — by most authors due to the lack of detail. The next 'orchid fossils' that came to light were the three species placed in the new genus *Orchidacites*. They were described by Strauss from the Upper Pliocene of Willershausen, Germany. The first two, *Orchidacites orchidioides* and *O. wegelei*, were published in 1954, and the third, *O. cypripedioides* in 1969. All three are made up of fossilised ellipsoidal fruit capsules, bearing longitudinal ribs and remnants of the corolla — which Strauss compared with those of extant orchids. The fossil fruits have received relatively little attention — in comparison to the other orchid fossils — but in general authors are skeptical given the number of ribs and the fused elongate corolla, lacking features that are diagnostic for the family Orchidaceae and which could definitively exclude other plant families. I am personally disinclined to believe that the first two species of *Orchidacites* are truly orchids, but I agree with Garay and Dressler that the third, *O. cypripedioides*, shows orchidaceous features and may deserve further consideration.

In 1977, Schmid and Schmid reviewed the fossil record of Orchidaceae and concluded that none of the four extinct orchid genera described on the basis of fossils — *Antholithes*, *Palaeorchis*, *Protorchis* and *Orchidacites* — were indisputably orchids. Objections had been raised against their orchidaceous nature and none could be assigned to the family with certainty. They concluded that similarities between the fossils and orchids were merely superficial and ruled that the Orchidaceae unfortunately lacked any positive or useful fossil record. The lack of a fossil record for orchids made understanding their

**Fig. 6.7.1.** Amber fossil of a female worker bee carrying the pollinarium of *Meliorchis caribea* on its back.
© Santiago Ramírez.

evolution a real challenge. This changed dramatically when, in 2007, a team led by Santiago Ramirez published a photograph of a bee fossilised in amber. The perfectly preserved specimen was a female worker bee carrying an orchid pollinarium on its back. The piece of amber, which was less than 1 cm in diameter, had been excavated in 2000 from lignite and sandy clay beds near Santiago in the Dominican Republic. The researchers concluded that the pollinarium belonged to an extinct, previously unknown orchid, which they called *Meliorchis caribea* (see Figure 6.7.1). With the discovery of *M. caribea*, scientists were finally able to infer the age of the whole orchid family, and for the first time it became clear that orchids were not as old as previously believed.

But how can a single fossilised orchid pollinarium tell us the age of orchids? To answer that question, we first need to talk about orchid pollinaria. Most plants produce dust-like pollen, made up of minute particles that are rather indistinctive and thus uninformative. As we have learned, in many orchid flowers pollen grains have been compacted into tetrads (discrete units composed of four pollen grains) and have a well-defined shape, size and composition, typically bearing specific accessory structures. They can be extremely complex and this complexity can be informative. Even though the grouping of four pollen grains into tetrads and the consolidation of those tetrads into pollinia is not exclusive to the orchid family — it is also found in the Asclepiadaceae (now part of the Apocynaceae) family — the cohesion of pollinia and their accessory structure into a single unit called a pollinarium is a feature that most orchids have in common. Darwin considered orchid pollinaria to be "as perfect as any of the most beautiful adaptations in the animal kingdom." They can be so characteristic that the pollinarium alone is enough to determine the orchid's identity. This highly specialised pollinarium morphology in orchids allowed researchers to determine what the closest living relatives of the fossilised *Meliorchis caribea* are. The fact that the fossilised pollen is compacted into discrete units that form a pollinarium excludes the possibility that it belongs to subfamilies Apostasioideae, Cypripedioideae or Vanilloideae. In fact, because the pollinarium of *M. caribea* is composed of two pollinia, and each one is made up of loose blocks of tightly-packed tetrads, we can confidently place it among the members of subtribe Goodyerinae — in the Orchidoideae subfamily. Pollinia that are made up of loosely packed blocks of pollen tetrads are known are 'sectile pollinia'. We can find sectile pollinia in many other species in the Orchidoideae subfamily alive today (see Figure 6.7.2).

Besides being able to identify orchids on the basis of their pollinarium morphology, there is another element that is highly informative when it comes to reconstructing the orchid's evolution. Fossils, like other rocks, can be dated. To establish the age of a fossil, radiometric dating methods are commonly used. These are based on the known rates of decay of certain radioactive isotopes. I will not go into the technical details here, but it suffices to say that some isotopes are unstable and change over time. This change is predictable, both in what the isotopes will decay into and how much time it takes for that to happen. Using such techniques, researchers can pinpoint the age of a fossil or rock directly. But because rocks appear in discrete layers, dating one rock allows us to establish the age of the whole layer. Once the age of a layer has been determined, we can infer how old any fossils appearing in that layer are. Establishing the age of a fossil is very important because it allows us to ascribe a minimum age to a biological group; tells us how old certain morphological features are; tells us when two or more organisms started to interact; and provides information about what their ancient geographical distribution was. The fossil of *M. caribea* was found in a layer of soil that has been established to be from the Miocene period, between 15 and 20 million years old. By combining these two key pieces

Fig. 6.7.2. Orchid pollinaria have a very particular morphology and composition, which allows researchers to identify them. With their sectile pollinia (**A, D**) and defined viscidium (**C, F**), *Habenaria monorrhiza* (**A–C**) and *Ludisia discolor* (**D–F**) are typical representatives of subtribes Orchidinae and Goodyerinae in the Orchidoideae subfamily.
© APK.

of information — which boils down to subtribe Goodyerinae being at least between 15 and 20 million years old — researchers were able to establish that the orchid family originated in the Late Cretaceous, between 76 and 84 million years ago.

The discovery of a fossilised orchid pollinarium on an insect in amber caused a sensation. The more dated fossils with identifiable orchid pollinaria we have, the more we can refine the evolutionary history of the family. When scientists started double-checking museum collections, more orchid fossils came to light. Five fossilised insects bearing orchid pollinaria were presented in publications by George Poinar between 2016 and 2017. Discovered in the Dominican Republic, the author proposed the fossil genera and species *Globosites apicola* and *Rudiculites dominicana* — placed tentatively in the Oncidiinae and Spiranthinae subtribes, respectively. From Mexican amber, the author described *Annulites mexicana* and *Cylindrocites browni* — assigned to the Cranichidinae and Spiranthinae subtribes, respectively, both in the Cranichideae tribe. Finally, *Succinanthera baltica* — assigned to the Epidendroideae subfamily — was described from the Samland Peninsula of the Baltic Sea, in Russia.

To scientists, fossils may be worth more than diamonds. Given their scarcity and the unique information they provide, well-preserved orchid fossils are invaluable. But like any other precious gem, fossils are illegally trafficked, falsified and, therefore, very zealously sought after and jealously guarded. In the 1990s, entomologists with large collections of insects in amber began attempting to extract DNA from insect fossils. This was very exciting! DNA from fossils could tell scientists a whole lot more about the evolution and extinction of organisms over time. A publication in the distinguished *Science* journal in 1992 by David Grimaldi and colleagues indicated that snippets of genetic material had been found in a termite in Dominican amber. The next year, a study led by Raúl Cano and George Poinar in the prestigious journal *Nature* published the results of DNA sequenced from a 120–135-million-year-old weevil trapped in Lebanese amber. The controversial paper was published on the very same day that Steven Spielberg's movie *Jurassic Park* came out. Grimaldi stated: "[It is] potentially the only time the scientific press has timed a paper's release to a pop-culture event." *Jurassic Park* — based on Michael Crichton's novel about scientists who recreate dinosaurs from DNA in blood sucked by a mosquito that was trapped in Dominican amber — was a gigantic success. The movie captivated the general public, interest in amber skyrocketed and prices soared.

"We speculate that on the basis of our investigation the majority of animal remains in amber have preserved DNA that can be extracted and studied, thus making amber a treasure chest for molecular paleontologists," stated Cano and his colleagues at the end of their 1993 paper. Unfortunately, these words did not age well. Time and time again, researchers failed to replicate the early studies that found DNA in amber fossils, and eventually the initial excitement began to fade. Some researchers have even suggested that the DNA extracted from the weevil in Lebanese amber was actually fungal contamination from the lab. The controversy over ancient DNA is still heatedly debated today. DNA in fossils is significantly degraded, and even if present it would be highly fragmented and in low quantities.

Therefore, a major issue with extracting ancient DNA from resin specimens is contamination with DNA in the environment. All around us — and in the laboratories where these ancient samples are analysed — are traces of DNA from modern organisms. Most studies have not been able to retrieve any DNA from amber-preserved specimens while those studies that claim to have succeeded have been subsequently deemed unreliable and/or disproven.

That being said, can DNA actually be extracted from amber specimens? Recent studies find that it is possible to study the genomics of organisms that have been artificially embedded in resin. But this has only been successful in very young specimens and under very strict laboratory procedures. DNA degradation is influenced by factors such as oxygen and water content, ambient temperature and time since the death of the organism. DNA fragments seem to be present in well-preserved geological material up to at least 100,000 years of age, and perhaps materials up to one million years old may yield DNA. But the pressure and heat generated through tectonic processes over millions of years reduces the likelihood of DNA preservation in amber specimens almost to zero. Unfortunately — contrary to what was promised by *Jurassic Park* — we are still far from recovering non-degraded dinosaur DNA from amber fossils, let alone cloning them!

Not being able to access genetic information from fossilised organisms means that their identification necessarily has to rely on morphological interpretation. This reinforces the need for careful documentation and detailed characterisation of orchid pollinaria so that putative fossils can be correctly and more accurately identified. Unfortunately, given the size of most orchid pollinaria, photographs and illustrations are often very schematic, and lack the necessary detail. However, the powerful micro-scanning and micro-photographic equipment that is available today is an extremely important improvement in the documentation of orchid pollinaria. At Lankester Botanical Garden we have created our own database of highly detailed photographs of orchid pollinaria. This database has become an important tool in distinguishing closely related taxa and identifying the pollinaria carried by insects. For example, it allowed genus-level identification of the pollinaria found on the Euglossini bees featured in the 'You're so vain' story in Chapter 3 (Fig. 3.7.2). Unfortunately, the database also gives us good reason to suspect that, contrary to Ramirez's

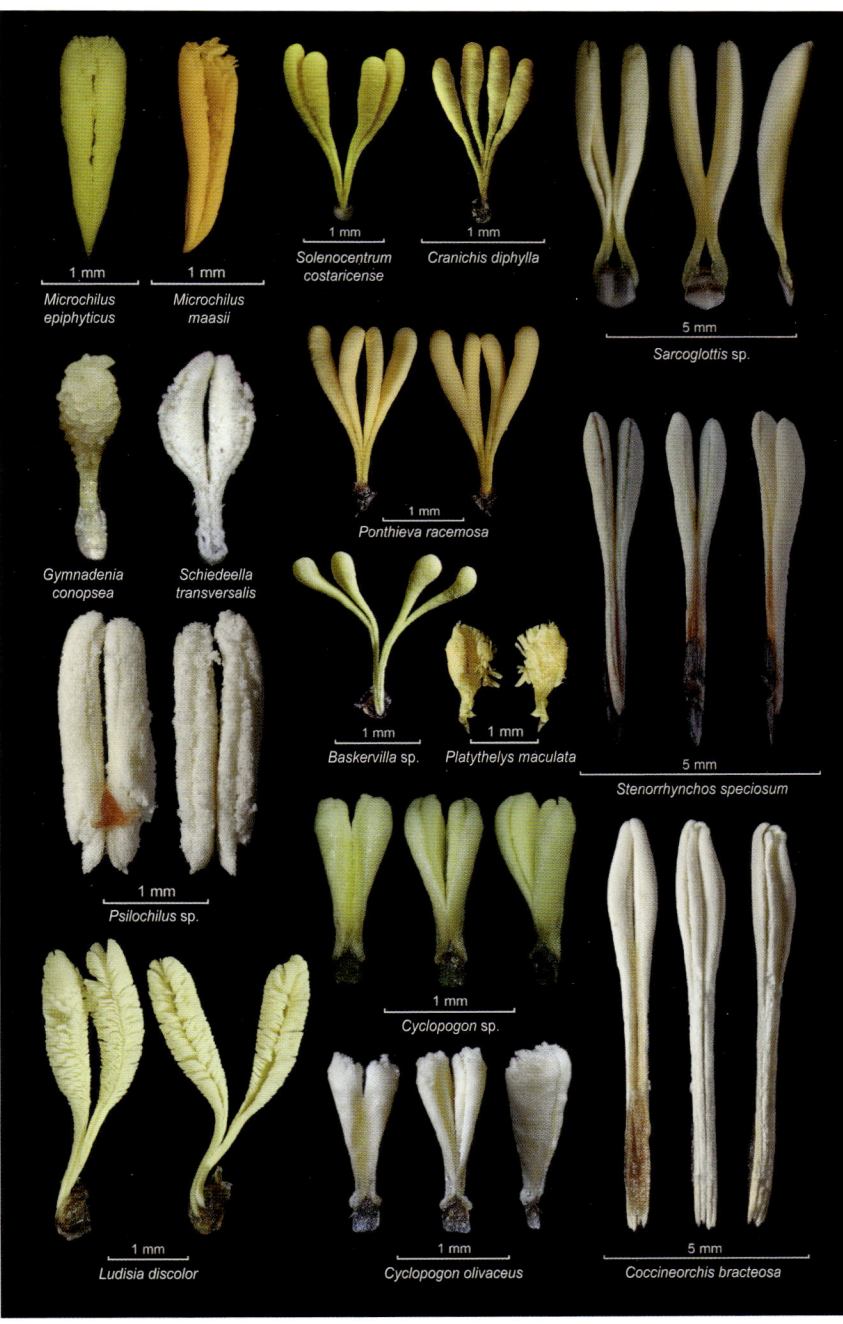

**Fig. 6.7.3.** High-resolution macrophotographs of pollinaria belonging to different members of the Cranichidinae, Goodyerinae, Orchidinae, and Spiranthinae subtribes in the Orchidoideae subfamily.
© Lankester Botanical Garden.

*Meliorchis caribea*, at least three of the Poinar fossils — *Annulites mexicana*, *Cylindrocites browni* and *Rudiculites dominicana*, are unlikely to be correctly identified* and are most probably not orchid pollinaria. But don't take my word for it. *Annulites*, *Cylindrocites*, *Meliorchis* and *Rudiculites* were all placed in subfamily Orchidoideae. You can carefully inspect the high-resolution photographs of pollinaria of different members of that subfamily shown here and compare them with the published photographs of the fossils (see Figure 6.7.3). Do any of them go beyond a superficial similarity? The best candidate is indeed *Meliorchis caribea*, but it is not easy to identify fossil records confidently. Even if the identities of the two remaining fossils — *Globosites apicola* and *Succinanthera baltica* — are confirmed, the fossil record of the Orchidaceae family remains extremely poor. When it comes to orchid evolution, we are not looking for the 'missing links' — virtually all the links are missing.

Understanding orchid pollination and documenting orchid pollinaria allows us to look for clues in amber fossils about the evolutionary history of orchids. Amber extraction is profitable, but amber mining is a dangerous business that leaves a trail of death. Many amber mines are found in conflict areas — conflicts that the mines themselves help to finance, fuelling bloodshed and violence. Poorly paid labourers typically work under substandard conditions in the mines, which results in many accidents and deaths. Consequently, the human cost of amber — like that of other precious gems — is high. Therefore, anyone working with amber fossils — especially in scientific institutions — should do this in a sustainable and socially conscientious manner.

More pollinaria-like objects may appear in amber fossils in the future, and they should certainly be brought forward — even if there is doubt regarding their identity. Given the overall lack of known orchid fossils, any specimen that could contribute to the record is highly desired. They may even turn out to be authentic. To quote Austrian writer Marie von Ebner-Eschenbach: "Even a stopped clock is right twice a day."

---

\* *Annulites mexicana* was compared with the pollinaria of genus *Solenocentrum* (tribe Cranichideae: subtribe Cranichidinae); *Cylindrocites browni* was compared with the pollinaria of the genera *Dichromanthus*, *Cyclopogon* and *Schiedeella* (tribe Cranichideae: subtribe Spiranthinae); *Rudiculites dominicana* was compared with a member of subtribe Spiranthinae without specifying a genus.

## 6.8 YOU ONLY LIVE ONCE

*"The world is so huge that people are always getting lost in it. There are too many ideas and things and people, too many directions to go. I was starting to believe that the reason it matters to care passionately about something is that it whittles the world down to a more manageable size. It makes the world seem not huge and empty but full of possibility."*
SUSAN ORLEAN in *The Orchid Thief* (1998)

As I lay there staring at the flowers of an unvisited bucket orchid, my mind started to unravel. This is pointless, I thought. What am I even doing here? I had driven eight hours to reach this particular spot the day before, successfully beating the grand opening of these exotic flowers, only to find out the next morning that the bee frenzy I was expecting did not take place. Instead, the opened flowers were mostly being ignored. I had been sleeping in a rundown cabin, accompanied only by bedbugs and the sounds of the forest — soothing at times, but mostly nerve-wrecking. At home, my pregnant wife patiently waited for me, probably wondering — as much as I was at this point — why I had suddenly decided to drop everything and head out to the middle of nowhere to film these orchids. This was a particularly large specimen of bucket orchid that produces several inflorescences at once, and the ant nest on which it thrived sat on a relatively low branch. It seemed an ideal candidate to get the work done without disturbing the plant. But I was isolated, with no phone or internet connection. The cameras were ready, fully charged, sitting on their tripods, but they aimed at a seemingly sterile flower. Was it all really worth it?

I have loved nature — and especially orchids — since I was a kid. But I come from a mixed agricultural and botanical background, and only became interested in ecology much later in my career. Botany typically entails going out to the field to collect samples or document plants, but there is not much sitting-still-and-waiting-for-something-to-happen involved. Somehow plants seem much more reliable and predictable than animals, and so when it comes to observing and documenting animal behaviour around orchids — specifically flowers — one may feel one's spirit crushed over and over again by feelings of disappointment and failure.

Perhaps the most celebrated study showcasing the challenge — and mystique — behind the hunt for an endangered orchid and the discovery and documentation of its pollinators is that of the ghost orchid. *Dendrophylax lindenii* is a rare, but very famous orchid (see Figure 6.8.1). It featured in Susan Orlean's bestseller *The Orchid Thief* and also in Spike Jonze's Oscar winner *Adaptation* — a movie based on Orlean's book, starring Nicolas Cage and Meryl Streep. This famed orchid has been the subject of studies involving fungal and host tree associations, seed germination and development, propagation and culture, fragrance composition, phylogenetic relationships, and conservation assessments and protection programmes. And yet, virtually nothing is known about the ecological interactions between the ghost orchid and its pollinating insects.

The ghost orchid is native to southern Florida and Cuba, where it is a rare and endangered epiphyte growing on woody trees that overhang standing water in remote swamps. The leafless plant forms a clump of roots that easily blends into the background, becoming virtually invisible when not in bloom. Its common name derives from the ghostly impression of the large white flowers hovering over the trunks and branches of trees in the dimly lit environments where it grows. Given its rarity and mysteriousness — Zettler called it "the most familiar and sought-after of all orchids in the Western Hemisphere" — the ghost orchid is commonly targeted by illegal poachers. It is a challenge to grow the ghost orchid under greenhouse conditions, and in the wild it is threatened by pests, periodic hurricanes, deforestation, and rising sea levels as a result of climate change. It's a perfect candidate for a good story combining mystery and drama. Given its elusiveness, studying the ecological interactions of this orchid has presented a true challenge. Hawkmoths have long been suspected to pollinate *Dendrophylax lindenii*, but definite proof was unavailable until very recently. In the past, based on the length of its proboscis, the ghost orchid had been assumed to be pollinated by one particular hawkmoth, the giant sphinx moth — *Cocytius antaeus*. It was speculated that the giant sphinx was the only species of moth with a tongue long enough to fit the long nectary spur of *D. lindenii*, and therefore it 'has to be' responsible for the orchid's pollination in Florida. However, south Florida and Cuba are particularly rich in hawkmoth species and several of them have a proboscis length that would also be sufficient to reach the nectar of

**Fig. 6.8.1.** The beautifully enigmatic ghost orchid, *Dendrophylax lindenii*. © Ron Parsons.

*D. lindenii* — and thus be effective as pollinators. So does the giant sphinx moth pollinate the ghost orchid?

To solve this longstanding enigma, two teams of nature photographers and researchers set out into the swamps in southern Florida. Now remember these plants are rather infrequent, and the inundated, dimly lit sites where they thrive are infested with mosquitoes, snakes and alligators. The first challenge is, of course, to locate an adequate site with mature plants that are about to bloom and will not be disturbed. To actually document the floral visitors, very fast and sophisticated camera trap systems were installed in front of the flowers. As the plant typically grows a few metres above the waterline, the camera trap systems had to be elevated into the trees — either with ropes or building platforms. Such a system includes cameras, flashes and triggers. Camera traps typically work by triggering when they detect a body passing in front of the sensors. They can detect subtle changes in both movement and heat. Using camera traps that trigger with movement seems to be a straightforward solution for studying the pollination mechanism of rare orchids — growing in inaccessible places — that are very occasionally visited by pollinators. Unfortunately, anyone who has tried using such cameras knows that the smaller your animal, the more sensitive you need to make the trigger, and the more sensitive your trigger, the more easily it will be triggered by anything moving in front of the lens — especially by the wind moving objects within the frame. This not only significantly increases the number of false positives — photos or video of nothing — that fill up your memory card, but it also seriously limits the number of hours the batteries will last. Powering your equipment — camera, sensors and flashes — is a real challenge out in the field, especially when you are deep in the wilderness.

The Florida Panther National Wildlife Refuge is a palustrine wetland, with a mix of cypress and hardwood-dominated swamps. The Fakahatchee Strand within the refuge is believed to support the highest diversity of native orchids and bromeliad species in North America, including more than 500 recorded individuals of *Dendrophylax lindenii*. At the strand, photographer Carlton Ward Jr. set up three digital camera traps during a three-year period between 2016 and 2018. Each camera was focused on a different specimen of flowering ghost orchid for several days at a time between June and October each year. Despite having three cameras running for several months amounting to almost 5,000 hours, only 21 instances of butterflies and moths visiting the flowers of *D. lindenii* were recorded. Similarly, conservation biologist Peter Houlihan and photographer Mac Stone also set out into the Corkscrew Swamp Sanctuary to catch pollinators of the ghost orchid in action. To increase their chances of documenting the pollinator, they concentrated their efforts on a 'super ghost' — multiple ghost orchid root masses clustered on top of one another. The super ghost, located 15 metres above the flooded forest, was accessed by tree climbing using ropes. Having a group of flowers has the advantage of allowing the monitoring of several flowers simultaneously. During their monitoring of the ghost orchid, the Houlihan and Stone team kept the camera active for 75 days, triggering almost 8,000 images, of which 23 captured visitations by hawkmoths. That means they had a successful trigger once every three days and just 0.3% of the triggered camera shots were usable!

Despite the challenges, both teams solved the ghost orchid mystery. *Dendrophylax lindenii* is indeed pollinated by hawkmoths. But contrary to popular belief, it is pollinated by more than a single species of large hawkmoth, and the giant sphinx moth is not among them. Instead, they discovered that it is the fig sphinx moth — *Pachylia ficus* (see Figure 6.8.2) — and the pawpaw sphinx moth — *Dolba hyloeus* — that transfer the ghost's pollinarium while reaching into the nectary tube with their extended proboscs. Interestingly, the flowers' smell attracts

**Fig. 6.8.2.** *Dendrophylax lindenii* pollinated by the fig sphinx moth *Pachylia ficus*. © Carlton Ward Jr courtesy of WILDPATH.

**Fig. 6.8.3.** *Dendrophylax lindenii* approached by the streaked sphinx moth *Protambulyx strigilis*.
© Carlton Ward Jr courtesy of WILDPATH.

several other large hawkmoths, including the giant sphinx moth, banded sphinx moth, and streaked sphinx moth (see Figure 6.8.3), and several much smaller nocturnal Lepidoptera. But all of them are either too large or too small to remove the pollinarium of the flower while reaching for the nectar. So it seems that the ghost orchid is adapted to pollination by a guild of moderately sized moths. Unlike the extreme case of Darwin's orchid in Madagascar, the nectar spur of *D. lindenii* falls within a spectrum of proboscis lengths of the moths native to Florida and Cuba.

Such studies not only answer longstanding questions and rectify enduring misconceptions, but they provide new insights into the evolution and conservation of ecological relationships. It is only thanks to these extraordinary efforts, in combination with the use of sophisticated equipment, that we can finally fill in the blanks in the ghost orchid thriller. But there are 14 other species of *Dendrophylax* — all rare epiphytes restricted to the Caribbean — and absolutely nothing is known about their ecology. This is merely one example, but it shows the crude reality of orchid pollination studies. Despite the big leaps in recent years, we are only just uncovering the tip of the iceberg.

There are good reasons why the pollination strategies and pollinators of most orchid species are still unknown or poorly understood — even in temperate regions with significantly more researchers, less biodiversity, and certainly less gruesome conditions than snake-infested swamps. Studying orchid ecology can be brutal. You may fully invest yourself into a project, be relentless, endure unsanitary and even risky working conditions, spend enormous amounts of time and resources, and at the end of the day still have nothing to show for it. Stories of success are often an obsessive combination of patience, hardship and incredible sacrifice. "I don't actually know the ingredients of obsession. Maybe it's a chance of trying something that seems impossible," admitted one of the authors of the ghost orchid pollination paper.

The study of orchid pollination is not easy, and typically becomes borderline obsessive. "I was a fool ever to touch orchids," Darwin wrote in a letter to Hooker. It took Maurice-Alexandre Pouyanne 20 years to finish his observations on the pseudocopulation of *Ophrys* in Algeria. Whereas in Denmark, entomologist Torben Wolff observed the flowers of this genus being pollinated just twice in an eight-year period. When asked if he had ever photographed orchids being pollinated in his many travels, Andre Schuiteman responded: "I have never seen any of the tropical species being pollinated in the wild, let alone managed to photograph it. There is simply not the time to do so." And he couldn't have been more accurate. Studying and documenting orchid pollination often requires a lot of patience and endurance. Even if you are able to obtain data, they may not be enough for a scientific journal to consider publishing it.

As academics we continue to use and celebrate the works of the early naturalists, but at the same time we underestimate and dismiss the scientific merit of such studies, even when rigorous. Darwin admitted that his readers needed "a strong taste for Natural History" to appreciate his books, and yet, journals, editors and senior researchers have made it clear that natural history is no longer acceptable within scientific literature. This policy may boost the journal and researcher's statistics, but it has unforeseen negative consequences to modern biology. A direct downside of the dismissal of careful observations on the natural world is that a young generation of botany and ecology students — who are required to publish quickly and plentifully in so-called first-tier journals — become indifferent and detached from the very biological object they study. They prefer the safety of laboratory tests and computational analyses — which secures high-impact publications — over the challenges and uncertainties of field observations — which become a nightmare to publish anywhere. Another consequence is that a significant amount of ecological data remain unpublished — and thus unknown — simply because it is not groundbreaking or apparently significant enough. We fail to realise that without the meticulous observations of naturalists such as Darwin and Wallace, Pouyanne and Coleman, Dodson and Dressler, we wouldn't have as many hypotheses to test today.

Academia tells students that modern biology is about answering 'the big questions', rather than producing relatively small observations. So we lament the fact that the vast majority of tropical orchids lack information on their reproductive strategies, that pollination information is only available for 10% of orchid species worldwide, and that much of the available basic ecological knowledge is not reliable. But at the same time we don't provide any stimulation or platform for people to collect and publish such data. We are forgetting that it is the accumulation of a multitude of observations on minor experiments — such as those featuring in his orchid book — that drove Darwin to formulate his big hypothesis on evolution through natural selection. To answer the so-called big questions we need high-quality, basic data, and lots of it. So I invite everyone with an interest in ecology to record and document species' interactions (even with a camera phone), reach out to scientists, or publish your findings in journals or magazines, social media or citizen science webpages. Be curious, experiment rigorously, but most importantly get the information out there!

It may never become easy or comfortable to do fieldwork. Try sitting in front of a flowering plant in your garden and observe pollination in action. Our increased knowledge and development of new technologies certainly facilitate the documentation process. We have come a long way since naturalists such as Sprengel, Darwin and Wallace patiently and untiringly observed and described nature in action. Sturdy setups with powerful cameras, highly sensitive triggers, long-lasting batteries, processing software and analytical capabilities are becoming more common in ecological studies. Experimentation, replication and analysis accompany and complement natural history in modern biology. But ecological problems cannot be solved without leaving the office. We need to engage our study subject and observe it first hand in its natural environment. Then, and only then — through continued and painstaking perseverance — will you, as researchers, as students, as naturalists, be able to formulate the big questions and take the first solid steps towards answering them. The hardships are not pointless. There are many questions that remain to be answered — big and small — and perhaps more interestingly, there are brand new questions to be asked. Life is short. We need to make the most of it.

> *"Life seemed to be filled with things that were just like the ghost orchid — wonderful to imagine and easy to fall in love with but a little fantastic, fleeting and out of reach."*
> SUSAN ORLEAN in *Adaptation* (2002)

*Dendrophylax lindenii* approached by Brazilian skipper *Calpodes ethlius*.
© Carlton Ward Jr courtesy of WILDPATH.

Confined to the mountains of Europe, *Gymnadenia (Nigritella) nigra* populations show severe decline. © Jean Claessens.

CHAPTER 7

# CHANGE

The Orchidaceae are considered to have the highest speciation rate, the highest rate of extinctions and the rarest species of any plant family in the world. Orchids face unprecedented levels of threat from climate change, habitat destruction and fragmentation, and over-collection. Conservation efforts are failing to avert extinctions. Unsurprisingly, orchids prominently feature on lists of threatened plant species. The number of endangered orchid species is likely to continue to grow, given that the state of the flora in biodiversity hotspots remains poorly known. These plants' unusual biology, patchy distribution and rarity also present unique conservation challenges, particularly given the specialised ecological preferences and complex interactions in the Orchidaceae.

A delicate balance exists between partner organisms that are involved in mutualistic relationships. From host trees to seed dispersers, from pollinators to patrolling protectors, in the preceding chapters we have reviewed a number of ecological interactions with other organisms on which orchids depend. Given the highly specific nature of these intricate relationships, biologists and ecologists have been concerned about the possible disruptive effects that habitat disturbance and climate change may have on these dynamics. The continued transformation of our planet has long been suspected to upset the balance between partner organisms and threatens to provoke the uncoupling of important species interactions. Unfortunately, we no longer need to theorise and speculate about it. Change is here and it's unfolding right before our eyes.

## 7.1 UNDER PRESSURE

*"We will respond to the threat of climate change, knowing that the failure to do so would betray our children and future generations."*
PRESIDENT BARACK OBAMA in his inaugural address (2013)

In 2020, Daniel Janzen and Winnie Hallwachs published a paper with a frightening conclusion. After five decades of studying insect abundance and diversity in the Guanacaste Conservation Area (*Area de Conservación Guanacaste* or ACG) in northwestern Costa Rica, they found a steep and continued decline in biomass and species richness. This is not very surprising given the all too familiar tendency of increased human onslaught on the surrounding nature, such as forest clearing and fragmentation, logging, slash-and-burn agriculture, indiscriminate hunting, and excessive use of pesticides, in addition to other impacts such as droughts, flooding and hurricanes. So what makes their account particularly alarming? In the 50 years since the authors began monitoring insect diversity, the extent of protected land belonging to the ACG has been steadily growing and forested areas have been actively recovered and recuperated. In the mid-1980s this national park was just 18% of its current area, so we would be perfectly justified in expecting a parallel increase in insect diversity and numbers. But the opposite has turned out to be true: while efforts carried out at ACG have managed to successfully save — and in fact increase — Costa Rica's tropical dry forest, its insect population and their ecological interactions are continuously and rapidly declining. The authors unequivocally attribute this to "the climate change excesses, speed, and erratic annual timing of temperature, wind, rain, cloud cover, and a plethora of combinations of these changes."

Climate change affects orchids and their pollinators much more than we might suspect. In Australia, one of the largest pollination studies ever undertaken showed that differences in the pollination success of orchid species was directly linked to factors such as density, flowering times and pollination syndromes. Pollination rates in these Australian Orchidaceae were severely impacted by drought and extreme temperatures at certain times of the year — occurring more frequently than in the past. Similarly, in the Tropics it has been established experimentally that cloud forest epiphytes transplanted to lower elevations — where there is less humidity — suffer higher mortality, lower development and reduced longevity. Epiphytic orchids, such as *Lepanthes* species, are expected to experience severe stress as a result of current warming and drying trends, especially given that they often live in habitats that are highly exposed to sunlight and wind, and where they benefit from substantial levels of moss cover to prevent desiccation. Increased stress leads to limited reproduction and lower survival rates. Studies on the reproductive success and spatial distribution of *L. rupestris* — a rare species endemic to Puerto Rico — has shown that there is a relationship between temperature and the colonisation-extinction dynamics. Temperature influences the orchid's establishment and reproduction. Fruit production increases with optimal lower temperatures as these lead to higher inflorescence and flower production, and as a consequence to the attraction of more pollinators. Even though little is known about the pollinators of these orchids, it is suspected that the tiny flies that pollinate *L. rupestris* are highly sensitive to fluctuations in temperature. Natural pollination of this orchid is already quite rare and fruiting is rather infrequent. Therefore, rising temperatures are likely to negatively affect pollination dynamics, given the specificity of plant–pollinator relationships in sexually deceptive orchids such as *Lepanthes*.

One way to determine the negative effects of climate change on orchids is to measure its potential to disrupt the synchrony of highly specific interactions between plants and insects. Specialist plants with few pollinator species — which include most orchids — are likely to be highly sensitive to this decoupling. Potentially, upsetting the delicate balances between partner organisms can destabilise entire ecosystems and can be a serious threat to human food security and ecosystem services in general. However, studies that show the effects of climate change on natural pollination continue to be rare, given how costly and time-consuming monitoring efforts can be. Accumulating hard evidence of the effects of climate change should concern not only biologists and ecologists but also the conservation community. Assessing the effects that global warming can have on complex mutualistic relationships continues to be a challenge.

In December 2014, a group of British scientists published their findings on the effects of higher temperatures on the pollination of an orchid species. Their hypothesis was that warmer springs would advance the flowering time of plants and the flight time of insects. By using herbarium and field records collected by scientists and dedicated

**Fig. 7.1.1.** The delicate synchrony between the emergence of *Andrena nigroaenea* males and the flowers of *Ophrys sphegodes* is starting to decouple as a consequence of a warmer climate, threatening to leave this specialised orchid without pollinators.
© Matteo Perilli.

laymen over 159 years, they were able to show that the peak flowering time of *Ophrys sphegodes*, commonly known as early spider orchid, has been advanced by the warmer springs. The orchid is pollinated through pseudocopulation by males of the solitary mining bee *Andrena nigroaenea*, one of the first bees to emerge in spring (see Figure 7.1.1). By examining museum specimens and field records of the flight date of the solitary mining bee, the authors were able to establish that this has also advanced with increasing temperatures. If both organisms show up earlier, what is the issue? Pollination by insects, an ecosystem service of immense economic and conservation importance, depends on synchrony between insect activity and flowering time. If plants and their pollinators show different responses to climate warming, pollination could fail. This is exactly what was discovered: the orchid and the bee respond differently to spring warming. Based on museum specimens and field records dating back to 1893, the researchers realised that the flight dates of *A. nigroaenea* is advanced more by warmer temperatures than is the flowering of *O. sphegodes*. On average, the flight date of male bees was significantly earlier than the peak flowering date of the orchid. An increase in temperature of as little as 2°C, results in both male and female *A. nigroaenea* emerging, flying and successfully copulating before the sexually deceptive flowers of *O. sphegodes* appear, effectively disrupting the orchid's pollination strategy.

One straightforward and widely used way to assess the consequence of continued warming is to estimate the extent of potentially suitable habitats based on current and future climatic characteristics. To model the ecological niches, studies use variables such as day-length, maximum, minimum and mean temperatures, and annual and seasonal rainfall, as well as the known distribution of a species. The effect that climate change will have on those variables can be estimated using robust climate scenarios, allowing an assessment of the species' future distribution based on its current climatic preferences. Studies on *Gymnadenia nigra* (also known as *Nigritella nigra*, see Figure 7.1.2) show that global warming will significantly reduce its suitable niches. By predicting future climatic niches for the orchid and its lepidopteran pollinators, the authors of this study were able to determine that *G. nigra* is facing both significant habitat loss and reduced pollinator availability. In this particular case, no evidence for desynchronisation with the pollinators was found, but the reduction in suitable habitats coupled with limited pollinator availability

**Fig. 7.1.2.** Even though uncoupling is not suspected in the case of *Gymnadenia* (*Nigritella*) *nigra*, both the orchid and its pollinator are facing significant habitat loss, with limited pollinator availability and increased isolation of the orchid's populations.
© Jean Claessens.

is thought to reduce gene transfer between distant populations. Climate change may therefore shape the geographic and genetic structure of G. *nigra* populations in the future, leading to increased isolation and the accumulation of genetic differences.

The effects that global warming can have on tropical orchids and their pollinators is notoriously understudied. In Brazil, research carried out in the Atlantic Forest Biodiversity Hotspot suggests that climate change will reduce the persistence of most species of orchid bees (Euglossini). Even though all species retain some suitable climatic areas, and certain species may gain additional areas, the abundance, distribution and diversity of orchid bees decreases overall. Given the importance and specificity of Euglossine bees as pollinators of many orchid species in Brazil — and elsewhere in the Neotropics — such reductions in their numbers and diversity are likely to disrupt plant–pollinator relationships.

Researchers in Southwestern China identified climate change as one of the major threats to the conservation of wild orchids. They identified a reduction in suitable habitats, extreme rainfall events, and disruption of synchrony with pollinators as key factors that potentially threaten the survival of orchids in the Yachang Orchid Nature Reserve in Guangxi. The region — which is an orchid biodiversity hotspot — has registered a 0.5°C increase in average temperatures over the past century. Unfortunately, long-term data on seasonal changes in the orchids at the reserve — and elsewhere — are rare. Fluctuations in flowering times have been observed, but it is unclear whether they affect these orchids and whether such differences in the start of blooming will cause an uncoupling from their pollinators. However, most of the orchids at the reserve are highly specialised in their pollination strategies, and have a single known insect acting as effective pollinator. Their reproductive success is therefore likely to be affected by mismatches between the orchids and the insects that are induced by climate change.

There appears to be no single general trend in how climate change affects the distribution and abundance of species. One study evaluating the impact of future climate characteristics on the distribution of orchids suggests that some are likely to go extinct, while others will see their niches reduced by up to 50%, and yet others may actually see an increase in their suitable climatic conditions of up to 74 times their current coverage. The latter result drove the authors to publish their study under the rather provocative — and slightly misleading — title

"Global warming not so harmful for all plants." Indeed, one can foresee — given how varied the habitat preferences for different orchids species are — that climate change is likely to decrease the extent of habitats suitable for some, while at the same time increasing the extent of habitats that are suitable for others. Unfortunately, as was pointed out in the study itself, ecological relationships are highly complex and even though the extent of habitats suitable for certain orchid species appears to increase, these very same orchid species may face other restrictions and threats that have yet to be accounted for. How the activity, distribution and abundance of orchid pollinators, and the distribution of mycorrhizal fungi that they require to germinate and develop, may be affected remains to be tested. "One wonders, does an increase in the extent of suitable habitat necessarily translate into an increase in orchid population size and reproductive success? Not necessarily.

It is well established that many orchid species are not currently occupying the whole extent of their — climatically speaking — suitable habitats. Limiting factors may include — but are not restricted to — geographical barriers, deforestation, agriculture, poaching, pollution, pollinator limitation, and a multitude of unaccounted ecological and biological factors. In other words, even though potentially suitable habitats increase in area, there is no guarantee that the orchid population will do the same, let alone that their ecological interactions will remain unaffected. This is exactly what a more recent study found when determining the effect of climate change on the distribution of two Greek orchids belonging to the genus *Ophrys*. Modelling of potential distributions shows that the spatial distribution of suitable habitat for these *Ophrys* will change due to global warming, though it will occupy the same overall extent. So one might infer that they are unaffected by climate change. However, the sexually deceptive O. *argolica* (see Figure 7.1.3) and O. *delphinensis* are known to be dependent on male *Anthophora plagiata* bees. When the authors accounted for this highly specialised interaction in their models, the resulting future scenario changed completely. *Ophrys argolica* suddenly becomes restricted to southern Greece, while O. *delphinensis* was expected to become extinct altogether due to the loss of its pollinator.

Studies on the effects of climate change on future potential distributions of orchids are a powerful tool. The first study teaches us that each orchid species may be affected very differently by a warming

**Fig. 7.1.3.** Climate change will directly affect the highly specialised relationship between *Ophrys argolica* and its pollinator, the male bee *Anthophora plagiata*, reducing the orchid's suitable habitat to restricted areas in southern Greece. © Matteo Perilli.

climate. Even closely related orchids currently occupying similar habitats are likely to be affected uniquely by environmental changes. However, both studies also reveal that we need much more information — high-quality data, as discussed at the end of the previous chapter — to estimate correctly how complex ecological interactions will cope with climate change. Without a clearer picture of the highly specialised and intricate ecological relationships of orchids, we may naively believe that global warming is 'not so harmful'. But once multiple organisms have been factored into our analyses, we may discover a much grimmer future scenario, such as that found in the *Area de Conservación Guanacaste*, where suitable habitats seem to be increasing, and yet, insect abundance and species richness are continuously declining. But such studies can only be done if we have adequately determined which organisms are involved in these interactions. In other words, we not only need to act now to reverse certain changes before it's too late, we also desperately need to get out there and carefully document orchids and their pollinators, seed dispersers, and hosts before they are lost forever.

*"The scientific literature provides numerous examples of climate-driven changes in species distribution ... However, when it comes to research on species interactions ... there is still a lack of information."*
KJØHL et al. in Potential Effects of Climate Change on Crop Pollination for the Food and Agriculture Organization of the United Nations (2011)

## 7.2  IT'S THE END OF THE WORLD AS WE KNOW IT

*"Orchids are at the front-line of extinction, with more species under threat globally than any other plant family."*
NIGEL D. SWARTS and KINGSLEY W. DIXON (2009)

There are of course many other threats to orchid conservation besides climate change. Some are not directly related to pollination and are therefore beyond the scope of this book. Others are known to impact both the orchid and its ecological interactions. Habitat loss, fragmentation and deterioration as a consequence of human activities are often interrelated processes resulting from landscape disturbance, and their individual effects on orchid pollination are commonly not separated in the literature. It is a challenge to determine which — either landscape composition or fragmentation — impacts ecological dynamics more severely, and therefore they are treated as a single issue under landscape disturbance.

Around 87% of all flowering plants, including 75% of the most productive crop plants, are dependent to some extent on animal pollination. Pollination is one of the most vital biodiversity-dependent ecosystem services in both wild plant communities and agriculture. It has been well-established that agriculture benefits from high biodiversity, because the presence of pollinating agents can boost a crop's productivity. Pollinators, therefore, are highly important both ecologically and economically. Agriculture depends on pollination services, but ironically it is the food production process itself that has the most negative impact on pollinators, as a result of habitat loss, land-use change and pesticide use. Plant species with the most specialised pollination requirements have the highest risk of suffering from pollinator limitation. Given that more than 80% of orchid species depend on animal pollination, and half of them have a single — known — species of pollinator, it may not be an exaggeration to state that orchids are naturally pollinator-limited. This has been confirmed by studies showing that supplementary hand pollination results in increased fruit set compared with natural populations in virtually every orchid species tested.

Even though the evidence for substantial and widespread decline of pollinator services has been questioned by some authors, the fact is that case studies often paint a worrying picture. There is clear regional evidence that domestic honey bee stocks are declining, with significant colony losses between 1947 and 2005 in both Europe and the USA. The honey bee scenario may not necessarily reflect what is going on with wild bees and other orchid pollinators, but it certainly marks a trend for what may be occurring with less well documented insects. Bumblebees have also shown significant declines in Belgium and the UK, with regional extinctions already occurring in the latter. The abundance and diversity of pollinators are likely to decline as natural landscapes are modified by human activities, but few studies address the effect of landscape disturbance on wild pollinator populations. Studies on the diversity of orchid bees — Euglossini — in the Brazilian Amazon show

**Fig. 7.2.1.** Populations of the main pollinator of *Platanthera ciliaris* are declining in North America. Pollination of the orchid species is supplemented in some areas by other large butterflies including other *Papilio* species.
© John Gange.

that oil palm plantations may have a quarter of the species richness of an adjacent forest reserve area. Interestingly, permanent protection areas on the farms showed a similar species richness to the forest reserve, which demonstrates that these can serve as effective biodiversity corridors, connecting fragmented populations. Forest fragments have indeed been shown to be important habitats for euglossine bees, and may therefore be critical in sustainable plantation management strategies as safe havens for pollinators. Conversely, in the case of La Réunion island studies on the pollination of sapromyiophilous orchids suggest that their fly pollinators are less abundant in disturbed habitats than in the native lowland rainforest.

A study by researchers in the Czech Republic found that orchid distribution is related to their pollination syndrome. By studying their abundance and distribution patterns, the authors found that deceiving and rewarding orchids occupy different geographical and altitudinal niches. A majority of orchids that offer nectar were found in forest habitats, while nectarless species were more common in open, sunny areas where there is higher pollinator activity and a greater diversity of bees and butterflies. At the same time, nectarless species appear to be more prone to local extinction than nectariferous ones. One report from Australia shows that the main threats to terrestrial orchid populations are grazing, low seed set and irregular population growth — independently of their pollination strategy. In North America, one of the major threats to native plant communities is the invasion by exotic plant pests and diseases. There are over 400 such cases known today, with at least one affecting every forested habitat. An example affecting an orchid species is that of the orange-fringed orchid *Platanthera ciliaris* (see Figure 7.2.1), which heavily relies on the Palamedes swallowtail butterfly *Papilio palamedes*. The Palamedes swallowtail is a common long-tongued butterfly in the southeastern USA, but a fungal disease spread by an exotic beetle has been seriously affecting the primary host for its larvae. The widespread loss of the redbay tree, on which the larvae of the butterfly depends, is already causing a decline in the population of *P. palamedes*. Even though the pollination services of the orange-fringed orchid may in part be supplemented by other butterflies — such as *Phoebis sennae*, the cloudless sulphur — the observed decline in the Palamedes swallowtail on account of fungal disease in redbay trees is expected to seriously limit its reproductive success.

We face the sixth mass extinction event in the history of life on Earth. The loss of rare and threatened plants confined to biodiversity hotspots and regions with high degrees of local endemism (having very narrow natural distributions) affects many orchids species directly. But the danger for orchids is compounded by the decline of associated organisms — such as pollinators — on which they depend. Many threatened orchid species have undergone significant population declines and conservation success stories are rare. Even reintroduction efforts seem to be ineffective because of the orchid's aversion to habitat disturbance and a lack of consideration for the availability of pollinators. A global review of transplanted threatened orchids showed that fruiting was recorded only in a quarter. Only 2.8% of the studies observed the actual establishment of new individuals in the orchid population. Therefore, the prospects of achieving self-sustaining populations are not good. Careful conservation planning for orchids requires taking into consideration pollinator needs and numbers, including their food plants, nest sites, and larval hosts.

> *"Extinction of species is perhaps the most fundamental assault that humans can inflict on the rest of the natural world."*
> JEFF OLLERTON in *Pollinators & pollination: nature and society* (2021)

## 7.3 BEAUTIFUL DAY

> *"From bitter searching of the heart we rise to play a greater part."*
> FRANK R. SCOTT in the poem *Villanelle For Our Time* (1944)

In August of 1883 one of the most powerful volcanic eruptions in recorded history occurred on Krakatoa (Krakatau), a small island east of Sumatra and west of Java. A series of huge explosions almost destroyed the island, splitting it into pieces. The blasts were so violent they were heard as far away as Perth in Western Australia. Pyroclastic flows, volcanic ashes and tsunamis wreaked havoc in the region and worldwide. The Dutch authorities recorded more than 30,000 dead, and it is believed that no living thing survived on the islands. Plants — including orchids — were wiped out completely. Tragic and traumatic as such events are, the documentation of step-by-step plant colonisation on Krakatau has provided a unique opportunity to learn about island recolonisation. The first higher plants were observed growing on the islands three years after the deadly volcanic explosions, and orchids were among the first vascular plants to colonise the remnants of the island. Some 13 years after the explosion, three terrestrial orchid species were registered. The orchids continued to accumulate as the vegetation structure became more complex. By 1935, the number had grown to 25 species, and by 1989 the number of orchid species growing on the islands had reached 59. A survey carried out in 1996 recorded 64 species of orchids — seven terrestrials and 58 epiphytes.

Despite their delicate flowers and intricate pollination strategies, orchids may be tougher than we think and capable of quick recovery after natural disturbance (see Figure 7.3.1). Some of them have the durability of weeds. Studies of the aftermath of hurricanes passing over islands in the West Indies have shown that orchids and their pollinators are much more resilient to storms than expected. On 18 September

Fig. 7.3.1. Cadavers still fixed to an ash covered tree at Turrialba Volcano National Park in Costa Rica. The active volcano has not been kind to the highly exposed epiphytic orchids.
© APK.

1989, Hurricane Hugo hit the island of Puerto Rico. The eye of the category 4 storm passed over the island's northeast corner, where a large population of the epiphytic *Epidendrum ciliare* thrived. Despite the dramatic visual effects and widespread destruction caused by Hugo, it had little effect on the reproductive success of *E. ciliare*, a food-deceptive orchid that is pollinated by the hawkmoth *Pseudosphinx tetrio*. After the hurricane hit, the population's flowering time was unaltered and it showed a similar number of pollinaria removals as in previous years. Pollinated flowers of *E. ciliare* did not have a higher failure rate of their fruits, and pollinator activity was found to be higher after the hurricane. These results are certainly encouraging, but such mild effects may not always be the case and observations of the impact of abiotic factors, such as climatic events, on the pollination of orchids are badly needed.

**Fig. 7.3.2.** Two *Prosthechea ochracea* plants reclaim an old cement wall in Santa Cruz, Costa Rica.
© APK.

The recovery of orchids from anthropogenic disturbances are even less well understood, but there are more opportunities for their assessment today (see Figure 7.3.2). Despite the alarming rates of deforestation in many parts of the world, the trend has seen reversals in both temperate and tropical regions. One well-documented case brings hope. In the 1940s, all but 5% of the island of Puerto Rico was deforested and converted to farmlands — decimating the orchids. Abandoned farmlands eventually formed secondary forests and orchid populations have mostly recovered. Very few species are thought to be lost forever, which suggests that small refuges were critical in this recovery process. Similarly, after experiencing some of the highest deforestation rates in Latin America through the 1960s and 1970s, Costa Rica has bounced back and has seen a dramatic and continued increase in forest cover. This was prompted by environmental awareness, which translated into strong policies that resulted in the establishment of numerous protected areas. Our ongoing studies on the orchid diversity at Guayabo National Monument — one of the country's smallest protected areas — demonstrates that half a century after declaring this a conserved site, the pastures of what was once a farm have made way for a lush, mature secondary forest that today hosts more than 140 species of epiphytic and terrestrial orchids. Two of them — *Isochilus latibracteatus* (see Figure 7.3.3) and *Maxillariella guareimensis* — had never been recorded before anywhere in the country.

Another stronghold for conservation are *ex situ* collections, especially when they are complemented with educational programmes for the general public. In January 2019, biologist Carlos Arrieta Quesada knocked on my door at Lankester Botanical Garden. Carlos was in charge of the environmental studies for the project '*South Ring Transmission Line*' of the Costa Rican Institute of Electricity and was in desperate need of some help. For the purpose of placing electricity towers to power the city, the institute had to remove trees from a large area at a site called El Tablón, on the mountains surrounding Cartago. Carlos noticed that the trees were full of epiphytic orchids. He wanted to know if we could take on the cultivation and care of both the rare and common plants that would be displaced. He was no expert, but thought some of them would be of conservation or research interest to us. As I glanced at the heap of orchids, ferns and bromeliads Carlos had rescued from the first clearing site, it immediately dawned on me

**Fig. 7.3.3.** *Isochilus latibracteatus*, never before recorded in Costa Rica, is one of the 140 species of orchids that have colonised Guayabo National Monument after it was declared a protected area.
© APK.

that we had hit the jackpot, but were also embarking on a gigantic task. He warned me about the number of plants and asked if we had the resources to take care of everything he would bring us. Without hesitation I said yes and we shook hands on it. Between January and November 2019, his team brought pickup-load after pickup-load of rescued plants from El Tablón. Most of the plants were small and could be easily transported, but others required two people to move them. By the end of the year the garden had received a few dozen cacti, aroids and pipers, 74 large bromeliads, 168 tall tree-ferns, and a whopping 817 epiphytic orchids. Today these plants are part of the living collections at Lankester Botanical Garden, where they now participate in its ongoing research, education and conservation programmes — both locally and in botanical collections around the world.

In 2005, the Ministry of Environment of Costa Rica formally recognised Lankester Botanical Garden as a 'National Epiphyte Sanctuary'. It is also the only institution in the country officially registered by CITES, the multilateral treaty to protect endangered plants and animals. Every year our permits for collecting and studying wild orchids are renewed by presenting the proper paperwork to the authorities at the National System of Conservation Areas. Much of our research is carried out in the country's extensive system of conservation areas, where we collaborate with park rangers and administration staff. We help to identify and cultivate confiscated plants, provide field guides for the national parks and organise workshops to train their staff. This coordination and collaboration between public institutions in Costa Rica not only allows Lankester to continue doing orchid research, but also facilitates the exchange of ideas, people and plants with other botanical institutions worldwide. Today, orchids that had been laying in the mud at El Tablón can be appreciated by the general public in places as far away as Gardens by the Bay in Singapore and the Royal Botanic Gardens, Kew, near London, where the *Dracula* and other rescued plants are now part of the general collections. This is a wonderful example of collaboration between public institutions for the conservation of flora in Costa Rica, and international collaboration to study and conserve biodiversity and educate visitors of all generations.

The conservation of orchids may present unique challenges, given their unusual life cycle, but it also opens up a whole new set of possibilities. Many orchids exhibit highly specialised ecological interactions and requirements, but these may continue to be effective even when population sizes are reduced. Studies show that organisms can persist after significant population decreases and genetic bottlenecks, which surely is encouraging for anyone working on the conservation of these plants. Studies on the reproduction of terrestrial orchids in South Africa show that pollination success was similar in large and small populations. They showed resilience in terms of both population size and density, with fruit set unaffected by the number of individuals. A single orchid seed capsule can potentially produce hundreds or thousands of plantlets, which under the right cultivation conditions could be an invaluable source of raw material for conservation programmes.

How pollination relates directly to the resilience of orchids is perhaps best explained with an example. The Hospital San Juan de Dios — Costa Rica's first hospital — was founded in 1845 in the very centre of the capital city San José. It can be found on the Paseo Colón, the main artery leading to the heart of the city. Today, the hospital is mostly surrounded by concrete and asphalt, but a row of scattered trees forms a lifeline of sorts to La Sabana Metropolitan Park — the country's largest urban park. The building has lost much of its original neogothic style, but it still conserves several of its small inner patios. There, the overpopulated hospital's hectic daily routine seems to catch a breath. Each patio has a few plants. In one of them I found two feeble individuals of the exotic *Arundina graminifolia*. This orchid is native to Asia, but it has become a very common garden plant in tropical areas around the world. Unlike the majority of invasive plants, *Arundina* is

Fig. 7.3.4. The exotic *Arundina graminifolia* somehow manages to flower and fruit in a small inner garden of the San Juan de Dios hospital in the center of San José, capital city of Costa Rica.
© APK.

not a prolific selfer. It has become common through human cultivation and asexual reproduction rather than through seed production. It is pollinator-dependent, but has been able to employ some of the local bees and therefore does set seed naturally outside of its native range. The two skimpy plants at the hospital had probably seen better days. They were thin and scant, barely hanging on to life, and were surrounded by cement. And yet, one of them had developed a fruit! I became emotional thinking about how — under the worst growing conditions possible — this non-native plant had somehow been able to reproduce. What are the odds of finding a bee — the right bee! — in the seemingly barren city centre? Inhospitable as conditions may appear, orchids do their utmost to strive and flourish (see Figure 7.3.4).

Pollination is not simply a consequence of biodiversity, it is a key force in the maintenance and promotion of ecosystem biodiversity. Pollinators are often shared, so different plants within a community benefit one another by supporting diverse pollinator communities. Ecological relationships are highly varied within a community, involving a spectrum of insects and other arthropods, which help to support the ecosystem's diversity. In essence, biodiversity maintains biodiversity. Therefore, the diversity of the plant community depends on the diversity of its pollinators. Orchids show great ecological resilience, having the capacity to recolonise a habitat that has recovered from disturbance. Wildlife refuges — small as they may seem — play a key role in this recovery, providing a source for both plants and animals. But recovery may no longer be possible if they cease to exist. There is hope for these extraordinary fruits of evolution and their faithful ecological partners, but we all need to take action today.

It's a beautiful summer day. The sky is clear blue, typical for early March, the air is lightly fragrant and the atmosphere buzzing with pollinators. As I glance at the tourists enjoying the spectacle of Costa Rica's national flower, the orchid *Guarianthe skinneri*, in full bloom at Lankester Botanical Garden, I am reminded of the responsibility we have not only to appreciate the grand diversity and complex ecology of these precious plants, but also to pass on their cautionary message to humanity:

*With great biodiversity comes great responsibility!*

A female *Euglossa* carrying a pollinarium of *Guarianthe skinneri*, the national flower of Costa Rica. © APK.

# ACKNOWLEDGEMENTS

I am grateful to my dear friend Rudolf Jenny, without whom this project would have been impossible. His database, Bibliorchidea, has been an invaluable source of literature and his sudden passing is a terrible loss to the orchid community. Important sources include the search engines of Google Scholar, the extensive access to literature granted through Sci-hub and ResearchGate, and the libraries at Lankester Botanical Garden, University of Costa Rica. My wife Miriam and my dad Jan were paramount in shaping and executing this project. I am thankful to numerous people whose photographs, videos and ideas are discussed and reproduced in this book, especially Jean Claessens, Ron Parsons, Rod Peakall, Charlotte Watteyn, Karen Gil and Jim Ackerman. Materials reproduced in this book were kindly provided by M. Ayasse, N. Belfort, H. Bellman, D. Bogarín, M. Brundrett, E. Carman, M. Cedeño, S. Dälstrom, M. Díaz, W. Driessen, J. Fournel, J. Gange, G. Gerlach, N. Gutiérrez, A. Hirtz, H. Jiang, A. Kocyan, H. Liu, C. Martel, L.E. Mejía, M. Méndez, K. Metzger, C. Micheneau, S. Moreno, J. M. Murillo, C.E.P. Nunes, J. O'Hanlon, H. Oakeley, P.T. Ong, M. Perilli, C. Peter, R. Phillips, F. Pupulin, R.-J. Quené, S. Ramírez, G. Rojas-Alvarado, C. Rose-Smyth, A. Salazar, G. Salguero, D. Scaccabarozzi, A. Schuiteman, R. Singer, J. Storey, M. Sunouchi, B. Valentine, J.J. Vermeulen, S. Vieira, D. Villalobos, R. van Vugt, C. Ward Jr, L.T. Wasserthal, L.E. Yupanki, C.-C. Zheng and J.J. Zúñiga. Thanks to Aleja, Carlos, Charlotte, Diego, Esteban, Ernesto, Francisco, Franco, Gerson, Grettel, Gustavo (both), Isler, Jorge, Jyotsna, Karen, Liz, Marco, Mari, Mario, Mauricio, Melania, Melissa, Miguel, Nelson, Noelia, Pablo, Paul, Priscilla, Ricardo, Sarah, Shanti and Steve. Scientific papers, books, documentaries, conferences, talks, interviews, songs, poems and late-night conversations have not only significantly enriched the ideas expressed throughout the text, but have been a constant source of inspiration and motivation. I am grateful to the countless authors whose words I have borrowed. Finally, I am very grateful to Lydia White, Gina Fullerlove, Georgie Hills, and editorial team, for their support, guidance and development of the book.

*Bombus pascuorum* with pollinaria of *Dactylorhiza praetermissa*.
© Jean Claessens.

# GLOSSARY

**abiotic**: not involving living organisms

**amber**: fossilised plant resin

**anemophily**: pollination by wind

**anther**: the part of the stamen containing pollen

**anthesis**: time when the flower opens

**anti-pollinator**: ambush predator, or an animal that feeds on pollinators

**autogamy**: self-fertilisation after pollination by pollen from the same individual

**Batesian floral mimicry**: where the species has evolved flowers that mimic the features of another species that rewards pollinators

**bilateral symmetry**: a basic body plan in which the organism can be divided along its midline into left and right sides that are approximate mirror images of each other

**bract**: a modified and specialised leaf in the inflorescence

**brood-site mimicry**: where flowers have evolved to attract insects that are looking for a place to lay their eggs

**callus**: (plural **calli**) a thickening, for example on the lip of some orchids

**campo rupestre**: rocky fields in Brazil characterised by a mosaic of herbs, grasses and shrubs

**cantharophily**: pollination by beetles

**caudicle**: a granular process, produced within the anther, to which the pollinia are attached

**chiropterophily**: pollination by bats

**cleistogamous**: self-fertilisation occurring within unopened flowers

**column**: the solid central body of a flower formed by the fused styles and stamens

**cross-fertilisation**: fertilisation by pollen from another individual of the same species

**dehiscent**: fruits and anthers that split open when ripe

**dichogamy**: when male and female parts develop at different times to prevent self-fertilisation

**dioecious**: having unisexual flowers, with male and female flowers on different plants

**diurnal**: occurring or flowering in the day-time (as opposed to nocturnal)

**diploid**: with two sets of chromosomes, one from each parent

**elaiophore**: an oil-secreting gland

**elaiosomes**: oily appendage on seeds that often serve as a food source for ants or other insects, which then disperse the seed

**electroantennography:** a technique for measuring the electrical output of an insect antenna to its brain when exposed to a given odour

**endemic:** restricted to, unique to, not naturally found elsewhere

**entomophagous:** feeding on insects

**environmental sex determination:** when the surrounding conditions influence whether males or females are produced

**epichile:** the terminal part of a lip that is divided into two or three distinct parts

**family:** higher taxonomic unit composed of one genus or several related genera, usually clearly separated from other families

**florivory:** of an animal that mainly eats the products of flowers

**grayanotoxins:** a group of closely related neurotoxins produced by rhododendrons and other species in the Ericaceae family

**generalised food deception:** a pollination strategy that exploits pollinators' innate foraging behaviour through general floral signals

**gland:** a secretory area or mass, either embedded in a surface or ending a hair

**gregarious flowering:** collective flowering of plants at a fixed interval after a climatic stimulus

**haplodiploid:** denoting a genetic system in which females develop from fertilised (diploid) eggs and males from unfertilised (haploid) ones

**haploid:** with one set of chromosomes

**hermaphrodite:** bisexual plant with stamens and pistil in the same flower

**hybridisation:** crossing between two different species

**hydrophily:** pollination that occurs through the dispersion of pollen in running water

**hypochile:** the basal part of a lip that is divided into two or three distinct parts

**indehiscent:** (of fruits) not splitting open

**inflorescence:** the part of the plant that bears the flowers, including all bracts, branches and flowers, but excluding unmodified leaves

**isotope:** atoms of the same element that have the same number of protons but a different number of neutrons

**keel:** narrow longitudinal ridge sticking out from a rounded surface, like that on the bottom of a boat

**kleptomyiophily:** strategy employed by fly-pollinated flowers that mimic the compounds released from freshly killed insects

**kleptoparasitism:** a form of parasitism by theft, where one animal takes resources from another

**melittophily:** bee pollination

**mesochile:** the mid-portion of a lip that is divided into three distinct portions

**mutualism:** an association between two organisms that is beneficial to both

**mycoheterotrophic:** obtaining food from decaying organisms through a symbiotic fungus

**myophilous:** being pollinated by flies

**nectar guides:** lines or blotches of colour that lead to the nectar-providing zones of the plant

**nectary:** (plural **nectaries**) organ(s) in which nectar is formed

**Neotropics:** the tropical part of the American continents, i.e. from central Mexico and the Caribbean islands to northern Chile, Paraguay and southern Brazil

**ombrophily:** rain-assisted self-pollination

**order:** (in nomenclature) a taxon below class and above family

**ornithophily:** pollination by birds

**outcrossing:** pollination that results from one plant fertilising another plant

**ovovivipary:** (of flies) that deposit hatched maggots instead of eggs on carrion, dung or the open wounds of animals

**palustrine:** vegetated, non-riverine, subcoastal wetland systems

**paramo:** a variety of alpine tundra found at high elevations in the Andes mountain range in South America

**peloric:** abnormally regular or symmetrical, when the usual condition is irregular

**perianth:** the outer part of the flowers, consisting of sepals and petals

**phalaenophily:** pollination by moths

**pheromone:** chemicals secreted by one individual animal capable of affecting the behaviour of other individuals

**pollinarium:** the complete set of pollinia from one anther, together with accessory structures such as caudicles, stipe and viscidium, if present

**pollinarium transformation:** a series of permanent changes that a pollinarium may undergo after removal from the flower

**pollination:** the transfer of pollen from anther to stigma

**pollination syndrome:** set of characters that together represent adaptations to particular types of pollinators

**pollinator:** an agent that effects pollination

**pollinium:** (plural **pollinia**) pollen grains cohering into a single group and distributed as such

**polylectic bees:** species of bee that collect pollen from a wide range of unrelated plants to feed their larvae

**protandry:** stamens releasing pollen before the stigma in the same flower is receptive

**pseudo-antagonism:** pollination strategy that exploits the territorial behaviour of certain animals

**pseudocopulation**: a type of mimicry in which the flowers resemble female insects and are pollinated by males that attempt to copulate with them

**pseudoparasitism**: a pollination strategy that relies on insects that attempt to sting and parasitise an apparent host

**pseudopollen**: a mealy deposit on the lip of some orchids

**psychophily**: pollination by butterflies

**radial symmetry**: symmetric around a central axis, as opposed to bilateral symmetry

**rostellum**: a shelf- or beak-like projection on the orchid column, derived from the median stigma lobe, that separates the fertile stigmatic surface from the anther, thereby preventing autogamy and aiding in gluing the pollinia to the pollinator

**sapromyiophily**: pollination by flies that visit decaying substances, dung or carrion

**saprophagous**: feeding on decaying organic matter

**scutellum**: a shield-like plate attached to the back end of the thorax of a fly

**self-compatible**: able to be fertilised by its own pollen.

**self-fertilising**: (or **selfer**) a flower that is fertilised by its own pollen

**speciation**: evolution into a new species

**species**: living organisms that share a similar morphology and constant, distinctive characters, and are thought to be capable of interbreeding to produce fertile offspring

**specimen**: (botany) dried plant or part of a plant in a herbarium, or any plant (part) collected for study

**sphingophily**: pollination by hawkmoths

**spur**: (in flowers) a slender, hollow extension (usually) of the perianth, often containing nectar

**stamen**: the male organ of a flower, consisting of a stalk (called the filament) and pollen-bearing anthers

**staminode**: a sterile or abortive stamen, having a different function than bearing pollen

**stipe**: (of pollinaria) a slender stalk often connecting a viscidium to the caudicles

**stigma**: (of an orchid flower) the pollen receptor on the column

**stoma**: (plural, **stomata**) a pore in the epidermis used primarily for transpiration (loss of water vapour)

**subfamily**: subdivision of family, ranked between family and tribe (subfamily names end in -ideae)

**subtribe**: taxonomic rank below tribe and above genus (subtribe names end in –inae)

**supergeneralist**: (of a flower) being pollinated by multiple different pollinator groups

**sympatric:** (of two or more taxa) living in the same area

**synchronous flowering:** flowering of plants together at a fixed interval

**tetrad:** discrete units composed of four pollen grains

**viscidium:** a sticky pad or viscid drop to which the pollinia are connected. Its main function is to attach the pollinarium to a pollinator

The butterfly *Eurema daira eugenia* carrying a pollinarium of *Epidendrum radicans*.
© APK.

# REFERENCES

## CHAPTER 1 PLOT
### 1.1 I PUT A SPELL ON YOU

Ackerman, J.D. (1986). Mechanisms and evolution of food-deceptive pollination systems in orchids. *Lindleyana* 1(2): 108–113.

Ackerman, J.D. & Tremblay, R.L. (2020). What hath he wrought: Dodson the divergent. *Icones Orchidacearum* 18(1): X–XV.

Ames, O. (1937). Pollination of orchids through pseudocopulation. *Bot. Mus. Leaf.* 5(1): 1–30.

Arditti, J. (1969). Annotated selected references on orchid pollination. *Orchid Rev.* 77: 249–251.

Argue, G.L. (2012a). *The Pollination Biology of North American Orchids, Vol. 1*. Springer, New York.

Argue, G.L. (2012b). *The Pollination Biology of North American Orchids, Vol. 2*. Springer, New York.

Berliocchi, L. (2004). *The Orchid in Lore and Legend*. Timber Press, Oregon.

Brown, R. (1833). On the organs and mode of fecundation in Orchideae and Asclepiadeae. *Trans. Linn. Soc.* 16: 685–733.

Candeias, M. (2021). *In Defense of Plants: An Exploration into the Wonder of Plants*. Mango Publishing Group, Coral Gables.

Darwin, C. (1862). *On the Various Contrivances by which British and Foreign Orchids are Fertilised by Insects, and on the Good Effects of Intercrossing*. John Murray, London.

Dressler, R.L. & Dodson, C.H. (1960). Classification and phylogeny in the Orchidaceae. *Ann. Missouri Bot. Gard.* 47: 25–68.

Endara, L. (2020). In memoriam. Calaway Homer Dodson (1928–2020). *Lankesteriana* 20(2): I–VII.

Jersáková, J. *et al.* (2006). Mechanisms and evolution of deceptive pollination in orchids. *Biol. Rev. Camb. Philos. Soc.* 81: 219–235.

Ossenbach, C. (2019). Obituary: Robert Dressler (1927–2019) — a botanist for all seasons. *Lankesteriana* 19(3): I–VIII.

Pridgeon, A. (2016). Robert L. Dressler: a biologist for all seasons. *Orchids* 85(2): 126–131.

Sprengel, C.K. (1793). *Das Entdeckte Geheimnis der Natur im Bau und in der Befruchtung der Blumen*. Friedrich Vieweg dem ältern, Berlin, p. 448.

Van der Cingel N.A. (1995). *An Atlas of orchid Pollination: European Orchids*. A.A. Balkema Publishers, Rotterdam.

Van der Cingel N.A. (2001). *An Atlas of Orchid Pollination: America, Africa, Asia and Australia*. A.A. Balkema Publishers, Rotterdam.

Van der Pijl, L. & Dodson, C.H. (1966). *Orchid Flowers: Their Pollination and Evolution*. University of Miami Press, Coral Gables.

### 1.2 US AND THEM

Christenhusz, M.J.M. & Byng, J.W. (2016). The number of known plants species in the world and its annual increase. *Phytotaxa* 261(3): 201–217.

Dressler, R.L. (1993). *Phylogeny and Classification of the Orchid Family*. Cambridge University Press, Cambridge.

Ollerton, J. (2021). *Pollinators & Pollination: Nature and Society*. Pelagic Publishing, Exeter.

### 1.3 SOMETHING TO BELIEVE IN

Jersáková, J. *et al*. (2006). Mechanisms and evolution of deceptive pollination in orchids. *Biol. Rev. Camb. Philos. Soc.* 81: 219–235.

Ollerton, J. *et al*. (2009). A global test of the pollination syndrome hypothesis. *Ann. Bot.* 103: 1471–1480.

Ollerton, J. *et al*. (2015). Using the literature to test pollination syndromes — some methodological cautions. *J. Pollinat. Ecol.* 16(17): 119–125.

Ollerton, J. (2021). *Pollinators & Pollination: Nature and Society*. Pelagic Publishing, Exeter.

Pauw, A. (2006). Floral syndromes accurately predict pollination by a specialized oil-collecting bee (*Rediviva peringueyi*, Melittidae) in a guild of South African orchids (Coryciinae). *Am. J. Bot.* 93: 917–926.

Rosas-Guerrero, V. *et al*. (2014). A quantitative review of pollination syndromes: do floral traits predict effective pollinators? *Ecology Letters* 17: 388–400.

Van der Cingel, N.A. (1995). *An Atlas of Orchid Pollination: European Orchids*. A.A. Balkema Publishers, Rotterdam.

Van der Cingel, N.A. (2001). *An Atlas of Orchid Pollination: America, Africa, Asia and Australia*. A.A. Balkema Publishers, Rotterdam.

Van der Pijl, L. & Dodson, C.H. (1966). *Orchid Flowers: Their Pollination and Evolution*. University of Miami Press, Coral Gables.

Waser, N.M. *et al*. (1996). Generalization in pollination systems, and why it matters. *Ecology* 77: 1043–1060.

Willmer, P. (2011). *Pollination and Floral Ecology*. Princeton University Press.

## CHAPTER 2 DECEIT

Candeias, M. (2021). *In Defense of Plants: An Exploration into the Wonder of Plants*. Mango Publishing Group, Coral Gables.

Darwin Correspondence Project, "Letter no. 3193," accessed on 19 July 2022, https://www.darwinproject.ac.uk/letter/?docId=letters/DCP-LETT-3193.xml

Darwin, C. (1877). *The Various Contrivances by which Orchids are Fertilised by Insects. 2nd edition.* John Murray, London.

Dodson, C.H. & Frymire, G.P. (1961). Natural pollination of orchids. *Missouri Bot. Gard. Bull.* 49: 133–139.

Endersby, J. (2016). Deceived by orchids: sex, science, fiction and Darwin. *BJHS* 49: 205–229.

Jersáková, J. *et al.* (2006). Mechanisms and evolution of deceptive pollination in orchids. *Biol. Rev. Camb. Philos. Soc.* 81: 219–235.

Johnson, S.D. & Schiestl, F.P. (2016). *Floral Mimicry.* Oxford University Press, Oxford.

Sprengel, C.K. (1793). *Das Entdeckte Geheimnis der Natur im Bau und in der Befruchtung der Blumen.* Friedrich Vieweg dem ældern, Berlin, p. 448.

Vogel, S. (1996). Christian Konrad Sprengel's theory of the flower: the cradle of floral ecology. In *Floral Biology*, Lloyd, D.G. & Barrett, S.C.H. (eds), Springer, Boston, pp. 44–62.

## 2.1 AIN'T TALKIN' 'BOUT LOVE

Ames, O. (1937). Pollination of orchids through pseudocopulation. *Bot. Mus. Leaf.* 5(1): 1–30.

Ayasse, M. *et al.* (2000). Evolution of reproductive strategies in the sexually deceptive orchid *Ophrys sphegodes*: how does flower-specific variation of odor signals influence reproductive success? *Evolution* 54(6): 1995–2006.

Baguette, M. *et al.* (2020). Why are there so many bee-orchid species? Adaptive radiation by intra-specific competition for mnesic pollinators. *Biol. Rev.* 95: 1630–1663.

Brown, R. (1833). On the organs and mode of fecundation in Orchideae and Asclepiadeae. *Trans. Linn. Soc.* 16: 685–733.

Correvon, H. & Pouyanne, M. (1916). Un curieux cas de mimétisme chez les ophrydées. *J. Soc. Nat. Hort. Fr.* 4(17): 29–31; 41–47.

Correvon, H. & Pouyanne, M. (1923). Nouvelles observations sur le mimétisme et la fécondation chez les *Ophrys speculum* et *lutea*. *J. Soc. Nat. Hort. Fr.* 4(24): 372–377.

Darwin Correspondence Project, "Letter no. 7228A," accessed on 19 July 2022, https://www.darwinproject.ac.uk/letter/?docId=letters/DCP-LETT-7228A.xml

Darwin, C. (1862). On the Various Contrivances by which British and Foreign Orchids are Fertilised by Insects, and on the Good Effects of Intercrossing. John Murray, London.

Gaskett, A.C. (2011). Orchid pollination by sexual deception: pollinator perspectives. *Biol. Rev.* 86: 33–75.

Paulus, H.F. (2019). Speciation, pattern recognition and the maximization of pollination: general questions and answers given by the reproductive biology of the orchid genus *Ophrys*. *J. Comp. Physiol. A* 205: 285–300.

Pasteur, G. (1982). A classificatory review of mimicry systems. *Ann. Rev. Ecol. Syst.* 13: 169–199.

Pouyanne, M. (1917). La fécondation des *Ophrys* par les insectes. *Bull. Soc. Hist. Nat. Afr. N.* 8: 6–7.

Schiestl, F.P. *et al.* (2000). Sex pheromone mimicry in the early spider orchid (*Ophrys sphegodes*): patterns of hydrocarbons and the key mechanism for pollination by sexual deception. *J. Comp. Physiol. A* 186: 567–574.

Vereecken, N.J. & McNeil, J.N. (2010). Cheaters and liars: chemical mimicry at its finest. *Can. J. Zool.* 88: 725–752.

Vereecken, N.J. & Schiestl, F.P. (2008). Evolution of imperfect floral mimicry. *PNAS* 105: 7484–7488.

Vereecken, N.J. *et al.* (2012). Pre-adaptations and the evolution of pollination by sexual deception: Cope's rule of specialization revisited. *Proc. R. Soc. Lond. B* 279: 4786–4794.

Wolff, T. (1950). Pollination and fertilisation of the Fly Ophrys, *Ophrys insectifera* L. in Allindelille Fredskov, Denmark. *Oikos* 2(1): 20–59.

## 2.2 FIRST DATE

Alcock, J. (1981). Notes on the reproductive behavior of some Australian thynnine wasps. *J Kans. Entomol. Soc.* 54: 681–693.

Ames, O. (1937). Pollination of orchids through pseudocopulation. *Bot. Mus. Leaf.* 5(1): 1–30.

Brunton Martin, A.L. (2020). Orchid sexual deceit affects pollinator sperm transfer. *Funct. Ecol.* 34: 1336–1344.

Burrell, R.W. (1935). Notes on the habits of certain Australian Thynnidae. *J. N. Y. Ent. Soc.* 43: 19–29.

Coleman, E. (1927). Pollination of the orchid *Cryptostylis leptochila*. *Vic. Nat.* 44: 20–22.

Coleman, E. (1928). Pollination of *Cryptostylis leptochila* F.v.M. *Vic. Nat.* 44: 333–340.

Coleman, E. (1928). Pollination of an Australian orchid by the male Ichneumonid *Lissopimpla semipunctata*, Kirby. *Trans. Ent. Soc. Lond.* 2: 533–539.

Gaskett, A.C. *et al.* (2008). Orchid sexual deceit provokes ejaculation. *Am. Nat.* 171(6): E206–E212.

Gaskett, A.C. (2011). Orchid pollination by sexual deception: pollinator perspectives. *Biol. Rev. Camb. Philos. Soc.* 86: 33–75.

Hopper, S.D. & Brown, A.P. (2007). Australia's wasp-pollinated flying duck orchids revised (Paracaleana: Orchidaceae). *Austral. Syst. Bot.* 20: 252–285.

Martel, C. *et al.* (2016). *Telipogon peruvianus* (Orchidaceae) flowers elicit premating behavior in *Eudejeania* (Tachnidae) males for pollination. *PLoS ONE* 11(11): e0165896.

Peakall, R. (1990). Responses of male *Zaspilothynnus trilobatus* Turner wasps to females and the sexually deceptive orchid it pollinates. *Funct. Ecol.* 4(2): 159–167.

Peakall, R. & Beattie, A.J. (1996). Ecological and genetic consequences of pollination by sexual deception in the orchid *Caladenia tentaculata*. *Evol.* 50: 2207–2220.

Phillips, R.D. *et al.* (2017). Evolutionary relationships among pollinators and repeated pollinator sharing in sexually deceptive orchids. *J. Evol. Biol.* 30: 1674–1691.

Stoutamire, W.P. (1974). Australian terrestrial orchids, thynnid wasps, and pseudocopulation. *Am. Orchid Soc. Bull.* 43: 13–18.

Stoutamire, W.P. (1975). Pseudocopulation in Australian terrestrial orchids. *Am. Orchid Soc. Bull.* 44: 226–233.

Wong, B.B.M. & Schiestl, F.P. (2002). How an orchid harms its pollinator. *Proc. R. Soc. Lond. B* 269: 1529–1532.

## 2.3 YOUR LATEST TRICK

Dodson, C.H. (1962). The importance of pollination in the evolution of the orchids of tropical America. *Am. Orchid Soc. Bull.* 31: 525–534; 641–649; 731–735.

Martel C. *et al.* (2016). *Telipogon peruvianus* (Orchidaceae) flowers elicit premating behavior in *Eudejeania* (Tachnidae) males for pollination. *PLoS ONE* 11: e0165896.

Martel, C. *et al.* (2019). The chemical and visual bases of the pollination of the Neotropical sexually deceptive orchid *Telipogon peruvianus*. *New Phytol.* 223: 1989–2001.

Van der Pijl, L. & Dodson, C.H. (1966). *Orchid Flowers: Their Pollination and Evolution*. University of Miami Press, Coral Gables.

## 2.4 PERFECT STRANGERS

Blanco, M.A. & Barboza, G. (2005). Pseudocopulatory pollination in *Lepanthes* (Orchidaceae: Pleurothallidinae) by fungus gnats. *Ann. Bot.* 95: 763–772.

Blanco, M.A. (2005). Flowers, sex and lies: the scandalous pollination of *Lepanthes*. *Pleurothallid News and Views* 17(4): 26–30.

Bogarín, D. *et al.* (2019). Phylogenetic comparative methods improve the selection of characters for generic delimitations in a hyperdiverse Neotropical orchid clade. *Sci. Rep.* 9: 15098.

Bogarín, D. *et al.* (2019). Floral anatomy and evolution of pollination syndromes in *Lepanthes* and close relatives. In *Proceedings of the 22nd World Orchid Conference, Guayaquil, Ecuador, 2017*, Pridgeon, A.M. & Arosema, A. (eds), Asociación Ecuatoriana de Orquideología, Guayaquil: 396–410.

Gaskett, A.C. (2011). Orchid pollination by sexual deception: pollinator perspectives. *Biol. Rev. Camb. Philos. Soc.* 86: 33–75.

Hodson, C.N. & Ross, L. (2021). Evolutionary perspectives on germline-restricted chromosomes in flies (Diptera). *Genome Biol. Evol.* 13(6): evab072.

Karremans, A.P. & Vieira-Uribe, S. (2020). *Pleurothallids: Neotropical Jewels, Vol 1*. Mariscal, Quito.

Kerr, W.E. & López, C.R. (1962). Biologia da reprodução de *Trigona* (*Plebeia*) *droryana*. F. Smith. *Revista Brasileira de Biologia* 22: 335–341.

Singer, R.B. (2002). The pollination mechanism in *Trigonidium obtusum* Lindl. (Orchidaceae: Maxillariinae): Sexual mimicry and trap-flowers. *Ann. Bot.* 89: 157–163.

Singer, R.B. *et al.* (2004). Sexual mimicry in *Mormolyca ringens* (Lindl.) Schltr. (Orchidaceae: Maxillariinae). *Ann. Bot.* 93: 755–762.

Steiner, K.E. *et al.* (1994). Floral and pollinator divergence in two sexually deceptive South African orchids. *Am. J. Bot.* 81(2): 185–194.

Streinzer, M. *et al.* (2009). Floral colour signal increases short-range detectability of a sexually deceptive orchid to its bee pollinator. *J. Exp. Biol.* 212: 1365–1370.

Tremblay, R.L. (1997). *Lepanthes caritensis*; an endangered orchid: no sex, no future? *Selbyana* 18: 160–166.

Tremblay, R.L. & Ackerman, J.D. (2001) Gene flow and effective population size in *Lepanthes* (Orchidaceae): a case for genetic drift. *Bot. J. Linn. Soc.* 72: 47–62.

Tremblay, R.L. *et al.* (1998). Host specificity and low reproductive success in a rare endemic Puerto Rican orchid *Lepanthes caritensis*. *Biol. Conserv.* 85: 297–304.

## 2.5 A MILLION MILES AWAY

Ackerman, J.D. (1981). Pollination of *Calypso bulbosa* var. *occidentalis* (Orchidaceae): a food deception system. *Madroño* 28: 101–110.

Aguiar, J.M. *et al.* (2020). A cognitive analysis of deceptive pollination: associative mechanisms underlying pollinators' choices in non-rewarding colour polymorphic scenarios. *Sci. Rep.* 10: 9476.

Alexandersson, R. & Ågren, J. (1996). Population size, pollinator visitation and fruit production in the deceptive orchid *Calypso bulbosa*. *Oecologia* 107: 533–540.

Aragon, S. & Ackerman, J.D. (2004). Does flower color variation matter in deception pollinated *Psychilis monensis* (Orchidaceae)? *Oecologia* 138: 405–413.

Boyden, T.C. (1982). The pollination biology of *Calypso bulbosa* var. *americana* (Orchidaceae): initial deception of bumblebee visitors. *Oecologia* 55: 178–184.

Gigord, L.D.B. *et al.* (2001). Negative frequency-dependent selection maintains a dramatic flower color polymorphism in the rewardless orchid *Dactylorhiza sambucina* (L.) Soò. PNAS 98(11): 6253–6255.

Gigord, L.D.B. *et al.* (2002). Experimental evidence for floral mimicry in a rewardless orchid. *Proc. R. Soc. Lond. B* 269: 1389–1395.

Groiß, A.M. *et al.* (2017). Pollen tracking in the food-deceptive orchid *Dactylorhiza sambucina* showed no predominant switching behaviour of pollinators between flower colour morphs. *Flora* 232: 194–199.

Jersáková, J. *et al.* (2006). Mechanisms and evolution of deceptive pollination in orchids. *Biol. Rev. Camb. Philos. Soc.* 81: 219–235.

Jersáková, J. *et al.* (2008). Effect of nectar supplementation on male and female components of pollination success in the deceptive orchid *Dactylorhiza sambucina*. *Acta Oecologica* 33: 300–306.

Jersáková, J. *et al.* (2009). Deceptive behavior in plants. II. Food deception by plants: from generalized systems to specialized floral mimicry. In Baluška, F. (ed.) *Plant-Environment Interactions, Signaling and Communication in Plants*. Springer, Berlin, pp. 223–246.

Juillet, N. & Scopece, G. (2010). Does floral trait variability enhance reproductive success in deceptive orchids? *Perspect. Plant Ecol. Evol. Syst.* 12: 317–322.

Koivisto, A.-M. *et al.* (2002). Pollination and reproductive success of two colour variants of a deceptive orchid, *Dactylorhiza maculata* (Orchidaceae). *Nord. J. Bot.* 22(1): 53–58.

Lammi, A. & Kuitunen, M. (1995). Deceptive pollination of *Dactylorhiza incarnata*: an experimental test of the magnet species hypothesis. *Oecologia* 101: 500–503.

Mosquin, T. 1970. The reproductive biology of *Calypso bulbosa* (Orchidaceae). *Canad. Field-Naturalist* 84: 291–296.

Ostorowiecka, B. *et al.* (2019). Pollinators and visitors of the generalized food-deceptive orchid *Dactylorhiza majalis* in North-Eastern Poland. *Biologia* 74: 1247–1257.

Pellegrino, G. *et al.* (2005). Evidence of post-pollination barriers among three colour morphs of the deceptive orchid *Dactylorhiza sambucina* (L.) Soó. *Sex. Plant Reprod.* 18: 179–185.

Tuomi, J. *et al.* (2015). Pollinator behavior on a food-deceptive orchid *Calypso bulbosa* and coflowering species. *Sci. World J.* 2015: 482161.

## 2.6 DOWN UNDER

Arditti, J. (1981). Orchid ramblings from U.C.I. *Orchid Rev.* 89(1056): 334.

Bernhardt, P. & Burns-Balogh, P. (1986). Floral mimesis in *Thelymitra nuda* (Orchidaceae). *Pl. Syst. Evol.* 151: 187–202.

Burns-Balogh, P. & Bernhardt, P. (1988). Floral evolution and phylogeny in the tribe Thelymitreae (Orchidaceae: Neottioideae). *Pl. Syst. Evol.* 159: 19–47.

Cropper, S.C. & Calder, D.M. (1990). The floral biology of *Thelymitra epipactoides* (Orchidaceae), and the implications of pollination by deceit on the survival of this rare orchid. *Pl. Syst. Evol.* 170: 11–27.

Dafni, A. & Calder, D.M. (1987). Pollination by deceit and floral mimesis in *Thelymitra antennifera* (Orchidaceae). *Pl. Syst. Evol.* 158: 11–22.

Edens-Meier, R & Bernhardt, P. (2014). *Darwin's Orchids: Then and Now*. University of Chicago Press, Chicago.

Edens-Meier, R. *et al.* (2014). Floral fraudulence: do blue *Thelymitra* species (Orchidaceae) mimic *Orthrosanthus laxus* (Iridaceae)? *Telopea* 17: 15–28.

Indsto, J.O. *et al.* (2006). Pollination of *Diuris maculata* (Orchidaceae) by male *Trichocolletes venustus* bees. *Austral. J. Bot.* 54: 669–679.

Indsto, J.O. *et al.* (2007). Generalised pollination of *Diuris alba* R. Br. (Orchidaceae) by small bees and wasps. *Austral. J. Bot.* 55: 628–634.

Scaccabarozzi, D. *et al.* (2018). Masquerading as pea plants: behavioural and morphological evidence for mimicry of multiple models in an Australian orchid. *Ann. Bot.* 122: 1061–1073.

Scaccabarozzi, D. *et al.* (2020). Ecological factors driving pollination success in an orchid that mimics a range of pea plant species. *Bot. J. Linn. Soc.* 20: 1–17.

## 2.7 NEVER ENOUGH

Bierzychudek, P. (1981). *Asclepias*, *Lantana*, and *Epidendrum*: a floral mimicry complex? *Biotropica (Suppl.)* 13: 54–58.

Boyden, T.C. (1980). Floral mimicry by *Epidendrum ibaguense* (Orchidaceae) in Panama. *Evol.* 34: 135–136.

Cardoso-Gustavson *et al.* (2018). Unidirectional transitions in nectar gain and loss suggest food deception is a stable evolutionary strategy in *Epidendrum* (Orchidaceae): insights from anatomical and molecular evidence. *BMC Plant Biol.* 18: 179.

Deacon, N. (2000). Pollinia removal and visitation in *Epidendrum radicans* (Orchidaceae) and *Asclepias curassavica* (Asclepiadaceae). CIEE Program. Report accessed at: https://digitalcommons.usf.edu/tropical_ecology/622/

Dupree, S. (2004). Evidence for floral mimicry in *Epidendrum radicans* (Orchidaceae) with *Asclepias curassavica* (Apocynaceae) and *Latana camara* (Verbenaceae). CIEE Program. Report accessed at: https://digitalcommons.usf.edu/tropical_ecology/385/

Dressler, R. (1989). Will the real *Epidendrum ibaguense* please stand up? *Am. Orchid Soc. Bull.* 58: 796–800.

Fuhro, D. *et al.* (2010). Are there evidences of a complex mimicry system among *Asclepias curassavica* (Apocynaceae), *Epidendrum fulgens* (Orchidaceae), and

*Lantana camara* (Verbenaceae) in Southern Brazil? *Revista Brasil. Bot.* 33(4): 589–598.

Karremans, A.P. (2021). With great biodiversity comes great responsibility: the underestimated diversity of *Epidendrum* (Orchidaceae). *Harvard Pap. Bot.* 26: 299–369.

Pansarin, E.R. & Pansarin, L.M. (2014). Reproductive biology of *Epidendrum tridactylum* (Orchidaceae: Epidendroideae): a reward-producing species and its deceptive flowers. *Pl. Syst. Evol.* 300: 321–328.

Pansarin, E.R. & Pansarin, L.M. (2017). Crane flies and microlepidoptera also function as pollinators in *Epidendrum* (Orchidaceae: Laeliinae): the reproductive biology of *E. avicula*. *Pl. Sp. Biol.* 32: 200–209.

Shrestha, M. *et al.* (2020). Rewardlessness in orchids: how frequent and how rewardless? *Pl. Biol.* 22: 555–561.

Stpiczyńska, M. *et al.* (2018). Nectar-secreting and nectarless *Epidendrum*: structure of the inner floral spur. *Front. Pl. Sci.* 9: 840.

Wolfe, L.M. (1987). Inflorescence size and pollinaria removal in *Asclepias curassavica* and *Epidendrum radicans*. *Biotropica* 19: 86–89.

## 2.8 ORIGINAL PRANKSTER

Brodmann, J. *et al.* (2009). Orchid mimics honey bee alarm pheromone in order to attract hornets for pollination. *Curr. Biol.* 19: 1368–1372.

Dawkins, R. (2009). *The Greatest Show on Earth: The Evidence for Evolution*. Free Press, New York.

Dodson, C.H. (1990). *Brassia R.Br.* In: Escobar. R. (ed.) *Native Colombian Orchids, Vol. 1. Acacallis-Dryadella*. Editorial Colina, Medellín, pp. 52–53.

Johnson, S.D. (2005). Specialized pollination by spider-hunting wasps in the African orchid *Disa sankeyi*. *Pl. Syst. Evol.* 251: 153–160.

Ollerton, J. *et al.* (2003). The pollination ecology of an assemblage of grassland asclepiads in South Africa. *Ann. Bot.* 92: 807–834.

Ospina-Calderón, N.H. *et al.* (2007). Observaciones de la polinización y fenología reproductiva de *Brassia* cf. *antherotes* Rchb.f. (Orchidaceae) en un relicto de selva subandina en la Reserva Natural la Montaña del Ocaso en Quimbaya, Quindío (Colombia). *Universitas Scientiarum* 12: 83–95.

Punzo, F. (2005) Experience affects hunting behavior of the wasp, *Pepsis mildei* Stål (Hymenoptera: Pompilidae). *J. N. Y. Entomol. Soc.* 113: 222–229.

Ramírez-Benavides W. & Jansen-González, S. (2018). Flower visitation of *Passiflora apetala*, *P. auriculata* and *P. holosericea* (Passifloraceae) by *Pepsis aquila* (Hymenoptera: Pompilidae). *Fragmenta Entomologica* 50: 57–59.

Restrepo-Giraldo, C. *et al.* (2012). Temporal activity patterns of the spider wasp *Pepsis montezuma* Smith (Hymenoptera: Pompilidae) in a disturbed lower montane rainforest (Manizales, Colombia). *Psyche* 2012: 1–4.

Shuttleworth A. & Johnson, S.D. (2009). The importance of scent and nectar filters in a specialized wasp-pollination system. *Funct. Ecol.* 23: 931–940.

Shuttleworth A. & Johnson, S.D. (2012). The *Hemipepsis* wasp-pollination system in South Africa: a comparative analysis of trait convergence in a highly specialized plant guild. *Bot. J. Linn. Soc.* 168: 278–299.

Van der Pijl, L. & Dodson, C.H. (1966). *Orchid Flowers: Their Pollination and Evolution*. University of Miami Press, Coral Gables.

Vardy, C.R. (2000). The New World tarantula-hawk wasp genus *Pepsis* Fabricius (Hymenoptera: Pompilidae). Part 3. The *P. inclyta* to *P. auriguttata*-groups. *Zoologische Verhandelingen* 332: 1–86.

Williams, N.H. (1972). A reconsideration of *Ada* and the glumaceous brassias (Orchidaceae). *Brittonia* 24: 93–110.

## 2.9 FEAR OF THE DARK

Dentinger, B.T.M. & Roy, B.A. (2010). A mushroom by any other name would smell as sweet: *Dracula* orchids. *McIlvainea* 19: 1–13.

Endara, L. *et al.* (2010). Lord of the flies: pollination of *Dracula* orchids. *Lankesteriana* 10: 1–11.

Humeau, L. *et al.* (2011). Sapromyiophily in the native orchid, *Bulbophyllum variegatum*, on Réunion (Mascarene Archipelago, Indian Ocean). *J. Trop. Ecol.* 27: 591–599.

Ong, P.T. (2012). Notes on the pollination of *Bulbophyllum mandibulare* Rchb.f. *Malayan Orchid Rev.* 46: 67–68.

Ong, P.T. & Tan, K.H. (2011). Fly pollination in four Malaysian species of *Bulbophyllum* (Section *Sestochilus*) – *B. lasianthum*, *B. lobbii*, *B. subumbellatum* and *B. virescens*. *Malesian Orchid Journal* 8: 103–110.

Ong, P.T. & Tam, S.M. (2019). Pollination notes for seven *Bulbophyllum* species (section *Ephippium*) from Peninsular Malaysia. *Malesian Orchid Journal* 23: 87–96.

Policha, T. *et al.* (2016). Disentangling visual and olfactory signals in mushroom-mimicking *Dracula* orchids using realistic three-dimensional printed flowers. *N. Phytol.* 210: 1058–1071.

Urru, I. *et al.* (2011). Pollination by brood-site deception. *Phytochemistry* 72: 1655–1666.

## 2.10 IT'S A LONG WAY TO THE TOP

Atwood, J.T. (1985). Pollination of *Paphiopedilum rothschildianum*: brood-site deception. *Nat. Geog. Res.* 1: 247–254.

Bänziger, H. (1996). The mesmerizing wart: the pollination strategy of epiphytic lady-slipper orchid *Paphiopedilum villosum* (Lindl.) Stein (Orchidaceae). *Bot. J. Linn. Soc.* 121: 59–90.

Darwin Correspondence Project, "Letter no. 3595," accessed on 19 July 2022, https://www.darwinproject.ac.uk/letter/?docId=letters/DCP-LETT-3595.xml

Darwin Correspondence Project, "Letter no. 3795," accessed on 19 July 2022, https://www.darwinproject.ac.uk/letter/?docId=letters/DCP-LETT-3795.xml

Darwin, C. (1862). *On the Various Contrivances by which British and Foreign Orchids are Fertilised by Insects, and on the Good Effects of Intercrossing*. John Murray, London.

Darwin, C. (1877). *The Various Contrivances by which Orchids are Fertilised by Insects. 2nd edition*. John Murray, London.

Díaz-Morales, M. *et al.* (2019). Reproductive biology of *Phragmipedium longifolium* and floral anatomy associated with pollinator attraction in the genus. In *Proceedings of the 22nd World Orchid Conference, Guayaquil, Ecuador, 2017*, Pridgeon, A.M. & Arosema, A. (eds), Asociación Ecuatoriana de Orquideología, Guayaquil, pp. 92–97.

Dodson, C. (1966). Studies in orchid pollination: *Cypripedium*, *Phragmipedium* and allied genera. *Am. Orch. Soc. Bull.* 35: 125–128.

Ferguson, C.S. *et al.* (2005). *Cypripedium fasciculatum* (Orchidaceae) anthesis and fruit set in relationship to diapriid activity. *Selbyana* 26(1,2): 103–113.

Gray, A. (1862). Dimorphism in the genitalia of flowers. *Am. J. Sci. Arts* 34(100–102): 419–429.

Jiang, H. *et al.* (2020). *Cypripedium subtropicum* (Orchidaceae) employs aphid colony mimicry to attract hoverfly (Syrphidae) pollinators. *New Phytologist* 227: 1213–1221.

Li, P. *et al.* (2012). Fly pollination in *Cypripedium*: a case study of sympatric C. *sichuanense* and C. *micranthum*. *Bot. J. Linn. Soc.* 170: 50–58.

Liu, Q. *et al.* (2020). Reproductive ecology of *Paphiopedilum spicerianum*: implications for conservation of a critically endangered orchid in China. *Glob. Ecol. Conserv.* 23: e01063.

Pemberton, R.W. (2011). Pollination studies in phragmipediums: flower fly (Syrphidae) pollination and mechanical self-pollination (autogamy) in *Phragmipedium* species (Cypripedioideae). *Orchids* 80: 364–367.

Pemberton, R.W. (2013). Pollination of slipper orchids (Cypripedioideae): a review. *Lankesteriana* 13(1–2): 65–73.

Ren, Z. *et al.* (2011). Flowers of *Cypripedium fargesii* (Orchidaceae) fool flat-footed flies (Platypezidae) by faking fungus-infected foliage. *PNAS* 108: 7478–7480.

Shi, J. *et al.* (2009). Pollination by deceit in *Paphiopedilum barbigerum* (Orchidaceae): a staminode exploits the innate colour preferences of hoverflies (Syrphidae). *Pl. Biol.* 11: 17–28.

Zheng, C.C. *et al.* (2022). *Cypripedium lichiangense* (Orchidaceae) mimics a humus-rich oviposition site to attract its female pollinator, *Ferdinandea cuprea* (Syrphidae). *Pl. Biol.* 24: 145–156.

# CHAPTER 3 REWARD

Ackerman, J.D. (1986). Mechanisms and evolution of food-deceptive pollination systems in orchids. *Lindleyana* 1(2): 108–113.

## 3.1 EVERYBODY KNOWS

Arditti, J. *et al.* (2012). 'Good Heavens what insect can suck it' – Charles Darwin, *Angraecum sesquipedale* and *Xanthopan morganii praedicta*. Bot. J. Linn. Soc. 169: 403–432.

Boberg, E. *et al.* (2014). Pollinator shifts and the evolution of spur length in the moth-pollinated orchid *Platanthera bifolia*. Ann. Bot. 113: 267–275.

Campbell, G.D. (1867). *The Reign of Law*. A. Strahan, London.

Darwin Correspondence Project, "Letter no. 3411," accessed on 18 July 2022, https://www.darwinproject.ac.uk/letter/?docId=letters/DCP-LETT-3411.xml

Darwin Correspondence Project, "Letter no. 3421," accessed on 18 July 2022, https://www.darwinproject.ac.uk/letter/?docId=letters/DCP-LETT-3421.xml

Darwin Correspondence Project, "Letter no. 3563," accessed on 18 July 2022, https://www.darwinproject.ac.uk/letter/?docId=letters/DCP-LETT-3563.xml

Darwin Correspondence Project, "Letter no. 3710," accessed on 18 July 2022, https://www.darwinproject.ac.uk/letter/?docId=letters/DCP-LETT-3710.xml

Darwin Correspondence Project, "Letter no. 5637," accessed on 18 July 2022, https://www.darwinproject.ac.uk/letter/?docId=letters/DCP-LETT-5637.xml

Darwin Correspondence Project, "Letter no. 5648," accessed on 18 July 2022, https://www.darwinproject.ac.uk/letter/?docId=letters/DCP-LETT-5648.xml

Darwin, C. (1862). On the Various Contrivances by which British and Foreign Orchids are Fertilised by Insects, and on the Good Effects of Intercrossing. John Murray, London.

Darwin, C. (1877). *The various contrivances by which orchids are fertilised by insects. 2nd edition*. John Murray, London.

Edens-Meier, R. & Bernhardt, P. (2014). *Darwin's Orchids: Then and Now*. University of Chicago Press, Chicago.

Micheneau, C. *et al.* (2009). Orchid pollination: from Darwin to the present day. *Bot. J. Linn. Soc.* 161(1): 1–19.

Nielsen, L.J. & Møller, B.L. (2015). Scent emission profiles from Darwin's orchid — *Angraecum sesquipedale*: investigation of the aldoxime metabolism using clustering analysis. *Phytochemistry* 120: 3–18.

Nilsson, L.A. *et al.* (1985). Monophily and pollination mechanisms in *Angraecum arachnites* Schltr. (Orchidaceae) in a guild of long-tongued hawk-moths (Sphingidae) in Madagascar. *Biol. J. Linn. Soc.* 26: 1–19.

Rothschild, L.W. & Jordan, K. (1903). A revision of the lepidopterous family Sphingidae. *Novitates Zoologicae Supplement* 9: 1–972.

Wallace, A.R. (1867). Creation by law. *Quart. J. Sci.* 4: 471–488.

Wasserthal, L.T. (1997). The pollinators of the Malagasy star orchids *Angraecum sesquipedale*, *A. sororium*, and *A. compactum* and the evolution of extremely long spurs by pollinator shift. *Botanica Acta* 110: 343–359.

Wasserthal, L.T. (2015). *Angraecum*-Orchideen und langrüsslige Schwärmer, Bestäubung und Evolution. *Die Orchidee* 66(3): 175–181.

## 3.2 LINGER

Chase, M.W. (1985). Pollination of *Pleurothallis endotrachys*. *Am. Orch. Soc. Bull.* 54: 431–434.

Karremans, A.P. *et al.* (2015). Pollination of *Specklinia* by nectar-feeding *Drosophila*: the first reported case of a deceptive syndrome employing aggregation pheromones in Orchidaceae. *Ann. Bot.* 116(3): 437–455.

Karremans, A.P. & Díaz-Morales, M. (2019). The Pleurothallidinae: extremely high speciation driven by pollinator adaptation. In *Proceedings of the 22nd World Orchid Conference, Guayaquil, Ecuador, 2017*, Pridgeon, A.M. & Arosema, A. (eds), Asociación Ecuatoriana de Orquideología, Guayaquil, pp. 363–388.

Pupulin, F. *et al.* (2012). A reconsideration of the empusellous species of *Specklinia* (Orchidaceae: Pleurothallidinae). *Phytotaxa* 63: 1–20.

Pupulin, F. *et al.* (2013). Taxonomy in watercolor: the *Specklinia endotrachys* group, part 1. *Die Orchidee* 64(5): 392–399.

## 3.3 MISUNDERSTOOD

Dressler, R.L. (1971). Dark pollinia in hummingbird-pollinated orchids or do hummingbirds suffer from strabismus? *Am. Nat.* 105(941): 80–83.

Mayr, G. (2004). Old World fossil record of modern-type hummingbirds. *Science* 304(5672): 861–864.

McGuire, J.A. *et al.* (2014). Molecular phylogenetics and the diversification of hummingbirds. *Curr. Biol.* 24: 910–916.

Neubig, K.M. *et al.* (2015). Nectary structure and nectar in *Sobralia* and *Elleanthus* (Sobralieae: Orchidaceae). *Lankesteriana* 15(2): 113–127.

Nunes, C.E.P. *et al.* (2016). Pollination ecology of two species of *Elleanthus* (Orchidaceae): novel mechanisms and underlying adaptations to hummingbird pollination. *Pl. Biol. (Stuttg.)* 18(1): 15–25.

Pansarin, E.R. & Amaral, M.C. (2008). Reproductive biology and pollination mechanisms of *Epidendrum secundum* (Orchidaceae). Floral variation: consequence of natural hybridization? *Pl. Biol. (Stuttg.)* 10(2): 211–219.

Pansarin, E.R. *et al.* (2015). Floral features, pollination biology, and breeding system of *Comparettia coccinea* (Orchidaceae: Oncidiinae). *Flora* 217: 57–63.

Pansarin, E.R. *et al.* (2018). Comparative reproductive biology reveals two distinct pollination strategies in Neotropical twig-epiphyte orchids. *Pl. Syst. Evol.* 304(6): 793–806.

Rodríguez-Robles, J.A. *et al.* (1992). The effects of display size, flowering phenology, and standing crop of nectar on the visitation frequency of *Comparettia falcata* (Orchidaceae). *Am. J. Bot.* 79: 1009–1017.

Siegel, C. (2011). Orchids and hummingbirds: sex in the fast lane. Part 1 of orchids and their pollinators. *Orchid Digest* 75: 8–17.

Singer, R.B. & Sazima, M. (2000). The pollination of *Stenorrhynchos lanceolatus* (Aublet) L.C.Rich. (Orchidaceae: Spiranthinae) by hummingbirds in Southeastern Brazil. *Pl. Syst. Evol.* 223: 221–227.

Stpiczyńska, M. *et al.* (2004). Nectary structure and nectar secretion in *Maxillaria coccinea* (Jacq.) L.O. Williams ex Hodge (Orchidaceae). *Ann. Bot.* 93: 87–95.

Stpiczyńska, M. *et al.* (2009). Nectary structure of *Ornithidium sophronitis* Rchb.f. (Orchidaceae: Maxillariinae). *Acta Agrobot.* 62(2): 3–12.

## 3.4 MADNESS

Aleva, J.F. *et al.* (1996). The history of the Herbarium Vadense. In *Herbarium Vadense, 1896–1996*, Breteler, F.J. & Sosef, M.S.M. (eds), Backhuys Publishers, Leiden, pp. 11–24.

Blume, C.L. (1825). *Bijdragen tot de Flora van Nederlandsch Indië. 8ste Stuk*. Ter Lands Drukkerij, Batavia, pp. 355–434.

De Vogel, E.F. (1969). Monograph of the tribe Apostasieae (Orchidaceae). *Blumea* 17: 313–350.

Gregg, K.B. (1991). Defrauding the deceitful orchid pollen collection by pollinators of *Cleistes divaricata* and *Cleistes bifaria*. *Lindleyana* 6(4): 214–220.

Inoue, K. *et al.* (1995). Pollination ecology of *Dendrobium setifolium*, *Neuwiedia borneensis*, and *Lecanorchis multiflora* (Orchidaceae) in Sarawak. *Tropics* 5(1/2): 95–100.

Kalkman, C. (1979). The Rijksherbarium, in past and present. *Blumea* 25: 13–26.

Kocyan, A. & Endres, P.K. (2001). Floral structure and development in *Apostasia* and *Neuwiedia* (Orchidaceae) and their relationships to other Orchidaceae. *Int. J. Plant Sci.* 162(4): 847–867.

Okada, H. *et al.* (1996). Pollination system of *Neuwiedia veratrifolia* Blume (Orchidaceae, Apostasioideae) in the Malesian wet tropics. *Acta Phytotax. Geobot.* 47(2): 173–181.

Pansarin, E.R. & Amaral, M.C.E. (2008). Pollen and nectar as a reward in the basal epidendroid *Psilochilus modestus* (Orchidaceae: Triphoreae): a study of floral morphology, reproductive biology and pollination strategy. *Flora* 203: 474–483.

Pansarin, E.R. & Pansarin, L.M. (2014). Floral biology of two Vanilloideae (Orchidaceae) primarily adapted to pollination by euglossine bees. *Plant Biology* 16: 1104–1113.

Pridgeon, A.M. *et al.* (1999). *Genera Orchidacearum, Volume 1: General Introduction, Apostasioideae, Cypripedioideae*. Oxford University Press, Oxford.

## 3.5 TWO OUT OF THREE AIN'T BAD

Davies, K.L. *et al.* (2000). Pseudopollen: its structure and development in *Maxillaria* (Orchidaceae). *Ann. Bot.* 85: 887–895.

Davies, K.L. *et al.* (2002). Pseudopollen and food-hair diversity in *Polystachya* Hook. (Orchidaceae). *Ann. Bot.* 90: 477–484.

Davies, K.L. (2009). Food-hair form and diversification in orchids. In *Orchid Biology: Reviews and Perspectives*, X., Kull, T., Arditti, J. & Wong, S.M. (eds), Springer Science + Business Media BV, Dordrecht, pp. 159–183.

Davies, K.L. & Turner, M.P. (2004). Pseudopollen in *Dendrobium unicum* Seidenf. (Orchidaceae): reward or deception? *Ann. Bot.* 94: 129–132.

Davies, K.L. & Turner, M.P. (2004). Pseudopollen in *Eria* Lindl. Section *Mycaranthes* Rchb.f. (Orchidaceae). *Ann. Bot.* 94: 707–715.

Davies, K.L. *et al.* (2013). Dual deceit in pseudopollen-producing *Maxillaria* s.s. (Orchidaceae: Maxillariinae). *Bot. J. Linn. Soc.* 173: 744–763.

Janse, J.M. (1886). Imitirte pollenkorner bei *Maxillaria* sp. *Ber. Deutsch. Bot. Ges.* 4: 277–283.

Krahl, A.H. *et al.* (2019). Study of the reproductive biology of an Amazonian *Heterotaxis* (Orchidaceae) demonstrates the collection of resin-like material by stingless bees. *Plant Syst. Evol.* 305: 281–291.

Leonhardt, S.D. *et al.* (2007). Foraging loads of stingless bees and utilisation of stored nectar for pollen harvesting. *Apidologie* 38: 125–135.

Pansarin, E.R. & Amaral, M.C.E. (2006). Biologia reprodutiva e polinização de duas espécies de *Polystachya* Hook. no Sudeste do Brasil: evidência de pseudocleistogamia em Polystachyeae (Orchidaceae). *Rev. Bras. Bot.* 26: 423–432.

Pansarin, E.R. & Maciel, A.A. (2017). Evolution of pollination systems involving edible trichomes in orchids. *AoB PLANTS* 9: 10.1093/aobpla/plx033.

Singer, R.B. & Koehler, S. (2004). Pollinarium morphology and floral rewards in Brazilian Maxillariinae. *Ann. Bot.* 93: 39–51.

## 3.6 DARK NECESSITIES

Bogarín, D. *et al.* (2018). Pollination of *Trichosalpinx* (Orchidaceae: Pleurothallidinae) by biting midges (Diptera: Ceratopogonidae). *Bot. J. Linn. Soc.* 186: 510–543.

Bogarín, D. *et al.* (2019). Floral anatomy and evolution of pollination syndromes in *Lepanthes* and close relatives. In *Proceedings of the 22nd World Orchid Conference, Guayaquil, Ecuador, 2017*, Pridgeon, A.M. & Arosema, A. (eds), Asociación Ecuatoriana de Orquideología, Guayaquil, pp. 396–410.

Heiduk, A. *et al.* (2015). Deceptive *Ceropegia dolichophylla* fools its kleptoparasitic fly pollinators with exceptional floral scent. *Front. Ecol. Evol.* 3: 1–13.

Karremans, A.P. & Díaz-Morales, M. (2019). The Pleurothallidinae: extremely high speciation driven by pollinator adaptation. In *Proceedings of the 22nd World Orchid Conference, Guayaquil, Ecuador, 2017*. Pridgeon, A.M. & Arosema, A. (eds), Asociación Ecuatoriana de Orquideología, Guayaquil, pp. 363–388.

Karremans, A.P. & Vieira-Uribe, S. (2020). *Pleurothallids Neotropical Jewels, Vol 1*. Mariscal, Quito.

Oelschlägel, B. *et al.* (2015). The betrayed thief – the extraordinary strategy of *Aristolochia rotunda* to deceive its pollinators. *New Phytol.* 206: 342–351.

## 3.7 YOU'RE SO VAIN

Ackerman, J.D. (1983). Specificity and mutual dependency of the orchid-euglossine bee interaction. *Biol. J. Linn. Soc.* 20: 301–314.

Crüger, H. (1865). A few notes on the fecundation of orchids and their morphology. *Bot. J. Linn. Soc.* 8: 129–135.

Dressler, R.L. (1968). Pollination by euglossine bees. *Evolution* 22: 202–210.

Hetherington-Rauth M. & Ramírez, S.R. (2015). Evolutionary trends and specialization in the euglossine bee-pollinated orchid genus *Gongora*. *Ann. Missouri Bot. Gard.* 100: 271–299.

Ramírez, S.R. *et al.* (2010). Phylogeny, diversification patterns and historical biogeography of euglossine orchid bees (Hymenoptera: Apidae). *Biol. J. Linn. Soc.* 100: 552–572.

## 3.8 THAT SMELL

Adachi, S.A. (2015). Structural and ultrastructural characterization of the floral lip in *Gongora bufonia* (Orchidaceae): understanding the slip-and-fall pollination mechanism. *Botany* 93: 759–768.

Allen, P.H. (1954). Pollination in *Gongora maculata*. *Ceiba* 4: 121–125.

Braga, P.I.S. (1976). Estudos da flora orquidológica do Estado do Amazonas. 1-Descrição e observação da biologia floral de *Stanhopea candida* Barb. Rodr. *Acta Amazon.* 6: 433–438.

Casique, J.V. *et al.* (2020). Novelties in the secretory structures of three species of *Gongora* (Orchidaceae: Stanhopeinae). *Biol. J. Linn. Soc.* XX: 1–21.

Dressler, R.L. (1966). Some observations on *Gongora*. *Orchid Digest* 30: 220–223.

Dressler, R.L. (1968). Pollination by euglossine bees. *Evolution* 22: 202–210.

Gerlach, G. (2021). Beobachtungen zur Gattung *Gongora* in Peru. *OrchideenJournal* 28(4): 137–142.

Hetherington-Rauth, M.C. & Ramírez, S.R. (2016). Evolution and diversity of floral scent chemistry in the euglossine bee-pollinated orchid genus *Gongora*. *Ann. Bot.* 118: 135–148.

Martini, P. *et al.* (2003). Pollination, flower longevity, and reproductive biology of *Gongora quinquenervis* Ruíz and Pavón (Orchidaceae) in an Atlantic Forest fragment of Pernambuco, Brazil. *Plant Biol.* 5: 495–503.

Nunes, C.E.P. *et al.* (2017). Two orchids, one scent? Floral volatiles of *Catasetum cernuum* and *Gongora bufonia* suggest convergent evolution to a unique pollination niche. *Flora* 232: 207–216.

Pansarin, E.R. & Amaral, M.C.E. (2009). Reproductive biology and pollination of southeastern Brazilian *Stanhopea* Frost ex Hook. (Orchidaceae). *Flora* 204: 238–249.

## 3.9 WHAT HAVE I DONE TO DESERVE THIS?

Allen, P.H. (1950). Pollination in *Coryanthes speciosa*. *Amer. Orch. Soc. Bull.* 19: 528–536.

Benzing, D.H. & Clements, M.A. (1991). Dispersal of the orchid *Dendrobium insigne* by the ant *Iridomyrmex cordatus* in Papua New Guinea. *Biotropica* 23: 604–607.

Chomicki, G. *et al.* (2017). The assembly of ant-farmed gardens: mutualism specialization following host broadening. *Proc. R. Soc. B* 284: 20161759.

Crüger, H. (1865). A few notes on the fecundation of orchids and their morphology. *Bot. J. Linn. Soc.* 8: 129–135.

Dodson, C.H. (1965). Studies in orchid pollination — the genus *Coryanthes*. *Amer. Orch. Soc. Bull.* 34: 680–687.

Gerlach, G. (2011). The genus *Coryanthes*: a paradigm in ecology. *Lankesteriana* 11: 253–264.

Morales-Linares, J. *et al.* (2018). Orchid seed removal by ants in Neotropical ant-gardens. *Plant Biol.* 20(3): 525–530.

Nazarov, V.V. & Gerlach, G. (1997). The potential seed productivity of orchid flowers and peculiarities of their pollination systems. *Lindleyana* 12(4): 188–204.

Orivel, J. & Leroy, C. (2011). The diversity and ecology of ant gardens (Hymenoptera: Formicidae; Spermatophyta: Angiospermae). *Myrmecol. News* 14: 73–85.

Schnepf, E. *et al.* (1983). On the fine structure of the liquid producing floral gland of the orchid, *Coryanthes speciosa*. *Nordic J. Bot.* 3: 479–491.

## 3.10 HURT

Darwin, C. (1862). On the three remarkable sexual forms of *Catasetum tridentatum*, an orchid in the possession of the Linnean Society. *Bot. J. Linn. Soc.* 6: 151–157.

Darwin, C. (1862). On the Various Contrivances by which British and Foreign Orchids are Fertilised by Insects, and on the Good Effects of Intercrossing. John Murray, London.

Darwin, C. (1877). The Various Contrivances by which Orchids are Fertilised by Insects. 2nd edition. John Murray, London.

Davis, W. (2005). Decadence and the organic metaphor. *Representations* 89: 131–149.

Franken, E.P. *et al.* (2016). Osmophore diversity in the *Catasetum* alliance (Orchidaceae: Catasetinae). *Lankesteriana* 16: 317–327.

Gregg, K.B. (1982). Sunlight-enhanced ethylene evolution by developing inflorescences of *Catasetum* and *Cycnoches* and its relation to female flower production. *Bot. Gaz.* 143: 466–475.

Lindley, J. (1826). *Catasetum cristatum*. *Edwards's Bot. Reg.* 12: t. 966.

Lindley, J. (1832). *Cirrhaea loddigesii*. *Edwards's Bot. Reg.* 18: t. 1538.

Lindley, J. (1837). *Monachanthi et Myanthi cristati proles biformis*. *Edwards's Bot. Reg.* 23: t. 1951A.

Lindley, J. (1853). *The Vegetable Kingdom*. Bradbury & Evans, London.

Milet-Pinheiro, P. *et al.* (2015). Pollination biology in the dioecious orchid *Catasetum uncatum*: how does floral scent influence the behaviour of pollinators? *Phytochemistry* 116: 149–161.

Pérez-Escobar, O.A. *et al.* (2016). Sex and the Catasetinae (Darwin's favourite orchids). *Mol. Phylogenet. Evol.* 97: 1–10.

Pérez-Escobar, O.A. *et al.* (2017). Multiple geographical origins of environmental sex determination enhanced the diversification of Darwin's favourite orchids. *Sci. Rep.* 7: 12878.

Rolfe, R.A. (1891). On the sexual forms of *Catasetum*, with special reference to the researches of Darwin and others. *Bot. J. Linn. Soc.* 27: 206–227.

Romero, G.A. & Nelson, C.E. (1986). Sexual dimorphism in *Catasetum* orchids: forcible pollen emplacement and male flower competition. *Science* 232: 1538–1540.

Schomburgk, R. (1837). On the identity of three supposed genera of orchidaceous epiphytes. *Trans. Linn. Soc.* 17: 551–552.

## 3.11 WHOLE LOTTA LOVE

Fletcher, B.S. (1987). The biology of dacine fruit flies. *Annu. Rev. Entomol.* 32: 115–144.

Nakashira, M. *et al.* (2018). Floral synomone diversification of *Bulbophyllum* sibling species (Orchidaceae) in attracting fruit fly pollinators. *Biochem. Syst. Ecol.* 81: 86–95.

Ong, P.T. & O'Byrne, P. (2015). *Bulbophyllum praetervisum*, a new record of Orchidaceae for the flora of Peninsular Malaysia. *Malesian Orchid Journal* 15: 53–60.

Ridley, H.N. (1890). On the method of fertilisation in *Bulbophyllum macranthum*, and allied orchids. *Ann. Bot.* 4: 327–336.

Shamshir, R.A. & Wee, S.L. (2019). Zingerone improves mating performance of *Zeugodacus tau* (Diptera: Tephritidae) through enhancement of male courtship activity and sexual signaling. *J. Insect. Physiol.* 119: 103949.

Shelly, T.E. (2000). Flower-feeding affects mating performance in male oriental fruit flies *Bactrocera dorsalis*. *Ecol. Entomol.* 25: 109–114.

Shelly, T.E. (2010). Effects of methyl eugenol and raspberry ketone/cue lure on the sexual behavior of *Bactrocera* species (Diptera: Tephritidae). *Appl. Entomol. Zool.* 45: 349–361.

Tan, K.H. & Nishida, R. (1998). Ecological significance of male attractant in the defence and mating strategies of the fruit fly pest, *Bactrocera papayae*. *Entomol. Exp. Appl.* 89: 155–158.

Tan, K.H. & Nishida, R. (2000). Mutual reproductive benefits between a wild orchid, *Bulbophyllum patens*, and *Bactrocera* fruit flies via a floral synomone. *J. Chem. Ecol.* 26: 533–546.

Tan, K.H. & Nishida, R. (2012). Methyl eugenol: its occurrence, distribution, and role in nature, especially in relation to insect behavior and pollination. *J. Insect Sci.* 12: 1–60.

Tan, K.H. *et al.* (2002). Floral synomone of a wild orchid, *Bulbophyllum cheiri*, lures *Bactrocera* fruit flies for pollination. *J. Chem. Ecol.* 28: 1161–1172.

Wei, D. *et al.* (2015). Female remating inhibition and fitness of *Bactrocera dorsalis* (Diptera: Tephritidae) associated with male accessory glands. *Fla. Entomol.* 98(1): 52–58.

## 3.12 GIMME SHELTER

Claessens, J. & Kleynen, J. (2011). *The Flower of the European Orchid: Form and Function*. Schrijen-Lippertz, Voerendaal.

Dafni, A. *et al.* (1981). Pollination of *Serapias vomeracea* Briq. (Orchidaceae) by imitation of holes for sleeping solitary male bees (Hymenoptera). *Acta Bot. Neerl.* 30: 69–73.

Godfery, M.J. (1920). The fertilisation of *Serapias cordigera* and *S. longipetala*. *Gard. Chron.* 1920: 70.

Godfery, M.J. (1928). A male bee *Anthidium septemdentatum*, Latr., bearing on its head the pollinia from two orchids of the genus *Serapias*. *Proc. Entomol. Soc. London* 3: 38.

Godfery, M.J. (1928). The fertilisation of orchids of the genus *Serapias* probably effected by bees seeking the flowers as a nocturnal shelter. *Proc. Entomol. Soc. London* 3: 60–61.

Godfery, M.J. (1929). Further evidence that the orchid *Serapias cordigera* is fertilised by bees which use the flowers as a nocturnal shelter. *Proc. Entomol. Soc. London* 4: 105–106.

Godfery, M.J. (1931). The pollination of *Coeloglossum*, *Nigritella*, *Serapias*, etc. *J. Bot.* 60: 129–130.

Kirchner, O. (1900). Mitteilungen über die Bestäubungseinrichtungen der Blüten. *Jahresh. Ver. Vaterl. Naturk. Württemberg* 56: 347–384.

Pellegrino, G. *et al.* (2005). Reproductive biology and pollinator limitation in a deceptive orchid, *Serapias vomeracea* (Orchidaceae). *Plant Species Biol.* 20: 33–39.

Pellegrino, G. *et al.* (2012). Comparative analysis of floral scents in four sympatric species of *Serapias* L. (Orchidaceae): clues on their pollination strategies. *Plant Syst. Evol.* 298: 1837–1843.

Pellegrino, G. *et al.* (2017). Functional differentiation in pollination processes among floral traits in *Serapias* species (Orchidaceae). *Ecol. Evol.* 7(18): 7171–7177.

Tamer, C.E. *et al*. (2006). A traditional Turkish beverage: salep. *Food Rev. Int.* 22: 43–50.

Vereecken, N.J. *et al*. (2012). Pre-adaptations and the evolution of pollination by sexual deception: Cope's rule of specialization revisited. *Proc. R. Soc. B* 279(1748): 4786–4794.

Vereecken, N.J. *et al*. (2013). A pollinators' eye view of a shelter mimicry system. *Ann. Bot.* 111: 1155–1165.

## CHAPTER 4  MISFITS

Ackerman, J.D. *et al*. 2023. Beyond the various contrivances by which orchids are pollinated: global patterns in orchid pollination biology. *Bot. J. Linn. Soc.* boac082, https://doi.org/10.1093/botlinnean/boac082

Grimaldi, D. & Engel, M.S. (2005). *Evolution of the Insects*. Cambridge University Press, Cambridge.

Gullan, P.J. & Cranston, P.S. (2014). *The Insects: An Outline of Entomology, 5th Edition*. Blackwell Publishing, Malden.

Mayhew, P.J. (2018). Explaining global insect species richness: lessons from a decade of macroevolutionary entomology. *Entomol. Exp. Appl.* 166(4): 225–250.

### 4.1 YOU CAN'T ALWAYS GET WHAT YOU WANT

Adit, A. *et al*. (2022). Breeding system and response of the pollinator to floral larceny and florivory define the reproductive success in *Aerides odorata*. *Front. Plant Sci.* 12: 767725.

Almeida, A.M. & Figueiredo, R.A. (2003). Ants visit nectaries of *Epidendrum denticulatum* (Orchidaceae) in a Brazilian rainforest: effects on herbivory and pollination. *Braz. J. Biol.* 63(4): 551–558.

Bates, R. (1979). Pollination of orchids. (10). *Leporella fimbriata* and its ant pollinators. *J. Native Orchid Soc. S. Australia* 3(11): 9–10.

Baumann, B. & Baumann, H. (2010). Pollination of *Chamorchis alpina* (L.) Rich. in the alps by worker ants of *Formica lemani* Bondroit: first record of ant pollination in Europe. *J. Eur. Orch.* 42: 3–20.

Beatie, A.J. *et al*. (1984). Ant inhibition of pollen function: a possible reason why ant pollination is rare. *Amer. J. Bot.* 71(3): 421–426.

Brantjes, N.B.M. (1981). Ant, bee and fly pollination in *Epipactis palustris* (L.) Crantz (Orchidaceae). *Acta Bot. Neerl.* 30: 59–68.

Claessens, J. & Seifert, B. (2017). Significant ant pollination in two orchid species in the Alps as adaptation to the climate of the alpine zone? *Tuexenia* 37: 363–374.

Damon, A. & Pérez-Soriano, M.A. (2005). Interaction between ants and orchids in the Soconusco region, Chiapas, Mexico. *Entomotropica* 20(1): 59–65.

De Vega, C. *et al*. (2014). Floral volatiles play a key role in specialized ant pollination. *Perspect. Plant Ecol.* 16: 32–42.

García-Gila, J. & Blasco-Aróstegui, J. (2021). First report on the pollination of *Neotinea maculata* (Orchidaceae) by minor worker ants of the *Temnothorax exilis* group (Hymenoptera: Formicidae). *Mediterr. Bot.* 42: e71171.

Koopowitz, H. & Marchant, T.A. (1998). Postpollination nectar reabsorption in the African epiphyte *Aerangis verdickii* (Orchidaceae). *Amer. J. Bot.* 85: 508–512.

Peakall, R. (1989). The unique pollination of *Leporella fimbriata* (Orchidaceae): pollination by pseudocopulating male ants (*Myrmecia urens*, Formicidae). *Pl. Syst. Evol.* 167: 137–148.

Peakall, R. & Beattie, A.J. (1989). Ecological and genetic consequences of pollination by sexual deception in the orchid *Caladenia tentactulata*. *Funct. Ecol.* 3: 515–522.

Peakall, R. *et al.* (1987). Pseudocopulation of an orchid by male ants: a test of two hypotheses accounting for the rarity of ant pollination. *Oecologia* 73: 522–524.

Rico-Gray, V. & Thien, L.B. (1989). Effect of different ant species on reproductive fitness of *Schomburgkia tibicinis* (Orchidaceae). *Oecologia* 81: 487–489.

Robert, D.L. (2007). Observations on the effect of nectar-robbery on the reproductive success of *Aeranthes arachnitis* (Orchidaceae). *Lankesteriana* 7(3): 509–514.

Schiestl, F.P & Glaser, F. (2012). Specific ant-pollination in an alpine orchid and the role of floral scent in attracting pollinating ants. *Alp Bot.* 122: 1–9.

Siegel, C. (2014). Orchids and Formicidae: ants in your plants. *Orchid Digest* 78: 150–161.

Subedi, A. *et al.* (2011). Pollination and protection against herbivory of Nepalese Coelogyninae (Orchidaceae). *Am. J. Bot.* 98: 1095–1103.

Sugiura, N. & Yamaguchi, T. (1997). Pollination of *Goodyera foliosa* var. *maximowicziana* (Orchidaceae) by the bumblebee *Bombus diversus diversus*. *Plant Species Biol.* 12: 9–14.

Sugiura, N. *et al.* (2006). A supplementary contribution of ants in the pollination of an orchid, *Epipactis thunbergii*, usually pollinated by hover flies. *Plant Syst. Evol.* 258: 17–26.

Zhongjian, L. *et al.* (2008). *Chenorchis*, a new orchid genus, and its eco-strategy of ant pollination. *Acta Ecologica Sinica* 28: 2433–2444.

## 4.2 WITH A LITTLE HELP FROM MY FRIENDS

Claessens, J. & Kleynen, J. (2011). *The Flower of the European Orchid: Form and Function*. Schrijen-Lippertz, Voerendaal.

Claessens, J. & Seifert, B. (2017). Significant ant pollination in two orchid species in the Alps as adaptation to the climate of the alpine zone? *Tuexenia* 37: 363–374.

Claessens, J. & Seifert, B. (2018). Ant pollination of *Dactylorhiza viridis*. *Orchid Digest* 154–158.

Cohen, C. *et al.* (2021). Sexual deception of a beetle pollinator through floral mimicry. *Curr. Biol.* 31(9): 1962–1969.E6.

Darwin Correspondence Project, "Letter no. 3956," accessed on 21 July 2022, https://www.darwinproject.ac.uk/letter/?docId=letters/DCP-LETT-3956.xml

Johnson, S.D. (2007). Specialization for pollination by beetles and wasps: the role of lollipop hairs and fragrance in *Satyrium microrrhynchum* (Orchidaceae). *Am. J. Bot.* 94: 47–55.

Nunes, C.E.P. *et al.* (2018). Parasitoids turn herbivores into mutualists in a nursery system involving active pollination. *Curr. Biol.* 28: 980–986.

Pedersen, H.Æ. *et al.* (2013). Pollination biology of *Luisia curtisii* (Orchidaceae): indications of a deceptive system operated by beetles. *Plant Syst. Evol.* 299: 177–185.

Peter, C.I., (2009). Pollination, floral deception and evolutionary processes in *Eulophia* (Orchidaceae) and its allies. PhD thesis, University of KwaZulu-Natal, South Africa.

Peter, C.I. & Johnson, S.D. (2006). Anther cap retention prevents self-pollination by elaterid beetles in the South African orchid *Eulophia foliosa*. *Ann. Bot.* 97: 345–355.

Peter, C.I. & Johnson, S.D. (2009). Pollination by flower chafer beetles in *Eulophia ensata* and *Eulophia welwitschii* (Orchidaceae). *S. Afr. J. Bot.* 75: 762–770.

Peter, C.I. & Johnson, S.D. (2014). A pollinator shift explains floral divergence in an orchid species complex in South Africa. *Ann. Bot.* 113: 277–288.

Singer, R.B. & Cocucci, A.A. (1997). Pollination of *Pteroglossaspis ruwenzoriensis* (Rendle) Rolfe (Orchidaceae) by beetles in Argentina. *Bot. Acta* 110(4): 338–342.

Steiner, K.E. *et al.* (1994). Floral and pollinator divergence in two sexually deceptive South African orchids. *Am. J. Bot.* 81: 185–194.

Steiner, K.E. (1998). The evolution of beetle pollination in a South African orchid. *Am. J. Bot.* 85(9): 1180–1193.

Stork, N.E. *et al.* (2015). New approaches narrow global species estimates for beetles, insects, and terrestrial arthropods. *PNAS* 112(24): 7519–7523.

Sugiura, N. *et al.* (2021). Beetle pollination of *Luisia teres* (Orchidaceae) and implications of a geographic divergence in the pollination system. *Pl. Species Biol.* 36(1): 52–59.

## 4.3 WHAT YOU'RE PROPOSING

Darwin, C. (1877). *The Various Contrivances by which Orchids are Fertilised by Insects. 2nd edition.* John Murray, London.

Dod, D. (1986). Afidos y trípidos polinizan orquídeas en las Pleurothallidinae (Orchidaceae). *Moscosoa* 4: 200–202.

Karremans, A.P. & Díaz-Morales, M. (2019). The Pleurothallidinae: extremely high speciation driven by pollinator adaptation. In *Proceedings of the 22nd World Orchid Conference, Guayaquil, Ecuador, 2017*, Pridgeon, A.M. & Arosema, A. (eds), Asociación Ecuatoriana de Orquideología, Guayaquil, pp. 363–388.

Matsui, K. *et al.* (2001). Pollinator limitation in a deceptive orchid, *Pogonia japonica*, on a floating peat mat. *Pl. Species Biol.* 16: 231–235.

Shigeta, K. & Suetsugu, K. (2020). Contribution of thrips to seed production in *Habenaria radiata*, an orchid morphologically adapted to hawkmoths. *J. Pl. Res.* 133: 499–506.

Suetsugu, K. *et al.* (2019). Thrips as a supplementary pollinator in an orchid with granular pollinia: is this mutualism? *Ecology* 100(2): e02535.

## 4.4 SECRET GARDEN

Balogh, P. (1982). *Rhizanthella* R. S. Rogers, a misunderstood genus (Orchidaceae). *Selbyana* 7(1): 27–33.

Bougoure, J.J. *et al.* (2010). Carbon and nitrogen supply to the underground orchid, *Rhizanthella gardneri. New Phytol.* 186: 945–956.

Brundrett, M. (2011). *Wheatbelt Orchid Rescue Project: Case Studies of Collaborative Orchid Conservation in Western Australia.* University of Western Australia, Crawley.

Clements, M.A. & Cribb, P.J. (1984). The underground orchids of Australia. *Curtis's Bot. Mag.* 1(2): 84–91.

Clements, M.A. & Cribb, P.J. (1985). The underground orchids of Australia. *The Orchadian* 8(4): 86–90.

Clements, M.A. & Jones, D.L. (2020). Notes on Australasian orchids 6: a new species of *Rhizanthella* (Diurideae, subtribe Prasophyllinae) from eastern Australia. *Lankesteriana* 20(2): 221–227.

Delannoy, E. *et al.* (2011). Rampant gene loss in the underground orchid *Rhizanthella gardneri* highlights evolutionary constraints on plastid genomes. *Molec. Biol. Evol.* 28: 2077–2086.

Dixon, K.W. (2003). *Rhizanthella gardneri* Orchidaceae. *Curtis's Bot. Mag.* 20(2): 94–100.

Dixon, K.W. & Christenhusz, M.J.M. (2018). Flowering in darkness: a new species of subterranean orchid *Rhizanthella* (Orchidaceae; Orchidoideae; Diurideae) from Western Australia. *Phytotaxa* 334(1): 75–79.

Dixon, K.W. *et al.* (1990). The Western Australian subterranean orchid *Rhizanthella gardneri* Rogers. In: Arditti, J. (ed) *Orchid Biology: Reviews and Perspectives Vol. 5.* Timber Press, Portland.

Rogers, R.S. (1928). A new genus of Australian orchid. *J. Roy. Soc. West. Aust.* 15: 1–8.

Thorogood, C.J. *et al.* (2019). *Rhizanthella*: orchids unseen. *Plants People Planet* 1(7): 153–156.

## 4.5 IMAGINE

Kevan, P.G. & Baker, H.G. (1983). Insects as flower visitors and pollinators. *Ann. Rev. Entomol.* 28: 407–453.

Micheneau, C. *et al.* (2010). Orthoptera, a new order of pollinator. *Ann. Bot.* 105: 355–364.

Pedersen, H. *et al.* (2018). Pollination-system diversity in *Epipactis* (Orchidaceae): new insights from studies of *E. flava* in Thailand. *Pl. Syst. Evol.* 304: 895–909.

Phillip, M. *et al.* (2006). Structure of a plant-pollinator network on a pahoehoe lava desert of the Galápagos Islands. *Ecography* 29(4): 531–540.

Suetsugu, K. (2018). Independent recruitment of a novel seed dispersal system by camel crickets in achlorophyllous plants. *New Phytol.* 217: 828–835.

Suetsugu K. (2018). Seed dispersal in the mycoheterotrophic orchid *Yoania japonica*: further evidence for endozoochory by camel crickets. *Plant Biol.* 20: 707–712.

Suetsugu K. (2020). A novel seed dispersal mode of *Apostasia nipponica* could provide some clues to the early evolution of the seed dispersal system in Orchidaceae. *Evol. Lett.* 4(5): 457–464.

Suetsugu, K. & Tanaka, K. (2014). Consumption of *Habenaria sagittifera* pollinia by juveniles of the katydid *Ducetia japonica*. *Ent. Sci.* 17 (1): 122–124.

Tan, M.K. *et al.* (2017). Overlooked flower-visiting *Orthoptera* in Southeast Asia. *J. Orthopt. Res.* 26(2): 143–153.

## 4.6 LADY IN BLACK

Ashton, K. (2015). *How to Fly a Horse: The Secret History of Creation, Invention, and Discovery.* Anchor Books, New York.

Ecott, T. (2005). *Vanilla: Travels in Search of the Ice Cream Orchid.* Penguin, London.

Fleming, T.H. (2009). The evolution of bat pollination: a phylogenetic perspective. *Ann. Bot.* 104: 1017–1043.

Karremans, A.P. *et al.* (2020). A reappraisal of Neotropical *Vanilla*. With a note on taxonomic inflation and the importance of alpha taxonomy in biological studies. *Lankesteriana* 20(3): 395–497.

Karremans *et al.* (2022). First evidence for multimodal animal seed dispersal in orchids, Current Biology, https:// doi.org/10.1016/j.cub.2022.11.041

Pansarin, E.R. & Ferreira, A.W.C. (2022). Evolutionary disruption in the pollination system of *Vanilla* (Orchidaceae). *Pl. Biol.* 24: 157–167.

Pansarin, E.R. & Pansarin, L.M. (2014). Floral biology of two Vanilloideae (Orchidaceae) primarily adapted to pollination by euglossine bees. *Pl. Biol.* 16: 1104–1113.

Ruschi, A. (1978). A atual fauna de mamíferos, aves e répteis da Reserva Biológica de Comboios. *Bol. Mus. Biol. Prof. Mello Leitão*, sér. Zool. 90:1–26.

Soto Arenas, M.A. & Dressler, R.L. (2010). A revision of the Mexican and Central American species of *Vanilla* Plumier ex Miller with a characterization of their ITS region of the nuclear ribosomal DNA. *Lankesteriana* 9: 285–354.

Watteyn, C. *et al.* (2022). Trick or treat? Pollinator attraction in *Vanilla pompona* (Orchidaceae). *Biotropica* 54: 268–274.

Watteyn, C. *et al. In review.* Sweet as *Vanilla hartii*: evidence for nectar rewards in *Vanilla* (Orchidaceae) flowers. *Flora.*

**4.7 WHAT IF**

Bègue, J-F. et al. (2014). New record of day geckos feeding on orchid nectar in Reunion Island: can lizards pollinate orchid species? *Herpetol. Notes* 7: 689–692.

Cocucci, A.A. & Sérsic, A.N. (1998). Evidence of rodent pollination in *Cajophora coronata* (Loasaceae). *Pl. Syst. Evol.* 211: 113–128.

Johnson, C.M. & Pauw, A. (2014). Adaptation for rodent pollination in *Leucospermum arenarium* (Proteaceae) despite rapid pollen loss during grooming. *Ann. Bot.* 113: 931–938.

Mayer, C. et al. (2011). Pollination ecology in the 21st century: key questions for future research. *J. Pollinat. Ecol.* 3(2): 8–23.

Rose-Smyth, M.C. (2019). Investigating the pollination biology of a long-lived island endemic epiphyte in the presence of an adventive alien pollinator. In *Proceedings of the 22nd World Orchid Conference, Guayaquil, Ecuador, 2017*, Pridgeon, A.M. & Arosema, A. (eds), Asociación Ecuatoriana de Orquideología, Guayaquil, pp. 80–91.

Rose-Smyth, M.C. (2019). Role of a sweet-toothed anole (*Anolis conspersus*) in orchid pollination. In *Anolis Newsletter VII*, Stroud, J.T. et al. (eds), Washington University, St. Louis MO, pp. 235–241.

Tang, G.-D. (2014). A review of orchid pollination studies in China. *J. Syst. Evol.* 52: 411–422.

Tsuji, K. & Kato, M. (2010). Odor-guided bee pollinators of two endangered winter/early spring blooming orchids, *Cymbidium kanran* and *Cymbidium goeringii*, in Japan. *Pl. Spec. Biol.* 25: 249–253.

Wang, Y. et al. (2008). The unique mouse pollination in an orchid species. *Nat Prec*. https://doi.org/10.1038/npre.2008.1824.1

Wester, P. et al. (2009). Mice pollinate the Pagoda Lily, *Whiteheadia bifolia* (Hyacinthaceae) — first field observations with photographic documentation of rodent pollination in South Africa. *S. African J. Bot.* 75: 713–719.

Willmer, P. (2011). *Pollination and Floral Ecology*. Princeton University Press, Princeton.

## CHAPTER 5 REDESIGN
### 5.1 COMFORTABLY NUMB

Adler, L.S. (2000). The ecological significance of toxic nectar. *Oikos* 91(3): 409–420.

Brodmann, J. et al. (2008). Orchids mimic green-leaf volatiles to attract prey-hunting wasps for pollination. *Curr. Biol.* 18: 740–744.

Claessens, J. & Kleynen, J. (2011). *The Flower of the European Orchid: Form and Function*. Schrijen-Lippertz, Voerendaal.

Claessens, J. & Kleynen, J. (2016). Many ways to get happy: pollination modes of European *Epipactis* species. *Orchid Digest* 80(3): 144–152.

Dalström, S. *et al.* (2020). *The Odontoglossum Story*. Koeltz Botanical Books, Oberreifenberg.

Ehlers, B.K. & Olesen, J.M. (1997). The fruit-wasp route to toxic nectar in *Epipactis* orchids? *Flora* 192: 223–229.

Jakubska, A. *et al.* (2005). Why do pollinators become "sluggish"? Nectar chemical constituents from *Epipactis helleborine* (L.) Crantz (Orchidaceae). *Appl. Ecol. Environ. Res.* 3(2): 29–38.

Jakubska, A. *et al.* (2005). Pollination ecology of *Epipactis helleborine* (L.) Crantz (Orchidaceae, Neottieae) in the south-western Poland. *Acta Bot. Sil.* 2: 131–144.

Jansen, S.A. *et al.* (2012). Grayanotoxin poisoning: 'mad honey disease' and beyond. *Cardiovasc. Toxicol.* 12: 208–215.

Jacquemyn, H. *et al.* (2013). Microbial diversity in the floral nectar of seven *Epipactis* (Orchidaceae) species. *MicrobiologyOpen* 4: 644–658.

Karremans, A.P. *et al.* (2015). Pollination of *Specklinia* by nectar-feeding *Drosophila*: the first reported case of a deceptive syndrome employing aggregation pheromones in Orchidaceae. *Ann. Bot.* 116(3): 437–455.

Kevan, P.G. *et al.* (1998). Yeast-contaminated nectar and its effects on bee foraging. *J. Apicult. Res.* 27: 26–29.

Løjtnant, B. (1974). Giftig nektar, fulde hvepse og orchideer. *Kaskelot* 15: 3–7.

## 5.2 THUNDERSTRUCK

Adams, P.B. & Lawson, S.D. (1993). Pollination in Australian orchids: a critical assessment of the literature 1882–1992. *Aust. J. Bot.* 41: 553–575.

Arditti, J. (1989). History of several important orchid research contributions by South East Asia scientists. *Malayan Orchid Rev.* 23: 64–80.

Brown, J. & York, A. (2017). Fire, food and sexual deception in the neighbourhood of some Australian orchids. *Austral. Ecol.* 42: 468–478.

Cannell, M.G.R. (1985). Physiology of the coffee crop. In *Coffee*, Clifford, M.N. & Willson, K.C. (eds), Springer, Boston, MA, pp. 108–134.

Gerlach, W.W. (1992). Flowering behaviour of ephemeral orchids of Western Samoa: 2. Mechanism of anthesis induction. *Gartenbauwissenschaft* 57(6): 288–291.

Goh, C.J. *et al.* (1982). Flower induction and physiology in orchids. In Arditti, J. (ed.) *Orchid Biology: Reviews and Perspectives, Vol. 2*. Cornell University Press, Ithaca, New York, pp. 213–241.

Goh, C.J. *et al.* (1985). Ethylene evolution and sensitivity in cut orchid flowers. *Scientia Hortic.* 26(1): 57–67.

Holttum, R.E. (1949). Gregarious flowering of the terrestrial orchid *Bromheadia finlaysoniana*. *Gard. Bull.* 12: 295–302.

Janzen, D.H. (1967). Synchronization of sexual reproduction of trees within the dry season in Central America. *Evol.* 21: 620–637.

Leong, T.M. & Wee, Y.C. (2013). Observations of pollination in the pigeon orchid, *Dendrobium crumenatum* Swartz (Orchidaceae) in Singapore. *Nat. Singap.* 6: 91–96.

Massart, J. (1895). Un botaniste en Malaisie. *Bull. Soc. Roy. Belgique* 34: 151–343.

Ong, P.T. (2010). A one-day affair. *Orchid Rev.* 118: 144–147.

Seifriz, W. (1923). The gregarious flowering of the orchid *Dendrobium crumenatum*. *Am. J. Bot.* 10(1): 32–37.

Steiner, K.E. *et al.* (1994). Floral and pollinator divergence in two sexually deceptive South African orchids. *Am. J. Bot.* 81(2): 185–194.

Treub, M. (1887). Quelques observations sur la végétation dans l'ile de Java. *Bull. Soc. Roy. Belgique* 26: 182–186.

Williams, S.A. (1994). Observations on reproduction in *Triphora trianthophora* (Orchidaceae). *Rhodora* 96(885): 30–43.

Yap, Y.-M. *et al.* (2008). Regulation of flower development in *Dendrobium crumenatum* by changes in carbohydrate contents, water status and cell wall metabolism. *Scientia Hortic.* 119: 59–66.

## 5.3 TRIGGER

Cheeseman, T.F. (1872). On the fertilisation of the New Zealand species of *Pterostylis*. *Trans. & Proc. New Zealand Inst.* 5: 352–357.

Darwin, C. (1877). *The Various Contrivances by which Orchids are Fertilised by Insects. 2nd edition*. John Murray, London.

Darwin Correspondence Project, "Letter no. 8955," accessed on 21 July 2022, https://www.darwinproject.ac.uk/letter/?docId=letters/DCP-LETT-8955.xml

Darwin Correspondence Project, "Letter no. 9048," accessed on 21 July 2022, https://www.darwinproject.ac.uk/letter/?docId=letters/DCP-LETT-9048.xml

Karremans, A.P. & Vieira-Uribe, S. (2020). *Pleurothallids Neotropical Jewels, Vol 1*. Mariscal, Quito.

Lindley, J. (1832). *An Introduction to Botany*. Longman, Rees, Orme, Brown, Green, & Longman, London.

Lindley, J. (1853). *The Vegetable Kingdom*. Bradbury & Evans, London.

Luer, C.A. (1987). Icones Pleurothallidinarum IV: systematics of *Acostaea*, *Condylago*, and *Porroglossum*. *Monogr. Syst. Bot.* 24: 1–20.

McDaniel, J.L. (2019). Molecular systematics of *Porroglossum* Schltr. (Orchidaceae): phylogenetics, floral snap-trap kinematics, and fragrance. Dissertation, University of Wisconsin-Madison, U.S.A.

Oliver, F.W. (1888). On the sensitive labellum of *Masdevallia muscosa*, Rchb.f. *Ann. Bot.* 1(3–4): 237–253.

Phillips, R.D. *et al.* (2014). Caught in the act: pollination of sexually deceptive trap-flowers by fungus gnats in *Pterostylis* (Orchidaceae). *Ann. Bot.* 113: 629–641.

Reiter, N. *et al.* (2019). Pollination by sexual deception of fungus gnats (Keroplatidae and Mycetophilidae) in two clades of *Pterostylis* (Orchidaceae). *Bot. J. Linn. Soc.* 190: 101–116.

Sargent, O.H. (1909). Notes on the life-history of *Pterostylis*. *Ann. Bot.* 23: 265–274.

## 5.4 PATIENCE

Ackerman, J.D. & Mesler, M.R. (1979). Pollination biology of *Listera cordata* (Orchidaceae). *Am. J. Bot.* 66(7): 820–824.

Borba, E.L. & Semir, J. (1999). Temporal variation in pollinarium size after its removal in species of *Bulbophyllum*: a different mechanism preventing self-pollination in Orchidaceae. *Pl. Syst. Evol.* 217: 197–204.

Brown, R. (1833). On the organs and mode of fecundation in Orchideae and Asclepiadeae. *Trans. Linn. Soc.* 16: 685–733.

Castro, J.B & Singer, R.B. (2019). A literature review of the pollination strategies and breeding systems in Oncidiinae orchids. *Acta Bot. Brasilica* 33: 618–643.

Catling, P.M. (1983). Pollination of northeastern North American *Spiranthes* (Orchidaceae). *Canad. J. Bot.* 61: 1080–1093.

Claessens, J. & Kleynen, J. (2011). *The Flower of the European Orchid: Form and Function*. Schrijen-Lippertz, Voerendaal.

Darwin, C. (1862). *On the Various Contrivances by which British and Foreign Orchids are Fertilised by Insects, and on the Good Effects of Intercrossing*. John Murray, London.

Dodson, C.H. (1962). Pollination and variation in the subtribe Catasetinae (Orchidaceae). *Ann. Missouri Bot. Gard.* 49: 35–56.

Hooker, J.D. (1854). On the functions and structure of the rostellum of *Listera ovata*. *Philos. Trans. R. Soc. Lond.* 144: 259–263.

Jersáková, J. & Johnson, S.D. (2007). Protandry promotes male pollination success in a moth-pollinated orchid. *Funct. Ecol.* 21: 496–504.

Johnson, S.D. & Edwards, T.J. (2000). The structure and function of orchid pollinaria. *Plant Syst. Evol.* 222: 243–269.

Peter, C.I. & Johnson, S.D. (2006). Doing the twist: a test of Darwin's cross-pollination hypothesis for pollinarium reconfiguration. *Biol. Lett.* 2: 65–68.

Romero-González, G.A. (2018). Charles Darwin on Catasetinae (Cymbidieae, Orchidaceae). *Harvard Pap. Bot.* 23(2): 339–379.

Singer, R.B. (2002). The pollination biology of *Sauroglossum elatum* Lindl. (Orchidaceae: Spiranthinae): moth-pollination and protandry in neotropical Spiranthinae. *Bot. J. Linn. Soc.* 138: 9–16.

Tremblay, R.L. *et al.* (2006). Flower phenology and sexual maturation: partial protandrous behavior in three species of orchids. *Caribb. J. Sci.* 42(1): 75–80.

## 5.5 NOTHING ELSE MATTERS

Ackerman, J.D. *et al.* 2023. Beyond the various contrivances by which orchids are pollinated: global patterns in orchid pollination biology. *Bot. J. Linn. Soc.* boac082, https://doi.org/10.1093/botlinnean/boac082

Bateman, R.M. (1985). Peloria and pseudopeloria in British orchids. *Watsonia* 15: 357–359.

Catling, P.M. (1990). Auto-pollination in the Orchidaceae. In: *Orchid Biology: Reviews and Perspectives, Vol V,* Arditti, J. (ed). Timber Press, Portland, pp. 121–158.

Karremans, A.P. *et al.* (2019). Nomenclatural notes in the Pleurothallidinae (Orchidaceae): miscellaneous. *Phytotaxa* 406(5): 259–270.

Liu, K.W. *et al.* (2006). Pollination: self-fertilization strategy in an orchid. *Nature* 441: 945–946.

Masters, M.T. (1865). On a peloria and semidouble flower of *Ophrys aranifera*, Huds. *Bot. J. Linn. Soc.* 8(31): 207–211.

Mondragón-Palomino, M. & Theißen, G. (2009). Why are orchid flowers so diverse? Reduction of evolutionary constraints by paralogues of class B floral homeotic genes. *Ann. Bot.* 104: 583–594.

Müller, H. (1883). *The Fertilisation of Flowers*. Macmillan, London.

Suetsugu, K. (2015). Autonomous self-pollination and insect visitors in partially and fully mycoheterotrophic species of *Cymbidium* (Orchidaceae). *J. Plant Res.* 128: 115–125.

Tałałaj, I. & Skierczyński, M. (2015). Mechanism of spontaneous autogamy in the allogamous lepidopteran orchid *Gymnadenia conopsea* (L.) R.Br. (Orchidaceae). *Acta Biol. Cracov. Ser. Bot.* 57: 1–11.

Tałałaj, I. *et al.* (2019). Spontaneous caudicle reconfiguration in *Dactylorhiza fuchsii*: a new self-pollination mechanism for Orchideae. *Pl. Syst. Evol.* 305: 269–280.

Tremblay, R.L. *et al.* (2005). Variation in sexual reproduction in orchids and its evolutionary consequences: a spasmodic journey to diversification. *Biol. J. Linn. Soc.* 84(1): 1–54.

## 5.6 RIDERS ON THE STORM

Aguiar, J.M.R.B.V. *et al.* (2012). Biotic versus abiotic pollination in *Oeceoclades maculata* (Lindl.) Lindl. (Orchidaceae). *Pl. Spec. Bot.* 27: 86–95.

Catling, P.M. (1980). Rain-assisted autogamy in *Liparis loeselii*. *Bull. Torrey Bot. Club* 107: 525–529.

Fan, X.-L. *et al.* (2012). Rain pollination provides reproductive assurance in a deceptive orchid. *Ann. Bot.* 110: 953–958.

González-Díaz, N. & Ackerman, J.D. (1988). Pollination, fruit set, and seed production in the orchid, *Oeceoclades maculata*. *Lindleyana* 3: 150–155.

Pansarin, L.M. *et al.* (2008). Facultative autogamy in *Cyrtopodium polyphyllum* (Orchidaceae) through a rain-assisted pollination mechanism. *Austr. J. Bot.* 56: 366–367.

Sheviak, C.J. (2001). A role for water droplets in the pollination of *Platanthera aquilonis* (Orchidaceae). *Rhodora* 103(916): 380–386.

Siegel, C. (2020). The role of rain in orchid pollination. *Orchid Dig.* 84(3): 162–168.

Suetsugu, L. (2019). Rain-triggered self-pollination in *Liparis kumokiri*, an orchid that blooms during the rainy season. *Ecology* 100(7): e02683.

## 5.7 BLOWIN' IN THE WIND

Bogarín, D. *et al.* (2018). Pollination of *Trichosalpinx* (Orchidaceae: Pleurothallidinae) by biting midges (Diptera: Ceratopogonidae). *Bot. J. Linn. Soc.* 186: 510–543.

Borba, E.L. & Semir, J. (1998). Wind-assisted fly pollination in three *Bulbophyllum* (Orchidaceae) species occurring in the Brazilian campos rupestres. *Lindleyana* 13: 203–218.

Claessens, J. & Kleynen, J. (2011). *The Flower of the European Orchid: Form and Function*. Schrijen-Lippertz, Voerendaal.

Claessens, J. & Kleynen, J. (2002). Investigations on the autogamy in *Ophrys apifera* Hudson. *Jahr. naturwiss Ver. Wuppertal* 55: 62–77.

Darwin, C. (1877). *The Various Contrivances by which Orchids are Fertilised by Insects. 2nd edition*. John Murray, London.

Jongejan, P. (1994). Specializations in ways of attracting insects for pollination in the genus *Bulbophyllum*. *Proceedings of 14th* WOC. HMSO, Glasgow, pp. 383–388.

Karremans, A.P. & Vieira-Uribe, S. (2020). *Pleurothallids Neotropical Jewels*, Vol 1. Mariscal, Quito.

Liu, Z.-J. *et al.* (2010). A floral organ moving like a caterpillar for pollinating. *J. Syst. Evol.* 48(2): 102–108.

Sazima, M. (1978). Polinização por moscas em *Bulbophyllum warmingianum* Cogn. (Orchidacae), na Serra do Cipó, Minas Gerais, Brasil. *Revista Brasil. Bot.* 1: 133–138.

Schuiteman, A. *et al.* (2011). Nocturne for an unknown pollinator: first description of a night-flowering orchid (*Bulbophyllum nocturnum*). *Bot. J. Linn. Soc.* 167: 344–350.

# CHAPTER 6 FALLACIES

Endersby, J. (2016). *Orchid: A Cultural History*. University of Chicago Press, Chicago.

## 6.1 STRAY CAT STRUT

Annandale, N. (1900). Observations on the habits and natural surroundings of insects made during the "Skeat expedition" to the Malay Peninsula, 1899–1900. *J. Zool. (Lond.)* 69: 837–869.

Hingston, J. (1879). *The Australian Abroad: Branches from the Main Routes Round the World*. Sampson Low, Marston, Searle and Rivington, London.

Mizuno, T. *et al.* (2014). "Double-trick" visual and chemical mimicry by the juvenile orchid mantis *Hymenopus coronatus* used in predation of the oriental honeybee *Apis cerana*. *Zool. Sci.* 31(12): 795–801.

O'Hanlon, J.C. *et al.* (2014). Predatory pollinator deception: does the orchid mantis resemble a model species? *Curr. Zool.* 60(1): 90–103.

O'Hanlon, J.C. *et al.* (2014). Pollinator deception in the orchid mantis. *Am. Nat.* 183: 126–132.

O'Hanlon, J.C. *et al.* (2015). Habitat selection in a deceptive predator: maximizing resource availability and signal efficacy. *Behav. Ecol.* 26: 194–199.

Svenson, G.J. *et al.* (2016). Selection for predation, not female fecundity, explains sexual size dimorphism in the orchid mantises. *Sci. Rep.* 6(37753): 1–9.

Wallace, A.R. (1877). The colors of animals and plants. *Am. Nat.* 11: 641–662.

Wallace, A.R. (1889). *Darwinism: An Exposition of the Theory of Natural Selection with Some of its Applications*. Macmillan, London.

## 6.2 EVERYWHERE

Arditti, J. *et al.* (2012). 'Good Heavens what insect can suck it' — Charles Darwin, *Angraecum sesquipedale* and *Xanthopan morganii praedicta*. *Bot. J. Linn. Soc.* 169: 403–432.

Gentry, A. (1978). Anti-pollinators for mass-flowering plants? *Biotropica* 10(1): 68–69.

Hopkins, H.C. & Hopkins, M.J.G. (1982). Predation by a snake of a flower-visiting bat at *Parkia nitida* (Leguminosae Mimosoideae). *Brittonia* 34(2): 225–227.

Luer, C.A. (1974). Algo mas sobre anti-polinizadores. *Orquideología* 9(3): 238.

Ohm, J.R. & Miller, T.E.X. (2014). Balancing anti-herbivore benefits and anti-pollinator costs of defensive mutualists. *Ecol.* 95(10): 2924–2935.

Ospina, M. (1969). Los antipolinizadores. *Orquideología* 4(1): 23–27.

Ospina, M. (1972). Los antipolinizadores. *Orquídea Mex.* 2(6): 163–167.

Ospina-Calderón, N.H. *et al.* (2007). Observaciones de la polinización y fenología reproductiva de *Brassia* cf. *antherotes* Rchb.f. (Orchidaceae) en un relicto de selva subandina en la Reserva Natural la Montaña del Ocaso en Quimbaya, Quindío (Colombia). *Univ. Sci.* 12: 83–95.

Quintero, C. *et al.* (2015). Weak trophic links between a crab-spider and the effective pollinators of a rewardless orchid. *Acta Oecologica* 62: 32–39.

Roberts, D.L. & Bateman, R.M. (2007). Do ambush predators prefer rewarding or non-rewarding orchid inflorescences? *Bot. J. Linn. Soc.* 92: 763–771.

## 6.3 WORTH FIGHTING FOR

Ackerman, J.D. (1986). Mechanisms and evolution of food-deceptive pollination systems in orchids. *Lindleyana* 1: 108–113.

Castro, J.B & Singer, R.B. (2019). A literature review of the pollination strategies and breeding systems in Oncidiinae orchids. *Acta Bot. Brasilica* 33: 618–643.

Chase, M.W. (2009). Subtribe Oncidiinae. In *Genera Orchidacearum, Volume 5: Epidendroideae (Part II)*, Pridgeon et al. (eds), Oxford University Press, pp. 211–394.

Coville, R.E. et al. (1986). Nesting and male bahavior of *Centris heithausi* in Costa Rica (Hymenoptera: Anthophoridae) with chemical analysis of the hind leg glands of males. *J. Kans. Entomol. Soc.* 59(2): 325–336.

Dod, D.D. (1976). *Oncidium henekenii*. Bee orchid pollinated by bee. *Am. Orchid Soc. Bull.* 45: 792–794.

Dodson, C.H. & Frymire, G.P. (1961). Natural pollination of orchids. *Missouri Bot. Gard. Bull.* 49: 133–139.

Frankie, G.W. et al. (1980). Territorial behavior of *Centris adani* and its reproductive function in the Costa Rican dry forest (Hymenoptera: Anthophoridae). *J. Kans. Entomol. Soc.* 53: 837–857.

Nierenberg, L. (1972). The mechanism for the maintenance of species integrity in sympatrically occurring equitant oncidiums in the Caribbean. *Am. Orchid Soc. Bull.* 41: 873–882.

Van der Pijl, L. & Dodson, C.H. (1966). *Orchid Flowers: Their Pollination and Evolution*. University of Miami Press, Coral Gables.

## 6.4 PARANOID

Bower, C.C. & Branwhite, P. (1993). Observations on the pollination of *Calochilus campestris* R.Br. *The Orchadian* 11: 68–71.

Fordham, F. (1946). Pollination of *Calochilus campestris*. *Vic. Nat.* 62: 199–201.

Van der Pijl, L. & Dodson, C.H. (1966). *Orchid Flowers: Their Pollination and Evolution*. University of Miami Press, Coral Gables.

## 6.5 NICE GUYS FINISH LAST

Ackerman, J.D. et al. 2023. Beyond the various contrivances by which orchids are pollinated: global patterns in orchid pollination biology. *Bot. J. Linn. Soc.* boac082, https://doi.org/10.1093/botlinnean/boac082

Darwin Correspondence Project, "Letter no. 3705," accessed on 22 July 2022, https://www.darwinproject.ac.uk/letter/?docId=letters/DCP-LETT-3705.xml

Karremans, A.P. & Davin, N. (2017). Genera Pleurothallidinarum: the era of Carl Luer. *Lankesteriana* 17(2): I–VIII.

Karremans, A.P. & Díaz-Morales, M. (2019). The Pleurothallidinae: extremely high speciation driven by pollinator adaptation. In *Proceedings of the 22nd World Orchid Conference, Guayaquil, Ecuador, 2017*, Pridgeon, A.M. & Arosema, A. (eds), Asociación Ecuatoriana de Orquideología, Guayaquil, pp. 363–388.

Van der Pijl, L. & Dodson, C.H. (1966). *Orchid Flowers: Their Pollination and Evolution*. University of Miami Press, Coral Gables.

## 6.6 SOMEBODY TOLD ME

Ackerman, J.D. *et al.* 2023. Beyond the various contrivances by which orchids are pollinated: global patterns in orchid pollination biology. *Bot. J. Linn. Soc.* boac082, https://doi.org/10.1093/botlinnean/boac082

Antonelli, A. *et al.* (2009). Pollination of the Lady's slipper orchid (*Cypripedium calceolus*) in Scandinavia — taxonomic and conservational aspects. *Nord. J. Bot.* 27: 266–273.

Brzosko, E. & Wróblewska, A. (2012). How genetically variable are *Neottia ovata* (Orchidaceae) populations in northeast Poland? *Bot. J. Linn. Soc.* 170: 40–49.

Dressler, R.L. (1993). *Phylogeny and Classification of the Orchid Family*. Cambridge University Press, Cambridge.

Kowalkowska, A.K. & Krawczyńska, A.T. (2019). Anatomical features related with pollination of *Neottia ovata* (L.) Bluff & T Fingerh. (Orchidaceae). *Flora* 255: 24–33.

Lehnebach, C.A. & Robertson, A.W. (2004). Pollination ecology of four epiphytic orchids of New Zealand. *Ann. Bot.* 93(6): 773–781.

## 6.7 DIAMONDS AND RUST

Cano, R.J. *et al.* (1993). Amplification and sequencing of DNA from a 120–135-million-year-old weevil. *Nature* 363: 536–538.

Conran, J.G. *et al.* (2009). Earliest orchid macrofossils: Early Miocene *Dendrobium* and *Earina* (Orchidaceae: Epidendroideae) from New Zealand. *Am. J. Bot.* 96: 466–474.

Desalle, R. *et al.* (1992). DNA sequences from a fossil termite in Oligo-Miocene amber and their phylogenetic implications. *Science* 257(5078): 1933–1936.

Penney, D. *et al.* (2013). Absence of ancient DNA in sub-fossil insect inclusions preserved in 'Anthropocene' Colombian copal. *PLoS ONE* 8(9): e73150.

Peris, D. *et al.* (2020). DNA from resin-embedded organisms: past, present and future. *PLoS ONE* 15(9): e0239521.

Poinar, G. (1999). DNA from fossils: the past and the future. *Acta Paediatr Suppl.* 88(433): 133–140.

Poinar, G. (2016). Orchid pollinaria (Orchidaceae) attached to stingless bees (Hymenoptera: Apidae) in Dominican amber. *Neues Jahrb. Geol. Palaontol.* 279: 287–293.

Poinar, G. (2016). Beetles with orchid pollinaria in Dominican and Mexican amber. *Am. Entomol.* 62: 172–177.

Poinar, G. (2017). Two new genera, *Mycophoris* gen. nov., (Orchidaceae) and *Synaptomitus* gen. nov. (Basidiomycota) based on a fossil seed with developing embryo and associated fungus in Dominican amber. *Botany* 95(1): 1–8.

Poinar, G. & Rasmussen, F.N. (2017). Orchids from the past, with a new species in Baltic amber. *Bot. J. Linn. Soc.* 183: 327–333.

Pupulin, F. & Karremans, A.P. (2008). The orchid pollinaria collection at Lankester Botanical Garden. *Selbyana* 29(1): 69–86.

Ramirez, S.R. *et al.* (2007). Dating the origin of the Orchidaceae from a fossil orchid with its pollinator. *Nature* 448: 1042–1045.

Schmid, R. & Schmid, M.J. (1973). Fossils attributed to the Orchidaceae. *Am. Orchid Soc. Bull.* 42: 17–27.

Schmid, R. & Schmid, M.J. (1977). Fossil history of the Orchidaceae. In *Orchid Biology: Reviews and Perspectives, Vol. 1*, Arditti, J. (ed) Cornell University Press, Ithaca, pp. 25–45.

Selosse, M. *et al.* (2017). Why *Mycophoris* is not an orchid seedling, and why *Synaptomitus* is not a fungal symbiont within this fossil. *Botany* 95(9): 1–4.

Singer, R.B. *et al.* (2008). The use of orchid pollinia or pollinaria for taxonomic identification. *Selbyana* 29(1): 6–19.

### 6.8 YOU ONLY LIVE ONCE

Danaher, M.W. *et al.* (2019). Pollinia removal and suspected pollination of the endangered ghost orchid, *Dendrophylax lindenii* (Orchidaceae) by various hawk moths (Lepidoptera: Sphingidae): another mystery dispelled. *Fla. Entomol.* 102: 1–13.

Darwin Correspondence Project, "Letter no. 3315," accessed on 22 July 2022, https://www.darwinproject.ac.uk/letter/?docId=letters/DCP-LETT-3315.xml

Endersby, J. (2016). *Orchid: A Cultural History*. University of Chicago Press, Chicago.

Houlihan, P.R. *et al.* (2019). Pollination ecology of the ghost orchid (*Dendrophylax lindenii*): a first description with new hypotheses for Darwin's orchids. *Sci. Rep.* 9: 12850.

Mújica, E.B. *et al.* (2018). A comparison of ghost orchid (*Dendrophylax lindenii*) habitats in Florida and Cuba, with particular reference to seedling recruitment and mycorrhizal fungi. *Bot. J. Linn. Soc.* 186: 572–586.

Zettler, L.W. *et al.* (2019). The ghost orchid demystified: biology, ecology, and conservation of *Dendrophylax lindenii* in Florida and Cuba. In *Proceedings of the 22nd World Orchid Conference, Guayaquil, Ecuador, 2017*, Pridgeon, A.M. & Arosema, A. (eds), Asociación Ecuatoriana de Orquideología, Guayaquil, pp. 136–148.

## CHAPTER 7 CHANGE

Phillips, R.D. *et al.* (2020). Orchid conservation: from theory to practice. *Ann. Bot.* 126: 345–362.

### 7.1 UNDER PRESSURE

Acevedo, M.A. *et al.* (2020). Local extinction risk under climate change in a neotropical asymmetrically dispersed epiphyte. *J. Ecol.* 108: 1553–1564.

Brundrett, M.C. (2019). A comprehensive study of orchid seed production relative to pollination traits, plant density and climate in an urban reserve in Western Australia. *Diversity* 11: 123.

Crain, B.J. (2012). On the relationship between bryophyte cover and the distribution of *Lepanthes* spp.. *Lankesteriana* 12: 13–18.

Crain, B.J. & Tremblay, R.L. (2017). Hot and bothered: changes in microclimate alter chlorophyll fluorescence measures and increase stress levels in tropical epiphytic orchids. *Int. J. Plant Sci.* 178(7): 503–511.

Faleiro, F.V. *et al*. (2018). Climate change likely to reduce orchid bee abundance even in climatic suitable sites. *Glob. Chang. Biol.* 24(6): 2272–2283.

Fay, M.F. (2018). Orchid conservation: how can we meet the challenges in the twenty-first century? *Bot. Stud.* 59: 16.

Janzen, D.H. & Hallwachs, W. (2020). To us insectometers, it is clear that insect decline in our Costa Rican tropics is real, so let's be kind to the survivors. *PNAS* 118: e2002546117.

Kolanowska, M. *et al*. (2017). Global warming not so harmful for all plants — response of holomycotrophic orchid species for the future climate change. *Sci. Rep.* 7: 12704.

Kolanowska, M. *et al*. (2021). Significant habitat loss of the black vanilla orchid (*Nigritella nigra* s.l., Orchidaceae) and shifts in its pollinators availability as results of global warming. *Glob. Ecol. Conserv.* 27: e01560.

Liu, H. *et al*. (2010). Potential challenges of climate change to orchid conservation in a wild orchid hotspot in southwestern China. *Bot. Rev.* 76: 174–192.

Nadkarni, N.M. & Solano, R. (2002). Potential effects of climate change on canopy communities in a tropical cloud forest: an experimental approach. *Oecologia* 131: 580–586.

Olaya-Arenas, P. *et al*. (2011). Demographic response by a small epiphytic orchid. *Am. J. Bot.* 98: 2040–2048.

Robbirt, K. *et al*. (2014). Potential disruption of pollination in a sexually deceptive orchid by climatic change. *Curr. Biol.* 24: 2845–2849.

Tsiftsis, S. & Djordjević, V. (2020). Modelling sexually deceptive orchid species distributions under future climates: the importance of plant–pollinator interactions. *Sci. Rep.* 10: 10623.

Wilmer, P. (2014). Climate change: bees and orchids lose touch. *Curr. Biol.* 24: R1133–R1135.

## 7.2 IT'S THE END OF THE WORLD AS WE KNOW IT

Brito, T.F. *et al*. (2017). Forest reserves and riparian corridors help maintain orchid bee (Hymenoptera: Euglossini) communities in oil palm plantations in Brazil. *Apidologie* 48: 575–587.

Brundrett, M.C. (2016). Using vital statistics and core-habitat maps to manage critically endangered orchids in the Western Australian wheatbelt. *Aus. J. Bot.* 64: 51–64.

Chupp, A.D. *et al.* (2015). Orchid–pollinator interactions and potential vulnerability to biological invasion. *AoB PLANTS* 7: plv099.

Hadley, A.S. & Betts, M.G. (2012). The effects of landscape fragmentation on pollination dynamics: absence of evidence not evidence of absence. *Biol. Rev.* 87: 526–544.

Humeau, L. *et al.* (2011). Sapromyiophily in the native orchid, *Bulbophyllum variegatum*, on Réunion (Mascarene Archipelago, Indian Ocean). *J. Trop. Ecol.* 27: 591–599.

Ollerton, J. (2021). *Pollinators & Pollination: Nature and Society*. Pelagic Publishing, Exeter.

Phillips, R.D. *et al.* (2020). Orchid conservation: from theory to practice. *Ann. Bot.* 126: 345–362.

Potts, S.G. *et al.* (2010). Global pollinator declines: trends, impacts and drivers. *Trends Ecol. Evol.* 25: 345–353.

Štípková, Z. *et al.* (2020). Pollination mechanisms are driving orchid distribution in space. *Sci. Rep.* 10: 850.

Swarts, N.D. & Dixon, K.W. (2009). Terrestrial orchid conservation in the age of extinction. *Ann. Bot.* 104: 543–556.

## 7.3 BEAUTIFUL DAY

Ackerman, J.D. (2007). Invasive orchids: weeds we hate to love? *Lankesteriana* 7(1–2): 19–21.

Ackerman, J.D. (2014). Rapid transformation of orchids floras. *Lankesteriana* 13(3): 157–164.

Ackerman, J.D. & Moya, S. (1996). Hurricane aftermath: resiliency of an orchid-pollinator interaction in Puerto Rico. *Caribb. J. Sci.* 32(4): 369–374.

Johnson, S.D. *et al.* (2009). Relationships between population size and pollen fates in a moth-pollinated orchid. *Biol. Lett.* 5: 282–285.

Partomihardjo, T. (2003). Colonisation of orchids on the Krakatau Islands. *Teleopea* 10: 299–310.

Phillips, R.D. *et al.* (2020). Orchid conservation: from theory to practice. *Ann. Bot.* 126: 345–362.

Travers, S.E. *et al.* (2011). The hidden benefits of pollinator diversity for the rangelands of the Great Plains: western prairie fringed orchids as a case study. *Rangelands* 33: 20–26.

# INDEX

Images are indicated by page numbers in **bold**.

**A**
*Acampe* 297, 298
*Acampe rigida* **297**, 298
*Acineta* 164
*Acostaea* 271, 276
*Ada* 98
*Aerangis* 203
*Aeranthes* 203
*Aerides* 203
Agaricales 104, 108
Allodapini 237
*Amorphophallus* 103
*Amorphophallus titanum* 103
*Anacamptis* 316, 317, 344
*Anacamptis morio* **316**
*Anacamptis pyramidalis* 316, 317, **344**
*Anathallis* 160, 301, 303
*Andrena* 52, 53, 345, 368, 369
*Andrena nigroaenea* 52, **368**, 369
Angraecinae 337
*Angraecum* 38, 123–129, 228–231, 250
*Angraecum arachnites* **127**
*Angraecum bracteosum* 250
*Angraecum cadetii* **228–231**, 250
*Angraecum sesquipedale* **123–129**, 228
*Annulites* 351, 354
*Annulites mexicana* 351, 354
*Anolis* 249–251, 318
*Anolis conspersus* 250, **251**
*Antholithes* 247
*Anthophora* 372, 373
*Anthophora plagiata* 372, **373**
*Apedium* 293

Apidae 247, 343
*Apis* 247, 266, 267, 308
*Apis cerana* **247**, 266, **267**
*Apis mellifera* **308**
Apocynaceae 91, 92, 157, 285, 349
*Apostasia* 147–152, 229, 244
Apostasioideae 34, 147–152, 349
*Aristolochia* 157
Aristolochiaceae 157
*Arpophyllum* 13, 137–147, 239
*Arpophyllum giganteum* **13**, 137–147
*Arpophyllum spicatum* **144**
*Artibeus* 234
*Artibeus jamaicensis* 234
*Artibeus jamaicensis planirostris* 234
*Arundina* 384
*Arundina graminifolia* **384**
Asclepiadaceae 285, 349
*Asclepias* 91–96
*Asclepias curassavica* **91–96**
Asparagaceae 246
Asteraceae 213
*Atrichelaphinis* 213
*Atrichelaphinis tigrina* **213**
*Autographa* 128
*Autographa gamma* **128**

**B**
*Bactrocera* 190–194
*Bactrocera papayae* 190–194
*Baskervilla* 353
Blattodea 200
*Bombus* 39, 280, 388
*Bombus pascuorum* **388**
*Bombus terrestris* **39**

*Bossiaea* 89
*Bossiaea eriocarpa* **89**
*Bothrops* 315
*Bothrops atrox* 315
*Brassia* 96–101, 201, 317
*Brassia allenii* 98
*Brassia antherotes* 317
*Brassia arachnoidea* 96
*Brassia arcuigera* **97**, **99**
*Brachionidium* 287
*Bromheadia* 263
*Bulbophyllum* 29, 46, 105, 106, 159, 160, 190–194, 268, 285, 300–307, 337, 339
*Bulbophyllum apertum* 193
*Bulbophyllum cheiri* 193
*Bulbophyllum cimicinum* **304**
*Bulbophyllum davidii* 303, **304**
*Bulbophyllum dayanum* **302**
*Bulbophyllum flavofimbriatum* **304**
*Bulbophyllum johannuli* 303
*Bulbophyllum jolandae* **302**
*Bulbophyllum lasianthum* **105**, 107
*Bulbophyllum macneiceae* 303
*Bulbophyllum macranthum* 190, **191**, 193
*Bulbophyllum macrorhopalon* 303, **304**
*Bulbophyllum mandibulare* **105**, 107
*Bulbophyllum nocturnum* **304**, 305
*Bulbophyllum patens* **192**, 193
*Bulbophyllum penicillium* 300
*Bulbophyllum praetervisum* 193, **194**
*Bulbophyllum renipetalum* **46**
*Bulbophyllum* section *Epicranthes* 303–305
*Bulbophyllum sicyobulbon* **337**
*Bulbophyllum tarantula* 303
*Bulbophyllum tremulum* **302**
*Bulbophyllum variegatum* 106
*Bulbophyllum virescens* 107

C
*Caladenia* 60, 61, 64, 265
*Caladenia ambusta* **61**
*Caladenia crebra* **61**

*Caladenia gardneri* **61**
*Caladenia longicauda* **61**
Caladeniinae 60
*Caleana* 268
Calliphoridae 107
*Calochilus* 332–335
*Calochilus campestris* **333**, 334
*Calochilus stramenicola* **334**
*Calpodes* 363
*Calpodes ethlius* **363**
*Calypso* 76–78
*Calypso bulbosa* 76, **77**
*Camaridium* 144
*Camaridium horichii* **144**
*Campsomeris* 334
*Campsomeris tasmaniensis* 334
*Cannabis* 182
*Cannabis sativa* 182
*Cantharis* 211, 343
*Cantharis rustica* **211**, **343**
*Cardiophorus* 212
Catasetinae 164, 166, 181–189, 285
*Catasetum* 39, 40, 162, 164, 181–189, 201, 285
*Catasetum maculatum* 39, 40, **183**
*Catasetum saccatum* **183**
*Cattleya* 38, 40, 201, 202
*Cattleya dowiana* **202**
Centridini 237, 322
*Centris* 321–332
*Cephalanthera* 316
*Cephalanthera rubra* **316**
*Ceratandra* 211, 213
*Ceratandra grandiflora* 211, 213
*Ceratochilus* 317
*Ceratochilus biglandulosus* 317
Ceratopogonidae 157–160, 303, 341–343
*Ceropegia* 157
Cetoniinae/cetoniid 210–217
*Chamorchis* 205, 206
*Chamorchis alpina* **205**, 206
*Cheilosia* 113
*Chelonistele* 201
*Chenorchis* 206
*Chiloglottis* 39, 59, 64

*Chiloglottis trapeziformis* **39**
*Chloraea* 317
*Chloraea alpina* 317
*Chrysomya* 105, 107
*Chrysomya megacephala* **105**, 107
*Coccineorchis* 353
*Coccineorchis bracteosa* **353**
*Cochleanthes* 83, 164
*Cochleanthes aromatica* 83
*Cocytius* 256
*Cocytius antaeus* 256
Coeliopsidinae 164, 166
*Coelogyne* 201, 203, 318, 320
*Coelogyne intermedia* **320**
*Coffea* 266, 277
*Coffea arabica* 266, 277
*Cohniella* 328, 329, 330
*Cohniella ascendens* 328, **329**, 330
Coleoptera 200, 210, 215
*Comparettia* 140, 144–146
*Comparettia coccinea* 146
*Comparettia falcata* 144, **145**, 146
*Coryanthes* 36, 41, 162, 164, 165, 175–181, 201, 236
*Coryanthes horichiana* **41**
*Coryanthes kaiseriana* **176**, 181
Coryciinae 40
Cranichideae 280, 351, 354
Cranichidinae 351, 353, 354
*Cranichis* 353
*Cranichis diphylla* **353**
*Cryptostylis* 60, 62, 63
*Cryptostylis erecta* **62**
*Cryptostylis ovata* **62**
*Cyanicula* 228
*Cyanicula ixioides* 228
*Cyclopogon* 353, 354
*Cyclopogon olivaceus* **353**
*Cycnoches* 164, 165, 182, 185, 188, 189, 201, 282, 285, 315
*Cycnoches chlorochilon* 315
*Cycnoches egertonianum* **188**
*Cycnoches warszewiczii* **188**
*Cylindrocites* 351, 354
*Cylindrocites browni* 351, 354
Cymbidieae 164

*Cymbidium* 201, 246, 247, 293
*Cymbidium cyperifolium* **247**
*Cymbidium serratum* 246, **247**
Cypripedioideae 34, 349
*Cypripedium* 108–118, 345, 346, 347
*Cypripedium calceolus* 345, 346
*Cypripedium fargesii* 113, **114**
*Cypripedium fasciculatum* **114**, 115
*Cypripedium lichiangense* 113, **114**
*Cypripedium micranthum* 115
*Cypripedium sichuanense* **114**, 115
*Cypripedium subtropicum* 115, **119**
*Cyrtopodium* 298
*Cyrtopodium polyphyllum* 298

D

*Dactylorhiza* 74–83, 205, 210, 211, 293, 317, 388
*Dactylorhiza fuchsii* 293, 317
*Dactylorhiza incarnata* 81
*Dactylorhiza maculata* 79, 81
*Dactylorhiza praetermissa* 388
*Dactylorhiza romana* 80
*Dactylorhiza sambucina* **79–82**
*Dactylorhiza viridis* **211**
*Dasyscolia* 49, 51, 54
*Dasyscolia ciliata* 49, 51, 54
*Daviesia* 89, 90
*Daviesia horrida* **89**
Dendrobiinae 337
*Dendrobium* 101, 154, 156, 177, 201, 261–267, 317–320, 337, 342
*Dendrobium crumenatum* **261–267**
*Dendrobium cymbidioides* **320**
*Dendrobium insigne* 177
*Dendrobium sinense* 101
*Dendrophylax* 355–363
*Dendrophylax lindenii* **355–363**
Diapriidae/diapriid 114, 115
*Dichaea* 164, 165, 215–217
*Dichromanthus* 354
*Didelphis* 243
*Didelphis marsupialis* 243
*Dideopsis* 110
*Dideopsis aegrota* 110
*Diplocaulobium* 263

Diptera 37, 96, 200, 299, 335–340, 343, 345
*Disa* 73, 100, 214, 215, 265
*Disa atricapilla* 214, **215**
*Disa bivalvata* 214, **215**
*Disa forficaria* 214
*Diuris* 42, 88–91, 265
*Diuris alba* 90
*Diuris brumalis* **89**
*Diuris maculata* 90
*Diuris magnifica* **89**
*Diuris punctata* **42**
Diurideae 56, 59
*Dodecatheon* 76
*Dodecatheon radicatum* 76
*Dolba* 359
*Dolba hyloeus* 359
*Draconanthes* 287
*Dracula* 16, 102–108, 112, 383
*Dracula erythrochaete* **104**, 106
*Dracula vinacea* **16**
*Drakaea* 56–60, 64, 268
*Drakaea glyptodon* 57
*Drakaea livida* 57
Drakaeinae 60
*Dressleria* 22, 23, 164, 283
*Dressleria dilecta* **23**
*Drosophila* 102, 129–137, 300
Drosophilidae/drosophilid 16, 17, 102, 107, 274, 337
*Ducetia* 228

**E**
*Earina* 342
Elateridae 212
*Elleanthus* 137–147, 239, 315
*Elleanthus cynarocephalus* **144**
*Elleanthus oliganthus* **141**, **144**
*Elleanthus xanthocomus* 315
*Encyclia* 201
Enterobacteriaceae 259
*Epicranthes* 303–305
Epidendroideae 34, 166, 351
*Epidendrum* 36, 39, 91–96, 140, 144, 177, 201, 202, 315, 318, 337, 339, 380, 394

*Epidendrum ciliare* 315, 380
*Epidendrum ibaguense* 91
*Epidendrum fulgens* 94
*Epidendrum piliferum* **39**
*Epidendrum radicans* **91–96**, 318, 394
*Epipactis* 206. 207, 218, 220, 255–260, 288, 344
*Epipactis helleborine* **256–259**
*Epipactis palustris* **206**, 207, 344
*Epipactis purpurata* **256–258**
*Epipactis thunbergii* 218, **220**
*Epistephium* 152
*Eria* 156
Ericaceae 255
Eucerine 195
*Eudejeania* 67
*Eufriesea* 186
Euglossini/euglossine 23, 161–189, 216, 231–245, 319, 352, 371, 375, 377
*Euglossa* 120, 163, 165, 171, 172, 180, 181, 186, 243, 245, 386
*Euglossa alleni* **171**
*Euglossa aureiventris* **171**
*Euglossa decorata* **171**
*Euglossa imperialis* **172**
*Eulaema* 39, 40, 171, 183, 188, 237, 238, 243, 245
*Eulaema cingulata* **39**, 40, **238**
*Eulophia* 210–217, 285
*Eulophia ensata* **213**
*Eulophia foliosa* 212
*Eulophia parviflora* 213, 214, 285
*Eulophia ruwenzoriensis* 212
*Eulophia welwitschii* 213
*Euphoria* 212
*Eurema* 394
*Eurema daira* **394**
*Exaerete* 188

**F**
Fabaceae 88–90
*Ferdinandea* 113–115
*Ferdinandea cuprea* **114**
*Flickingeria* 263

## G

*Globosites* 350, 354
*Globosites apicola* 350, 354
*Glomeremus* 229–231
*Glomeremus orchidophilus* 229–231
*Glossophaga* 234
*Glossophaga soricina* 234
*Gongora* 36, 162, 164, 167–175, 201, 236, 285
*Gongora cruciformis* **171**
*Gongora latisepala* **171**
*Gongora quinquenervis* **171**
*Gongora unicolor* **171**
*Goodyera* 39, 203
*Goodyera foliosa* 203
*Goodyera repens* **39**
Goodyerinae 280, 349, 350, 353
*Grammatophyllum* 28, 29, 262
*Grammatophyllum speciosum* **28**
Gryllacridinae 229
*Guarianthe* 385, 386
*Guarianthe skinneri* 385, **386**
*Gymnadenia* 294, 317, 353, 365, 369
*Gymnadenia conopsea* 294, 317, 353
*Gymnadenia nigra* **365**, 369, 370

## H

*Habenaria* 218, 219, 228, 288, 289, 337, 339, 350
*Habenaria epipactidea* **339**
*Habenaria monorrhiza* 288, **289**, 350
*Habenaria radiata* 218, **219**
*Habenaria sagittifera* 228
Halictidae/halictid 154, 156, 264
*Halictus* 345
*Heterotaxis* 154, 155
*Heterotaxis maleolens* **155**
*Heterotaxis sessilis* **155**
*Holcoglossum* 206, 294
*Holcoglossum amesianum* **294**
Hymenoptera 35, 49, 59, 200, 210, 335–340, 343, 345
*Hymenopus* 310–314
*Hymenopus coronatus* **310**, 311–314

## I

Ichneumonidae/ichneumonid 60, 343, 345,
*Idiarthron* 229, 244
*Isochilus* 381, 382
*Isochilus latibracteatus* 381, **382**

## K

*Kefersteinia* 163, 165
*Kefersteinia orbicularis* **163**
Keroplatidae 271

## L

*Laelia* 201
*Lankesteriana* 160, 301, 303
*Lantana* 91–96
*Lantana camara* 91, **92**, 93
*Lasioglossum* 154, 156, 345
*Lepanthes* 6, 48, 68–74, 159, 219, 222, 281, 287, 337, 339, 367
*Lepanthes rupestris* 367
*Lepanthes sabinadaleyana* **69**
*Lepanthes telipogoniflora* 70
*Lepanthopsis* 219, 222
Lepidoptera [Chap 4 / Intro; Chap 6 / St 44, 45]
*Leporella* 207, 208
*Leporella fimbriata* **208**
*Leucocelis* 213
*Leucocelis haemorrhoidalis* **213**
*Liopygia* 105
*Liopygia ruficornis* **105**
*Liparis* 298, 299
*Liparis kumokiri* 299
*Liparis loeselii* 298
*Lissopimpla* 60, 62
*Lissopimpla excelsa* 60, **62**
*Listera* 277, 278
*Ludisia* 31, 33, 350, 353
*Ludisia discolor* **33**, **350**, **353**
*Luisia* 214–216,
*Luisia curtisii* 215
*Luisia teres* 214, **216**
*Lycaste* 163, 164, 165, 228
*Lycaste bradeorum* **163**
*Lycaste candida* **163**

*Lycaste desboisiana* **228**
*Lycaste xytriophora* **163**

# M

*Macradenia* 120, 163, 164, 281
*Macradenia brassavolae* **120**, **163**
*Macroclinium* 164, 281
Malaxidinae 337
*Malaxis* 277
Malpighiaceae 321–332
Mantodea 311
*Masdevallia* 17, 38, 39, 135, 136, 140, 144, 204, 274, 291, 292
*Masdevallia bicolor* **136**
*Masdevallia cupularis* 291, **292**
*Masdevallia driesseniana* 291, 292
*Masdevallia ignea* 136
*Masdevallia guttulata* **204**
*Masdevallia muscosa* 274
*Masdevallia rosea* 136
*Masdevallia rostriflora* 292
*Masdevallia smallmaniana* 291, **292**
*Masdevallia veitchiana* 39
*Masdevallia zahlbruckneri* 17
*Massonia* 246
*Maxillaria* 153–156
*Maxillaria crassifolia* **155**
*Maxillaria maleolens* **155**
*Maxillariella* 381
*Maxillariella guareimensis* 381
Maxillariinae 65, 164, 166, 285
*Megaselia* 225
*Melaleuca* 222–224
*Meliorchis* 348, 349, 354
*Meliorchis caribea* **348**, 349, 354
Meliponini 154
*Melitaea* 344
*Melitaea parthenoides* **344**
*Mexipedium* 109
*Microchilus* 353
*Microchilus epiphyticus* **353**
*Microchilus maasii* **353**
*Microtis* 207, 208, 265
*Microtis media* **208**
Milichiidae 301
*Monachanthus* 181, 184

*Montella* 215, 217
*Mormodes* 31, 36, 164, 165, 181–189, 281–286
*Mormodes andicola* **186**
*Mormodes colossus* **186**, **283**
*Mormodes fractiflexa* **283**
*Mormodes horichii* **186**, **283**
*Mormolyca* 48, 71, 72, 74
*Mormolyca ringens* **72**
*Muscarella* 305, 307
*Muscarella strumosa* 305
*Myanthus* 181, 184
*Myanthus cristatus* 184
Mycetophilidae 115, 271
*Mycaranthes* 156
*Myoxanthus* 263, 264, 266
*Myoxanthus parahybunensis* 263, **264**, 266
*Myrmecia* 207, 208
*Myrmecia urens* **208**
*Myrmecophila* 84, 201, 202, 248–251
*Myrmecophila thomsoniana* **248–251**

# N

*Neotinea* 206
*Neotinea maculata* 206
*Neottia* 218, 276–286, 308, 343, 344, 346
*Neottia cordata* 278
*Neottia nidus-avis* 218
*Neottia ovata* 277, 278, **308**, **343**, 344, 346
*Neuwiedia* 147–152
*Neuwiedia veratrifolia* **148**, **151**
*Neuwiedia zollingeri* **151**
*Nigritella* 364, 369, 370
*Nigritella nigra* **364**, **369**, **370**
*Notylia* 164, 201, 281

# O

*Octomeria* 263
*Octomeria valerioi* 263
*Ocyptamus* 116
*Oeceoclades* 254, 298

*Oeceoclades maculata* **254**, 298
*Oligographa* 339
*Oligographa juniperi* **339**
Oncidiinae 40, 65, 164, 166, 281, 321–332
*Oncidium* 145, 146, 201, 318, 319, 321–332
*Oncidium ascendens* 328
*Oncidium henekenii* 327
*Oncidium hyphaematicum* **323**, 330
*Oncidium ornithorhynchum* 330, **331**
*Oncidium pardothyrsus* 323, **324**
*Oncidium planilabre* 323, **324**
*Oncidium stenotis* 318, **319**, 328
*Oncidium strictum* **145**, 146
*Odontoglossum* 256
*Ophrys* 29, 47–55, 56, 64, 306, 361, 368, 372, 373
*Ophrys apifera* 306
*Ophrys argolica* 372, **373**
*Ophrys delphinensis* 372
*Ophrys fusca* 49, **50**
*Ophrys lojaconoi* **53**
*Ophrys lutea* 49, **50**
*Ophrys speculum* **49–52**
*Ophrys sphegodes* 52, **53**, **368**, 369
*Ophrys splendida* **53**
*Opuntia* 320
*Orchidacites* 347
*Orchidacites cypripedioides* 347
*Orchidacites orchidioides* 347
*Orchidacites wegelei* 347
Orchidoideae 34, 349, 350, 353, 354
*Orchis* 11, 29, 78, 81, 210, 211, 282, 309, 317
*Orchis anthropophora* 210, **211**
*Orchis fusca* 81
*Orchis italica* 317
*Orchis latifolia* 81
*Orchis maculata* 81
*Orchis mascula* 282
*Orchis morio* 75
*Orchis purpurea* 81
*Ornithidium* 142–145
*Ornithidium coccineum* 142
*Ornithidium fulgens* **143–145**

*Ornithidium sophronitis* 142
*Ornithocephalus* 306
*Ornithocephalus lankesterii* **306**
Orthoptera 200, 227–231
*Orthrosanthus* 87, 88
*Orthrosanthus laxus* 87, **88**
*Otochilus* 201

P
*Pachylia* 359
*Pachylia ficus* **359**
*Palaeorchis* 346, 347
*Palmorchis* 262, 263
*Panogena* 127
*Panogena lingens* **127**
*Paphinia* 171
*Paphiopedilum* 108–118
*Paphiopedilum barbigerum* 113
*Paphiopedilum insigne* **111**, 113
*Paphiopedilum rothschildianum* 110, **111**
*Paphiopedilum spicerianum* 112
*Paphiopedilum villosum* **111**, 112
*Papilio* 376, 377
*Papilio palamedes* 377
*Parkia* 316
*Parkia nitida* 316
*Pendusalpinx* 160
*Pepsis* 97–99
*Pepsis atalanta* **99**
*Phalaenopsis* 29, 262, 312, 313
*Phalaenopsis deliciosa* **313**
*Phelsuma* 250
*Phelsuma borbonica* 250
*Phoebis* 377
*Phoebis sennae* 377
*Pholeomyia* 301
*Pholidota* 201
Phoridae/phorid 225
*Phragmipedium* 108–118, 292
*Phragmipedium caudatum* 116
*Phragmipedium lindenii* 292
*Phragmipedium longifolium* 116, **117**, 118
*Phragmipedium pearcei* 116
Phymatinae 317

*Platanthera* 128, 129, 288, 290, 298, 299, 376, 377
*Platanthera aquilonis* 288, 298, **299**
*Platanthera bifolia* **128**
*Platanthera ciliaris* **376**, 377
*Platanthera huronensis* 288, **290**
*Platystele* 28, 29
*Platystele tica* **28**
Platystomatidae 106
*Platythelys* 353
*Platythelys maculata* **353**
*Plebeia* 72
Pleurothallidinae/pleurothallid 37, 219, 221, 222, 271, 276, 287, 303, 307, 337, 339
*Pleurothallis* 203, 204, 278, 279, 281, 341, 342
*Pleurothallis eumecocaulon* 278, **279**
*Pleurothallis helleri* **341**, 342
*Pleurothallis homalantha* **204**
*Pogonia* 218, 220
*Pogonia japonica* 218, **220**
Polygonaceae 112
*Polygonum* 112
*Polystachya* 154–156
*Polystachya masayensis* 154
*Polystachya foliosa* **155**
Pompilidae/pompilid 96–101
*Ponthieva* 353
*Ponthieva racemosa* **353**
*Porroglossum* 274–276
*Porroglossum hystrix* 276
*Porroglossum mordax* **275**
*Porroglossum muscosum* **274–276**
*Prasophyllum* 265
Prescottinae 280
*Proechimys* 242, 243
*Proechimys semispinosus* **242**, 243
*Prosthechea* 201, 380
*Prosthechea ochracea* **380**
*Protaetia* 214
*Protambulyx* 360
*Protambulyx strigilis* **360**
*Protea* 246
Proteaceae 246
Protorchidaceae 347

*Protorchis* 346, 347
*Pseudosphinx* 380
*Pseudosphinx tetrio* 380
*Psilochilus* 152, 263, 353
*Pterostylis* 268–276
*Pterostylis barbata* **270**
*Pterostylis picta* **270**
*Pterostylis sanguinea* 269, **270**
*Pterygodium* 265

R
*Rafflesia* 103
*Rediviva* 38
*Rediviva peringueyi* 38
Reduviidae 317
*Rhizanthella* 222–227
*Rhizanthella gardneri* **222–227**
*Rhizanthella johnstonii* **226**
*Rhizanthella omissa* 226
*Rhizanthella slateri* 223, 226
*Rhizanthella speciosa* 226
*Rhododendron* 255
*Rodriguezia* 142, 143
*Rodriguezia lanceolata* 143
*Rudiculites* 350, 354
*Rudiculites dominicana* 350, 354

S
*Sacoila* 142–144
*Sacoila lanceolata* 143
*Sarcoglottis* 354
*Satyrium* 214
*Satyrium microrrhynchum* 214
*Scaptotrigona* 72
Sciaridae/sciarid 68–74, 115
*Schiedeella* 353, 354
*Schiedeella transversalis* **353**
Scoliidae/scoliid 49, 54, 334, 335
*Selenipedium* 293
*Serapias* 194–197, 317
*Serapias cordigera* **197**
*Serapias lingua* **197**
*Serapias neglecta* 317
*Sobralia* 24, 40, 262–264, 266
*Sobralia labiata* **264**
*Solenocentrum* 353, 354

*Solenocentrum costaricense* 353
*Specklinia* 129–137, 159, 256, 271–276
*Specklinia caulophryne* 272, **273**
*Specklinia colombiana* 272, **273**
*Specklinia endotrachys* 130, 133, **134**, 137
*Specklinia pfavii* **131**
*Specklinia remotiflora* **131**
*Specklinia spectabilis* **131**
*Specklinia trilobata* 272, **273**
Sphingidae/sphynx 126, 343
*Spiculae* 268
*Spiranthes* 204, 280
*Spiranthes lacera* **280**
*Spiranthes tuberosa* **204**
Spiranthinae 280, 350, 351, 353
*Stanhopea* 83, 162, 164, 165, 172–174, 285
*Stanhopea cirrhata* **172**
*Stanhopea wardii* 83
Stanhopeinae 164, 166, 174, 285
*Stelis* 219, 221, 222, 252, 282, 293, 318, 337–339
*Stelis deregularis* 293
*Stelis gelida* 318
*Stelis kefersteiniana* **221**
*Stelis villosa* **252**
*Stenorrhynchos* 354
*Stenorrhynchos speciosum* **354**
*Succinanthera* 351, 354
*Succinanthera baltica* 351, 354
Syrphid 108–118
*Syrphus* 116

T
Tachinidae/tachinid 64–67
*Telipogon* 4, 48, 64–67
*Telipogon costaricensis* **66**
*Telipogon glicensteinii* **4**
*Telipogon peruvianus* **67**
Tephritidae/tephritid 191
*Tetragonula* 151
*Tetragonula laeviceps* **151**
*Tetragonula melina* **151**
Tettigoniidae 229, 244

*Thelymitra* 14, 83–91, 198, 265
*Thelymitra antennifera* **86**, 87
*Thelymitra crinita* **86**, 87
*Thelymitra epipactoides* **86**, 87
*Thelymitra ixioides* **86**
*Thelymitra jacksonii* **198**
*Thelymitra macrophylla* 87
*Thelymitra nuda* **86**, 87
*Thelymitra rubra* **86**
*Thelymitra speciosa* **14**
*Theobroma* 160
*Theobroma cacao* 160
Thomisidae 317
*Thrixspermum* 263
Thynnine/thynnid 56, 58, 64
Thysanoptera 200, 218
*Tolumnia* 326, 327
*Tolumnia henekenii* **326**, 327
*Trichoceros* 64, 65
*Trichocolletes* 90
*Trichopilia* 163, 164
*Trichopilia fragrans* **163**, 164
*Trichosalpinx* 157–160, 301, 303
*Trichosalpinx blaisdellii* **158**, 159
*Trichosalpinx reflexa* 159
*Trigona* 151, 154, 156, 330
*Trigonidium* 48, 65, 71–74, 285
*Trigonidium egertonianum* **72**
*Trigonidium obtusum* **72**
*Triphora* 263, 264
*Triphora trianthophoros* **264**

U
*Uropedium* 292
*Uropedium lindenii* 292

V
*Vanda* 32, 201, 262, 282, 284
*Vanda tricolor* **284**
*Vanilla* 18, 20, 29, 34, 39, 40, 164, 166, 201, 228, 229, 231–245, 262, 263, 319, 320
*Vanilla chamissonis* 234, 235, 239, 244
*Vanilla chamissonis* var. *longifolia* 234

*Vanilla costaricensis* **18**
*Vanilla dressleri* 233
*Vanilla hartii* 237, 239, 245
*Vanilla odorata* 237, 241, 243
*Vanilla palmarum* 238, 239
*Vanilla planifolia* 164, 229, **231–245**
*Vanilla pompona* **39**, 40, **231–245**
*Vanilla hartii* 237, 239, 245
*Vanilla trigonocarpa* 235, 244, 263
*Vanilla* × *tahitensis* 243
Vanilloideae 34, 166, 349
Verbenaceae 91, 92
*Vespula* 258, 259
*Vespula germanica* 259
*Vespula vulgaris* **258**, 259

## X
*Xanthopan* 123–129
*Xanthopan morganii* 125, 126
*Xanthopan morganii praedicta* 125

## Y
*Yoania* 229, 244

## Z
*Zaspilothynnus* 56, 57
*Zaspilothynnus nigripes* **57**
Zygopetalinae 39, 164, 166
*Zygaena* 344
*Zygaena purpuralis* **344**
*Zygothrica* 102–108
*Zaspilothynnus* 56, 57
*Zaspilothynnus nigripes* **57**